THI

Practice in PHYSICS

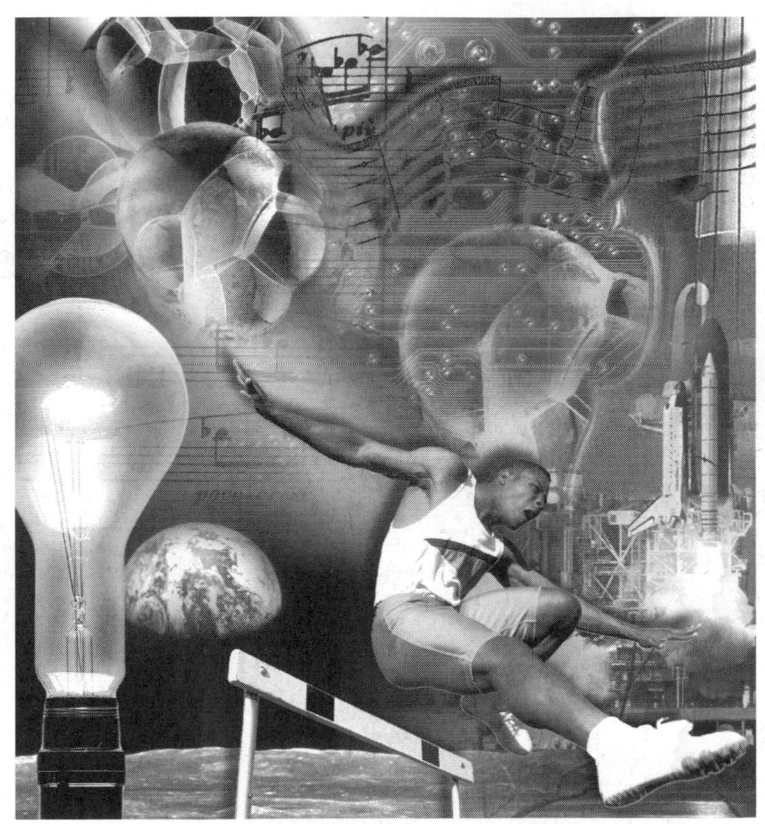

Akrill, Bennet and Millar

Hodder & Stoughton

A MEMBER OF THE HODDER HEADLINE GROUP

Orders: please contact Bookpoint Ltd, 78 Milton Park, Abingdon, Oxon OX14 4TD. Telephone: (44) 01235 827720; Fax: (44) 01235 400454. Lines are open from 9.00–6.00, Monday to Saturday, with a 24 hour message answering service. Email address: orders@bookpoint.co.uk

British Library Cataloguing in Publication Data
A catalogue record for this title is available from The British Library

ISBN 0 340 75813 9

First published 2000

Impression number	10 9 8 7 6 5 4 3 2
Year	2006 2005 2004 2003 2002 2001 2000

Cover design by Blue Pig Design Co.
Illustrated by Jeff Edwards.
Typeset by Multiplex Techniques Ltd, Brook Industrial Park, Mill Brook Road, St. Mary Cray, Kent BR5 3SR.
Printed in Great Britain by Hodder & Stoughton Educational, a division of Hodder Headline Plc, 338 Euston Road, London NW1 3BH by JW Arrowsmith, Bristol.

Photo acknowledgments
The publishers would like to thank the following individuals, institutions and companies for permission to reproduce photographs in this book. Every effort has been made to trace ownership of copyright. The publishers would be happy to make arrangements with any copyright holder whom it has not been possible to contact.

Action Plus (14,15, 21, 33, 87); Andrew Lambert (57); BI Technologies (59); Corbis (221); Heatrae Sadia (102); Imperial War Museum (202); KWC Watson (129, 193); Melcor (120); *Physics Education* 7 no. 6, July 1972 (178); *PSSC Physics 4th Edition* © DC Heath & Co 1976 (174); Private Collection (8, 41); Science Photo Library (84, 134, 162); Techni Measure, TML (241); Vauxhall Motors Ltd. (31); The Wellcome Trust (90).

Contents

About this book . . .

This is a book of questions to help you understand Physics at AS or A Level. None of the questions are from previous examination papers; questions in examination papers are meant to test you at the end of your course or module. What you need during the course is to do questions which will help you to check whether you have understood what you are being taught. That is why we have called this book *Practice in Physics.*

The first edition of this book was originally published in 1979 and it has been in print ever since. We have revised it to take account of the new syllabuses (specifications) for Advanced Level courses from September 2000. Some of the questions test whether you have understood the principles, and a few (very few) should make you think quite hard. These harder questions are indicated by an asterisk (*). At the end of the book there are answers to all the questions (except for one or two where to give the answer would take away any need to think). You will not need to do all the questions! But we hope you enjoy doing most of them because part of the pleasure of doing Physics is to discover that you can get the right answers, showing that you understand the ideas.

Chapter 22 contains synoptic questions, i.e. questions which need knowledge and skills from more than one section. Above each question are the numbers of the sections to which the questions relate. These are questions you should attempt near the end of the A Level course. We have also included a chapter of questions (Chapter 21) which will help you elsewhere in the book when you need certain mathematical techniques. Some of these are very general – others will help you with particular chapters.

Throughout the book answers are given to the same number (usually 2) of significant figures as the data in the question, but when answers to the later parts of a question depend on the answers to earlier parts, you should use the unrounded figures for the later parts.

Tim Akrill
George Bennet
Chris Millar

1 Linear and circular motion

Data free fall acceleration at the Earth's surface = $9.81\ \text{m s}^{-2}$

1.1 Speed and velocity

In this section you will need to

- use the equation average velocity = $\Delta s/\Delta t$
- understand that displacement and velocity are vector quantities
- remember how to measure speed
- understand that the gradient of a displacement–time graph is the velocity
- understand that the area between a velocity–time graph and the time axis represents the displacement.

1.1 Starting from home, a jogger runs 4.0 km (about 2.5 miles). She returns home after 20 minutes. What is **(a)** her average speed **(b)** her average velocity?

1.2 The diagram shows the oil spots left on a road by a motorbike with a leaky sump as the bike travels from A to G. Describe the journey, assuming that the drips come at regular time intervals.

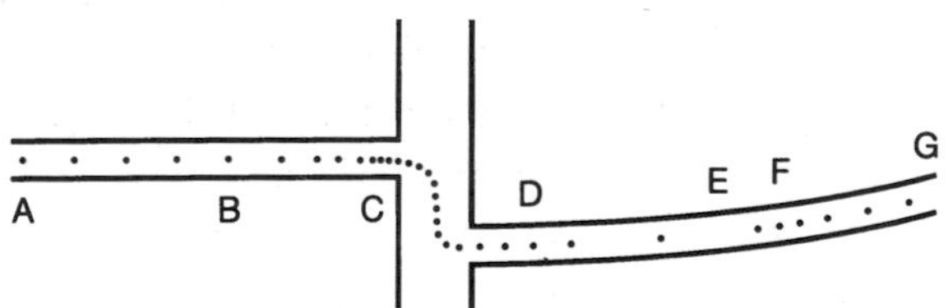

1.3 The diagram shows the movement of a smoke particle in a Brownian motion experiment.

(a) Use a ruler to find **(i)** the total distance moved by the smoke particle in going from A to B **(ii)** the displacement AB.

(b) If it took 1.20 s to travel from A to B, calculate **(i)** the average speed **(ii)** the average velocity of the smoke particle.

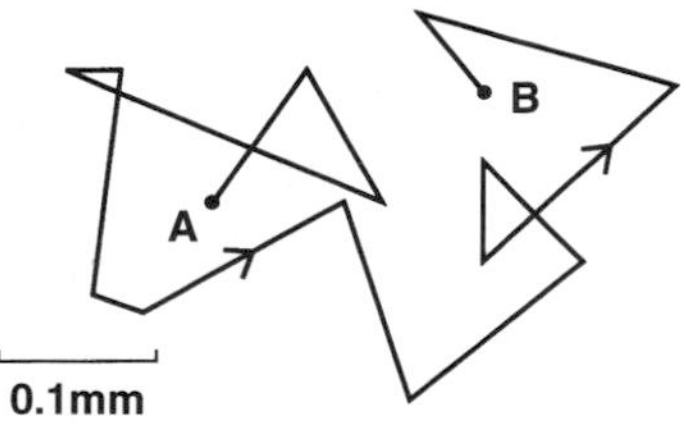

1.4 What is the change of velocity when

(a) $+6.0\ \text{m s}^{-1}$ becomes $+15\ \text{m s}^{-1}$ **(b)** $+6.0\ \text{m s}^{-1}$ becomes $-15\ \text{m s}^{-1}$

(c) $+6.0\ \text{m s}^{-1}$ becomes $-\ 6.0\ \text{m s}^{-1}$ **(d)** $5.0\ \text{m s}^{-1}$ east becomes $15\ \text{m s}^{-1}$ west?

1.5 A skier moves at $11\ \text{m s}^{-1}$ down a 16° slope. What is the skier's

(a) vertical velocity **(b)** horizontal velocity?

1.6 The graph shows displacement against time. Describe the motion in as much detail as possible.

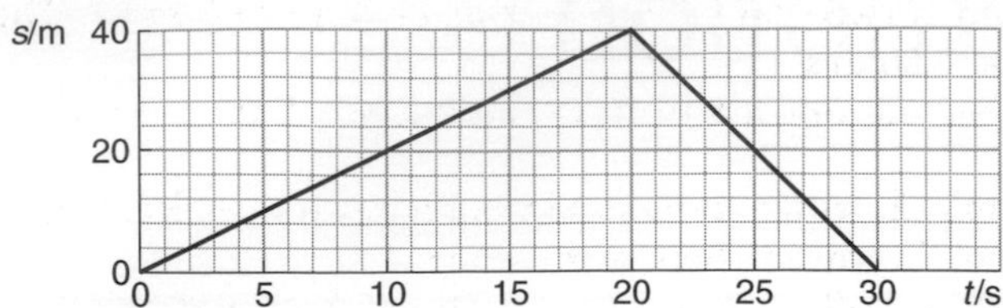

1.7 A swimmer dives in and is timed at various points in a 50 m race. The record is as shown in the table:

s/m	0	5	10	15	20	25	30	35	40	45	50
t/s	0.0	2.5	5.5	11.0	16.0	22.0	26.5	32.0	39.5	47.5	56.0

(a) Plot a graph of s (up) against t (along) and draw a smooth line through the points.
(b) Work out the average speed during **(i)** the first 10 s **(ii)** from 20 s to 35 s **(iii)** the last 10 s. Suggest why the average speeds are different.
(c) Draw a tangent to your curve at $t = 20$ s and deduce the swimmer's instantaneous speed then.

1.8 The diagram has been drawn after studying a stroboscopic photograph of a golf swing. The stroboscope was flashing 50 times per second.
(a) Sketch a graph showing how the speed of the head of the golf club varies with time from A to D.
(b) Using the scale from the 1.00 m ruler at the bottom of the diagram, estimate the speed of the club-head from **(i)** B to C **(ii)** P to Q.
(c) Estimate the speed of the golf ball between X and Y.

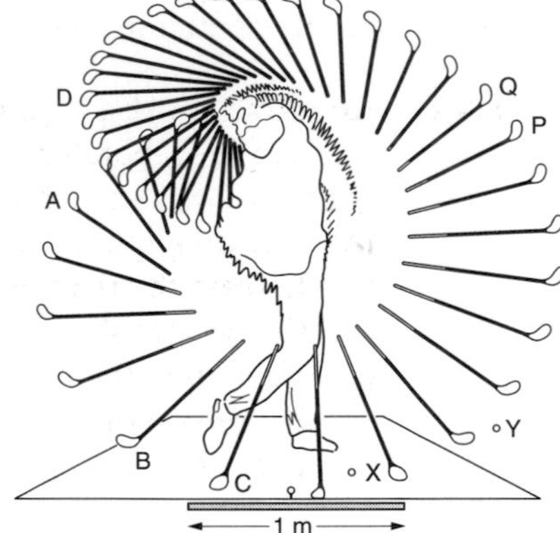

1.9 A starting pistol is fired in front of a microphone and the sound is arranged to start an electronic timer which counts in milliseconds. When the sound reaches another microphone 12.0 m away the timer is stopped. If the speed of sound in air is 340 m s^{-1}, calculate the time which would be recorded on the timer.

1.10 In an emergency, you have a reaction time of about 0.6 s. How far would you travel in this time on a bicycle moving at 12 m s^{-1} (just over 25 m.p.h.)?

1.11 A speed skier registers an average speed of 233.7 km h^{-1} over a distance of exactly one kilometre.
(a) How long did he take to cover the kilometre?
(b) Express his average speed in m s^{-1}.

1.12 How would you use laboratory apparatus to measure the speed of
(a) a snail **(b)** a model train **(c)** an air rifle pellet?

1.13 Approaching Terminal 3 at Heathrow Airport, passenger P uses the walkway and, having heavy luggage, allows it to take him along. Passenger Q walks alongside the walkway and passenger R walks on it, both walking briskly at 1.2 m s^{-1}. The walkway is 40 m long and moves at 0.80 m s^{-1}.
Calculate how long it takes P, Q and R to reach the other end of the runway.

1.14 The graph shows the motion of a stone thrown vertically upward. Calculate the maximum height reached by the stone
(a) by first finding the average velocity of the stone
(b) by finding the area under the graph.

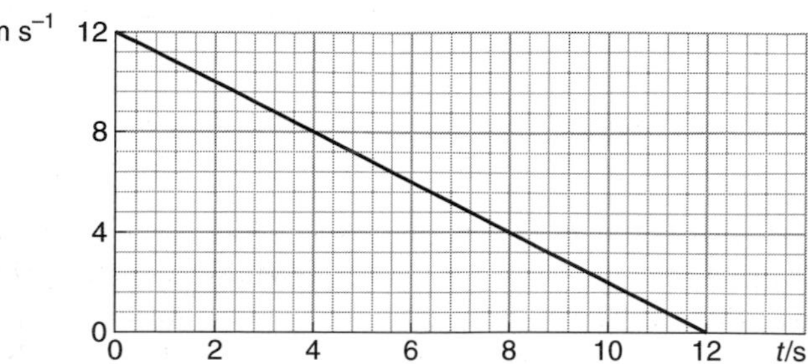

1.15 Draw a velocity–time graph for a tennis ball which is being volleyed backwards and forwards by two players close to the net. Assume that the ball travels horizontally and perpendicular to the net but that the players hit it so that it travels at a variety of speeds.

1.16 An ultrasonic displacement sensor is used to study the motion of a trolley sliding down a ramp in the laboratory. The displacement against time data is presented on screen as a graph and then converted to a velocity–time graph as shown.
(a) Which graph is s–t and which is v–t?
(b) Explain the relationship between the two graphs.
(c) Verify that the computer software has drawn the v–t graph correctly at $t = 2$ s.

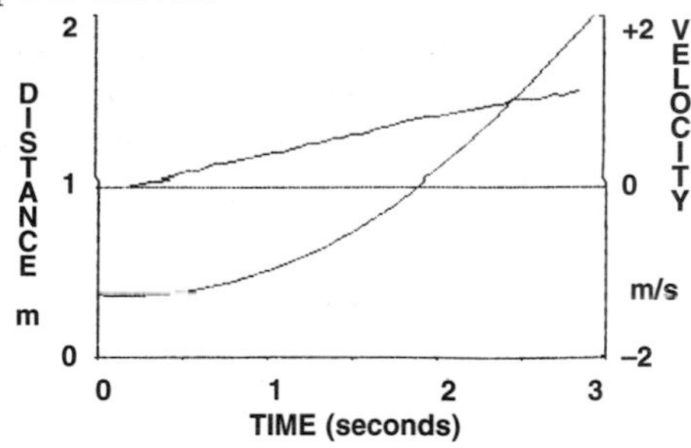

1.2 Acceleration along a line

In this section you will need to

- use the equation average acceleration = $\Delta v/\Delta t$
- understand that the gradient of a velocity–time graph is the acceleration
- remember how to measure acceleration
- understand how to draw graphs for displacement, velocity or acceleration against time when given only one of them
- use the following equations for uniform acceleration:
  $v = u + at, \quad s = \frac{1}{2}(u + v)/t, \quad s = \frac{1}{2}at^2$ when $u = 0$

1.17 A man, John L. Stapp, travelling in a rocket-powered sledge, accelerated from 0 to 284 m s^{-1} (about 630 m.p.h.) in 5.0 s and then came to a stop in only 1.5 s. Calculate his acceleration
(a) while he is speeding up **(b)** while he is slowing down.

1.18 The graph shows, in idealised form, a velocity–time graph for a typical short journey.
(a) Calculate the acceleration at each stage of the journey and display your answers on an acceleration–time graph.
(b) Sketch a displacement–time graph for this journey.

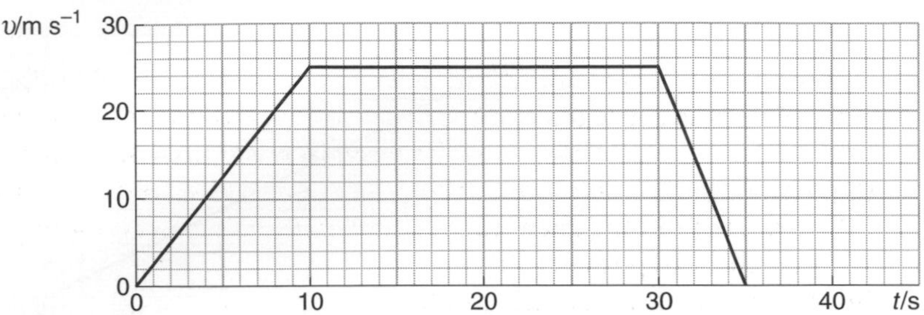

1.19 A baby buggy rolls down a ramp which is 15 m long. It starts from rest, accelerates uniformly, and takes 5.0 s to reach the bottom.
(a) Calculate its average velocity as it moves down the ramp.
(b) What is its velocity at the bottom of the ramp?
(c) What is its acceleration down the ramp?

1.20 One type of aeroplane has a maximum acceleration on the ground of 3.5 m s^{-2}.
(a) For how many seconds must it accelerate along a runway in order to reach its take-off speed of 115 m s^{-1}?
(b) What is the minimum length of runway needed to reach this speed?

1.21 A particle moves in a straight line. Its motion can be described as follows:
at $t = 0$, $v = 0$
$0 < t < 20$ s, $a = 4.0$ m s^{-2}
10 s $< t < 20$ s, $a = -4.0$ m s^{-2}.
Sketch the velocity–time graph and use it to find the change of displacement of the particle between $t = 0$ and $t = 20$ s.

1.22 Draw displacement–time, velocity–time and acceleration–time graphs for the following situations. Use the same time axes for all three graphs in each situation. [Hint: you will probably find it easiest to begin with the velocity–time graph.]
(a) An electrically-powered milk float moving from one house to another on a straight road.
(b) A ball, attached by an elastic cord to a fixed point on the ground, and hit horizontally away from that point.

1.23 The graph shows the horizontal speed of a long jumper from the start of his run-up to the moment when he takes off.
(a) What is his maximum acceleration?
(b) Estimate the distance he runs before he takes off.
(c) Sketch the general shape of his acceleration against time.

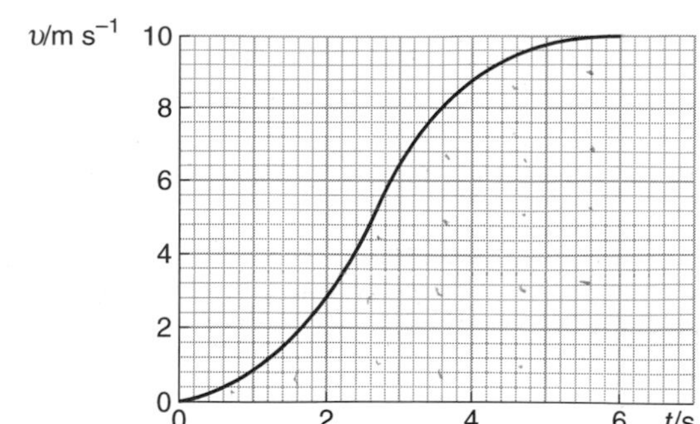

1.24 The diagram shows the positions of a line drawn on an accelerating air track glider every 100 milliseconds. The scale of the diagram is one twentieth that of the real situation.

(a) How far did the glider move between **(i)** positions 4 and 6 **(ii)** positions 10 and 12?

(b) Write down the average velocities of the glider between these two sets of positions.

(c) How long did the glider take to accelerate from position 4 to position 10 (or from position 6 to position 12)?

(d) What was the acceleration of the glider?

1.25 **(a)** Slow motion photography shows that a jumping flea pushes against the ground for about 0.001 s during which time it accelerates upwards to a maximum speed of 0.8 m s^{-1}. What is its upward acceleration during this 'take-off'?

(b) It then moves upwards with an acceleration (assumed constant) of –12 m s^{-2}. Calculate **(i)** how long it takes from leaving the ground to the top of its jump **(ii)** how high it jumps.

1.26 Electrons in a particle accelerator are moving at 8.0×10^5 m s^{-1} when they enter a tube where they are accelerated to 6.5×10^6 m s^{-1}.

(a) What is their acceleration in the tube?

(b) What is the length of the tube?

1.27 A road test report gives the following data for a standing start acceleration test on a car:

t/s	0	5	10	15	20	25	30	35	40
v/m s^{-1}	0	14	24	30	34	37	39	40	40

(a) Draw a velocity–time graph for the test.

(b) What is the displacement of the car when it has reached a speed of **(i)** 25 m s^{-1} **(ii)** 35 m s^{-1}?

(c) Find **(i)** the acceleration of the car when its speed is 30 m s^{-1} **(ii)** the car's maximum acceleration.

1.28 The graph on the next page shows the result of studying a sprint start.

(a) What, in m.p.h., was the maximum velocity reached? Take 1 m s^{-1} to equal 2.24 m.p.h.

(b) Calculate the acceleration of the sprinter **(i)** as she leaves her blocks **(ii)** after 2.0 s.

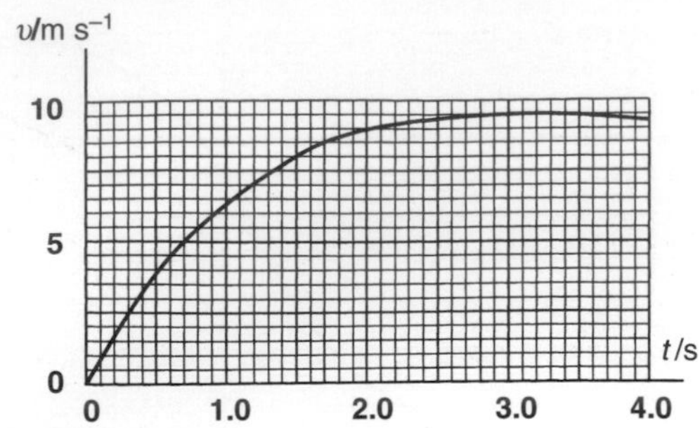

1.29 Suppose you are travelling on a bicycle with brakes which, at best, can produce an acceleration of -2.5 m s^{-2}.

(a) Draw, on the same axes, velocity–time graphs showing how you and the bicycle will slow down, with the brakes full on, from initial speeds of 15 m s^{-1}, 10 m s^{-1} and 5 m s^{-1}. [Hint: draw the 15 m s^{-1} case first.]

(b) Calculate the distances you travel in coming to rest in each case.

1.30 Use the first two equations for uniform acceleration, given at the bottom of page 3, to produce

(a) an equation linking u, v, a and s

(b) an equation linking s, u, a and t.

Check that each term of equations **(a)** and **(b)** has the same unit.

1.31 A high speed train can slow down smoothly from a speed of 190 km h^{-1} (120 m.p.h.) to rest within a distance of 1500 m.

(a) What is the average speed of the train, in m s^{-1}, as it slows down?

(b) Calculate how long it takes to come to rest from 120 m.p.h.

(c) Hence find the average acceleration of the train.

1.32 Find the acceleration of the train in the previous question using equation **(a)** from question 1.30.

1.33 The best throwers in the world are baseball pitchers. They can release a ball travelling at 40 m s^{-1}. In so doing they accelerate the baseball through a distance of 3.6 m. Calculate

(a) the average speed of the ball during the throw

(b) the time during which the ball is being accelerated

(c) the average acceleration of the baseball.

1.34 Calculate the acceleration of the ball in the previous question using equation **(a)** from question 1.30.

1.35 The graph describes the motion of a train moving in a speed-restricted area and then accelerating as it clears the area.

You are to calculate the total distance travelled by the train in the 40 s shown in three different ways.

(a) Use the average velocity of the train during each 20 s interval to calculate two separate distances and add them together.

(b) Use equation **(b)** from question 1.30.

(c) Find the number of squares under the graph and the distance represented by one square.

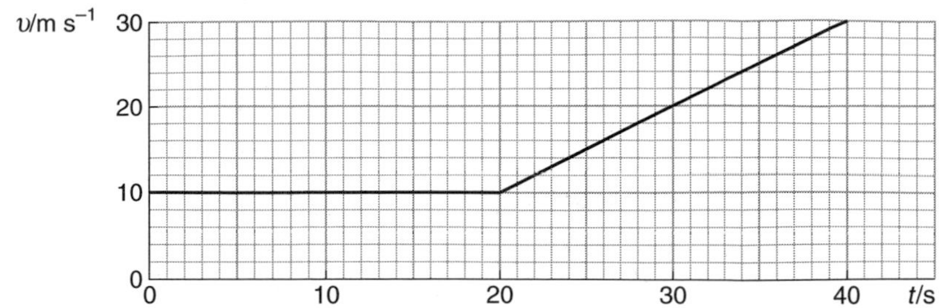

1.3 Free fall and projectile motion

In this section you will need to

- remember that the free fall acceleration at the Earth's surface is 9.8 m s^{-2}
- remember how to measure the free fall acceleration in the laboratory
- use the equations $v^2 = 2gs$ and $s = \frac{1}{2}gt^2$ for free fall from rest
- understand that when an object is falling freely its vertical motion is independent of its horizontal motion
- remember that velocity vectors can be resolved into two perpendicular components, $v_x = v\cos\theta$ and $v_y = v\sin\theta$, where θ is the angle between v and the x-axis.

1.36 A ball is thrown vertically upwards at 19.6 m s^{-1}.
(a) Make a table showing its velocity after 1.0 s, 2.0 s, 3.0 s and 4.0 s.
(b) What is its displacement after 2.0 s and 4.0 s?
(c) How far does it travel in the first 4.0 s?

1.37 **(a)** Explain how the apparatus shown can be used to measure g.
(b) What sources of error are there likely to be in such an experiment?

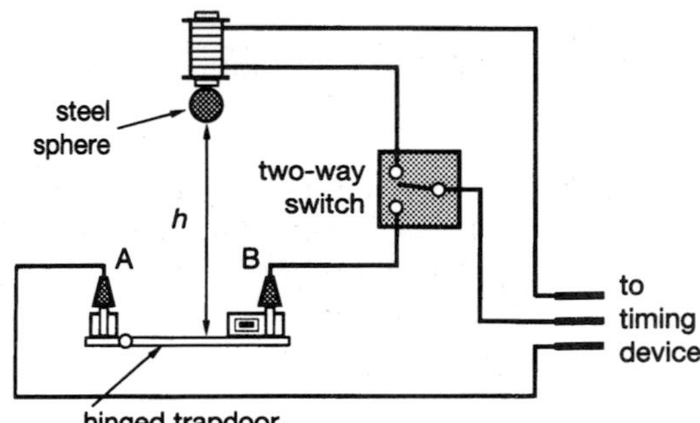

1.38 In an experiment with this apparatus a steel sphere is found to fall a distance 456 mm in 301 ms. Calculate
(a) the average velocity of the sphere as it falls
(b) the velocity with which the steel sphere hits the trap door
(c) the acceleration of the steel sphere.

1.39 A man throws a ball straight up into the air and catches it again. Take the upward direction to be positive and sketch a velocity–time graph for the ball, assuming air resistance to be negligible. How could you use your graph to find the height reached by the ball?

1.40 **(a)** Ignoring air resistance, how long does an object take to fall a distance of **(i)** 1.0 m **(ii)** 2.0 m?
(b) Why is the answer to **(ii)** not twice the answer to **(i)**?

1.41 In a cartoon two characters are standing by a well. One drops a stone down the well and starts to count. He stops counting when he hears the stone hit the water. He then announces proudly, 'Your well is exactly three seconds deep'. How deep is the well really?

1.42 The diagram shows a velocity–time graph for a ball bouncing vertically on a hard surface.
(a) Explain the shape of the graph.
(b) Use the graph to calculate three separate values for the acceleration of free fall.
(c) Calculate the height from which the ball was dropped and the height to which it bounced on **(i)** its first bounce **(ii)** its second bounce.

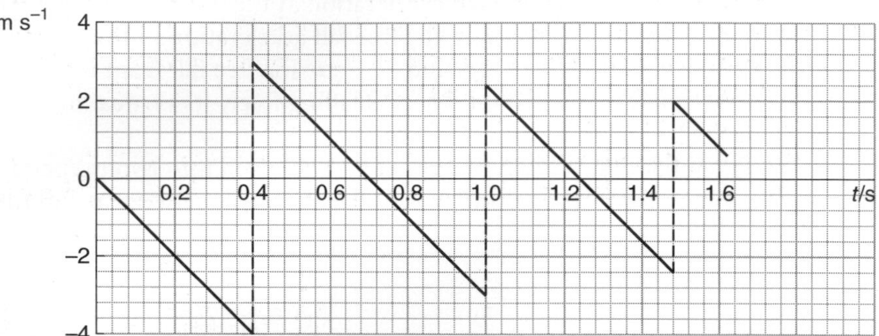

1.43 Parachutists hit the ground at about 6 m s^{-1}. How high a platform is needed for them to jump off in order to give them practice at hitting the ground at this speed?

1.44 A kangaroo was seen to jump to a vertical height of 2.8 m. For how long was it in the air?

1.45 A ball was photographed stroboscopically as it was released and fell freely. Eight images appear (the first two are overlapping) on the photo which is reduced in scale by a factor of five and shown below. Use the fact that the acceleration of the ball is 9.8 m s^{-2} to discover the time interval between the flashes of the stroboscope.

1.46 A bullet is fired horizontally at a speed of 200 m s^{-1} at a target which is 100 m away.

(a) Ignoring air resistance, calculate **(i)** how far the bullet has fallen when it hits the target **(ii)** the angle that it then makes with the horizontal.

(b) Explain whether air resistance would cause the bullet to fall a greater or lesser amount than the distance you have calculated.

1.47 A body is projected horizontally from ground level with a speed of 24 m s^{-1} at an angle of 30° above the horizontal. Neglect air resistance and calculate

(a) the vertical resolved part of its velocity

(b) the time taken to reach its highest point

(c) the greatest height reached

(d) the horizontal range of the body.

1.48* The long jumper in the diagram is shown at the instant he leaves the ground, at three positions during his flight and at the instant he first touches the sand. His long jump measures 7.5 m and he is recorded as being in the air for 0.80 s. His centre of gravity falls 0.95 m between his take-off and landing.

(a) Calculate his horizontal velocity at take-off.

(b) Show that his vertical velocity at take-off is 2.7(3) m s^{-1}.

(c) Hence calculate the angle at which he projects himself at take-off.

1.49 The cliff divers of Acapulco, Mexico, take off from a rocky cliff face 26.5 m above the surface of the water. In the course of their flight, of which all but the last 1.5 m of their horizontal motion is above rock, they travel 8.0 m forward.

(a) Make a sketch of the cliffs and water, and draw on the flight path of the diver.

(b) For how long are they in the air (ignore air resistance)?

(c) For how long are they over the water during their flight?

(d) What is their vertical velocity on entry?

(e) At what angle is their path to the vertical at entry?

1.50 You drop a heavy stone from a high suspension bridge and one second later you drop a second stone.

(a) Draw two v–t graphs for the two stones on the same graph axes.

(b) Explain how the distance between the two stones changes as they fall towards the water.

1.4 Motion in a circle

In this section you will need to

- use the equations $s = \theta / t$, $T = 1/f$, $v = r\omega$ and $\omega = 2\pi f$
- understand that a body moving in a circle at a constant speed is accelerating towards the centre of the circle
- use the expressions $a = v^2/r$ and $a = r\omega^2$ for centripetal acceleration

---

1.51 Calculate the angular velocity **(i)** in degrees **(ii)** in radians of
(a) a fan blade rotating at 2.5 r.p.m.
(b) the minute hand of a clock.

1.52 All points on the Earth rotate with an angular velocity of 7.3×10^{-5} rad s^{-1}. What is the speed of a point on the equator? Take the radius of the Earth to be 6.4×10^3 km.

1.53 A helicopter blade is designed to rotate at such an angular velocity that the tip of the blade is less than the speed of sound in air, which is 340 m s^{-1}. The blades rotate at 260 r.p.m. Calculate the maximum possible length of the blades.

1.54 A compact disc (CD) player varies the rate of rotation of the disc in order to keep the track from which the music is being reproduced moving at a constant linear speed of 1.30 m s^{-1}. Calculate the rates of rotation of a 12.0 cm disc when the music is being read from **(a)** the outer edge of the disc **(b)** a point 2.55 cm from the centre of the disc. Give your answers in rad s^{-1} and in rev min^{-1}.

1.55 The diagram shows a particle moving in a circle ABCDA of radius 8.0 m at a constant speed. It completes one revolution in 5.0 s. What is
(a) its average speed for one revolution
(b) its average speed from A to B
(c) its average velocity for one revolution
(d) its average velocity from A to C
(e) its average velocity from A to B
(f) its change of velocity from A to C
(g) its change of velocity from A to B?

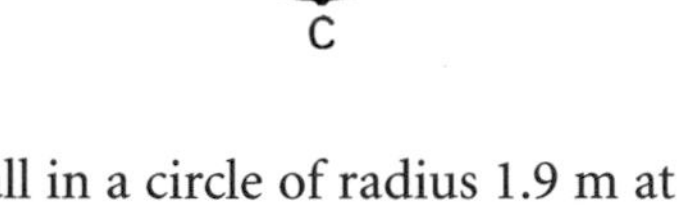

1.56 In throwing the hammer an athlete whirls a steel ball in a circle of radius 1.9 m at 2.4 rev s^{-1}. Calculate the acceleration of the ball.

1.57 A geosynchronous satellite makes a complete circular orbit once every 24 hours. The radius of the orbit is 4.2×10^7 m.
(a) What is its angular velocity? **(b)** What is its speed?
(c) What is its centripetal acceleration?

1.58 The designer of a loop-the-loop ride in an amusement park wants to create an acceleration of 6*g* at the bottom of a loop of radius 8.0 m.
At what speed should the car be moving to produce this acceleration?

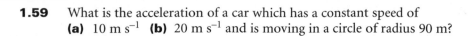

1.59 What is the acceleration of a car which has a constant speed of
(a) 10 m s^{-1} **(b)** 20 m s^{-1} and is moving in a circle of radius 90 m?

1.60 A space station is made roughly in the shape of the inner tube of a motor car tyre. The diagram shows a sectional view across a diameter of the station.

(a) With what period must the space station be rotated about the axis AA' in order that someone living inside it may experience a centripetal acceleration equal to *g*?

(b) What are the problems likely to be experienced by an astronaut living and working in such a rotating space station?

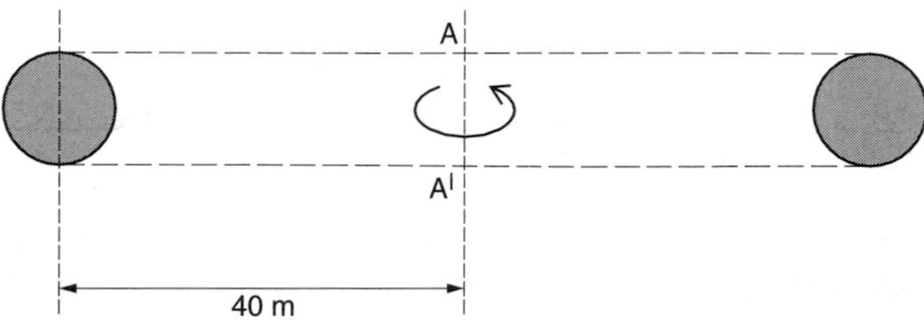

2 Balanced and unbalanced forces

Data: gravitational field strength $g = 9.81\ \text{m s}^{-2} = 9.81\ \text{N kg}^{-1}$

2.1 Forces in equilibrium

In this section you will need to

- understand that all forces are pushes or pulls of one body on another
- use the phrase 'the push or pull of A on B' when describing any particular force
- understand the meaning of the words weight and tension and of frictional and normal contact forces
- draw free-body force diagrams when analysing problems about bodies which are in equilibrium
- understand that when a body is in equilibrium the sum of the forces acting on it, resolved in any direction, is zero
- understand that forces occur in pairs which act on different bodies and that the push or pull of A on B is always equal in size to the push or pull of B on A
- understand how to resolve forces into two mutually perpendicular components and how to add two forces which are perpendicular to one another.

2.1 A new-born baby is said to be a healthy 7.8 pounds. What is the baby's weight in newtons? Take 2.2 lb = 1.0 kg.

2.2 The body shown in each of the following free-body force diagrams is in equilibrium. Write down the value(s) of the unknown force(s) in each case.

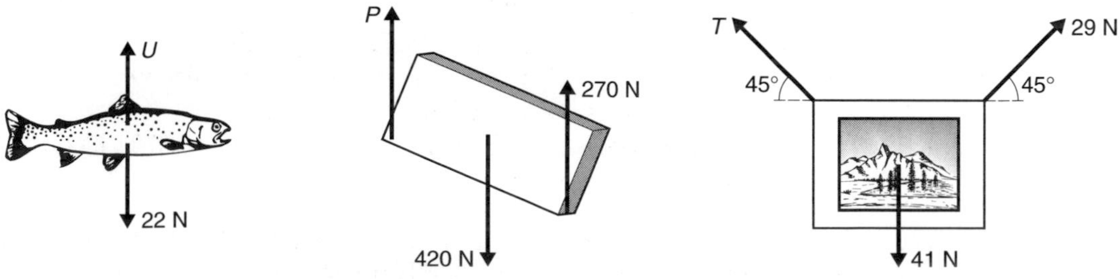

2.3 The diagram shows a man pulling a briefcase across a table. Free-body force diagrams for the briefcase and for the man are also drawn.

(a) Describe each of the forces acting on the briefcase with phrases like '*B* is the frictional contact push of the table on the briefcase'.

(b) Describe each of the forces acting on the man with phrases like '*F* is the normal contact push of the floor on the man'.

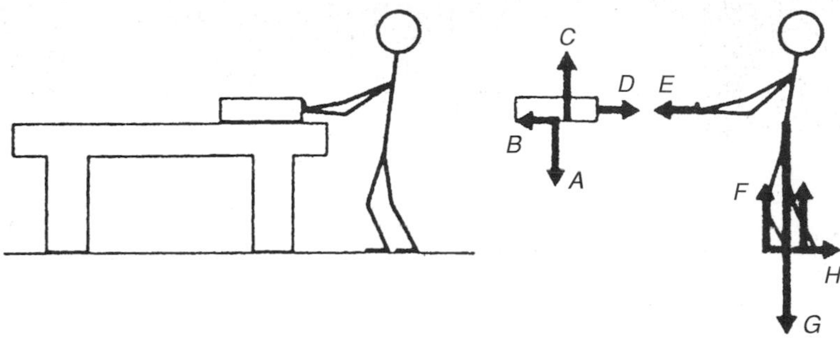

2.4 A child sits at rest on a swing. The figure shows free-body force diagrams for **(i)** the child **(ii)** the swing seat.

(a) For each of the forces *P*, *W*, *T*, *w* and P', identify the body which is producing the force. [P' is *not* produced by the Earth.]

(b) Write a phrase describing each force as the push or pull of the identified body on **(i)** the child **(ii)** the seat.

(c) In this situation $P = W$, $T = w + P'$ and $P = P'$. Explain each equation in terms of Newton's laws.

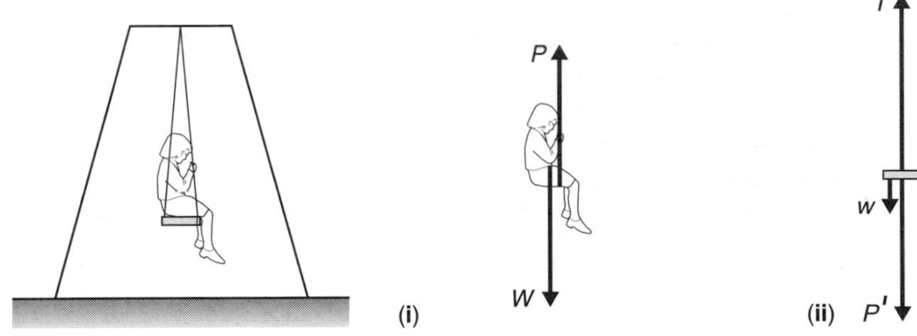

2.5 A child learning to swim is supported by her instructor who stands on the side of the pool. The forces acting on the child are:
the pull *W* of the Earth on the child, 300 N
the pull *P* of the harness on the child
the push *U* of the water on the child, 250 N.

(a) Draw a free-body force diagram for the child. How big is *P*?

(b) Newton's third law tells us that there are other forces equal in size to *W*, *P* and *U*. On which bodies do each of these forces act?

(c) Draw a free-body force diagram for the instructor who weighs 800 N. Deduce the normal contact push of the floor on him.

2.6 The photo on the next page shows two skaters. At the moment shown the woman is exerting a horizontal force on the man who is moving at a constant velocity.

(a) Draw a free-body force diagram for the man.

(b) If the man weighs 700 N and his partner's horizontal pull on him is 50 N, what are the sizes of **(i)** the normal contact push of the ice on the man **(ii)** the frictional contact push of the ice on the man **(iii)** the total push of the ice on the man?

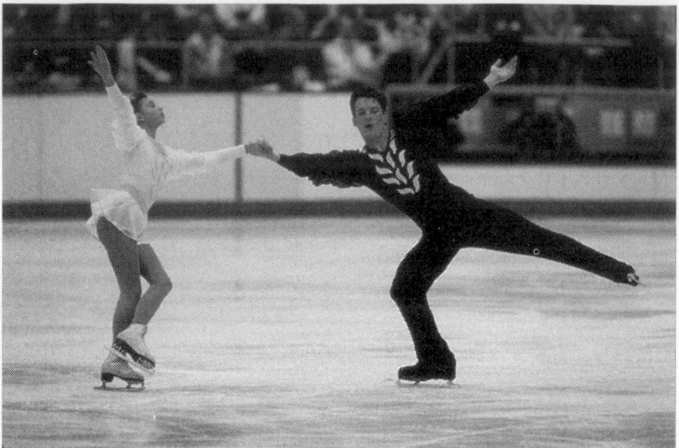

2.7 A racing car is shown in the diagram together with a free-body force diagram describing the forces acting on it.

(a) Copy the free-body force diagram and list the forces using phrases which end with 'on the car'. [Q is *not* the push or pull of the engine on the car.]

(b) How big are the forces P and Q?

(c) What would happen to the size of **(i)** W **(ii)** F, if the car was moving more slowly, e.g. at 70 m.p.h.?

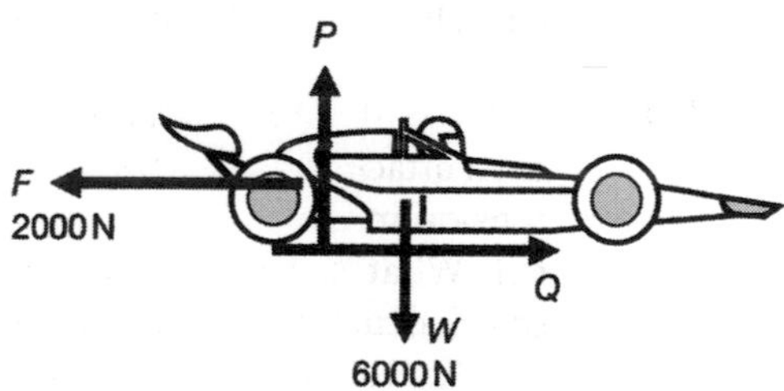

2.8 A man pushes a lawnmower with a force of 80 N directed along the handle which makes an angle of 40° to the vertical.
What is the size of the resolved part of this force
(a) in the horizontal direction **(b)** in the vertical direction?

2.9 A force P is 20 N in a direction N 60° E. What is the resolved part of the force
(a) north **(b)** east **(c)** N 30° E?

2.10 An empty sledge of mass 5.0 kg slides at a constant speed down a slope which makes an angle of 20° to the horizontal.

(a) Draw a free-body diagram for the sledge and add the forces acting on the it.

(b) By resolving the weight of the sledge parallel and perpendicular to the slope calculate the size of each of the forces in your free-body force diagram.

2.11 In designing a supermarket there is sometimes a need for connecting ramps between different parts of the shop. It is decided to limit the push required by shoppers to push a 24 kg trolley to less than 25 N. What is the maximum angle at which the ramps can be built?

2.12 A gymnast of weight 720 N is holding himself in the cross position on the high rings. He is quite still. A free-body force diagram for the gymnast shows the two upward pulls of the rings on his hands, each of size 380 N. Calculate the angle between the wires supporting the rings and the vertical

2.13 A very heavy sack is hung from a rope and pushed sideways. When the sideways push is 220 N the rope supporting the sack is inclined at 18° to the vertical.

(a) Calculate the tension in the rope.

(b) Hence find the mass of the sack.

2.14 A stone of weight 32 N is attached to a wire and hung from a rigid support. A string is then attached to the bob and is pulled sideways with a horizontal force P until the tension in the wire is 44 N.

(a) What is the angle between the wire and the vertical?

(b) Calculate the size of the force P.

(c) Repeat your calculations for the case when the tension reaches 140 N.

2.15 A framed picture of weight 15 N is to be hung on a wall, using a piece of string. The ends of the string are tied to two points, 0.60 m apart on the same horizontal level, on the back of the picture. Draw a free-body force diagram for the picture, and find the tension in the string if

(a) the string is 1.0 m long **(b)** the string is 0.66 m long.

2.2 Forces and moments

In this section you will need to

- use the equation: moment of a force about an axis = the force times the perpendicular distance from the axis to the line of action of the force
- draw free-body force diagrams for extended bodies in equilibrium when solving problems involving the principle of moments
- use the principle of moments: for a body in equilibrium the sum of the moments of the forces acting on it, about any axis, is zero
- understand that couples and torques exert moments on extended bodies.

2.16 The diagram shows a boy and a girl on a seesaw plus a free-body force diagram for the seesaw beam.

(a) Calculate the moment, about O, of the 400 N push of the girl on the beam.

(b) What is the moment, about O, of the 300 N pull of the Earth on the beam?

(c) Deduce a value for *F*, the push of the boy on the beam.

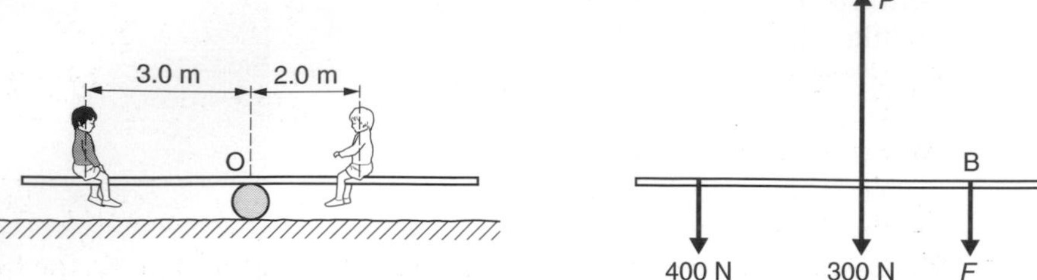

2.17 Using the diagram from the previous question:

(a) Calculate the moments, about B, of the 400 N and the 300 N forces.

(b) Deduce a value for *P*.

2.18 Two men are walking carrying a ladder of length 12 m and weight 200 N: it may be considered to be uniform. The first man is 2.0 m from the end of the ladder; the second man is at the end.

(a) Draw a free-body force diagram for the ladder and calculate the force which each man exerts on it.

(b) If the second man gradually slides his point of support forward until he is 4.0 m from the back, suggest how, in general terms, the force he exerts changes.

(c) How far from the back of the ladder is he when the men exert equal forces?

2.19 The diagram shows a lorry of mass 45 000 kg crossing a bridge. The span from A to B is 32 m.

(a) When the lorry is at the centre of the bridge, what is the increase in the upward push of the bridge supports A and B on the bridge structure?

(b) Where on the bridge will the lorry be when the additional upward push of the bridge support at A is 120 000 N?

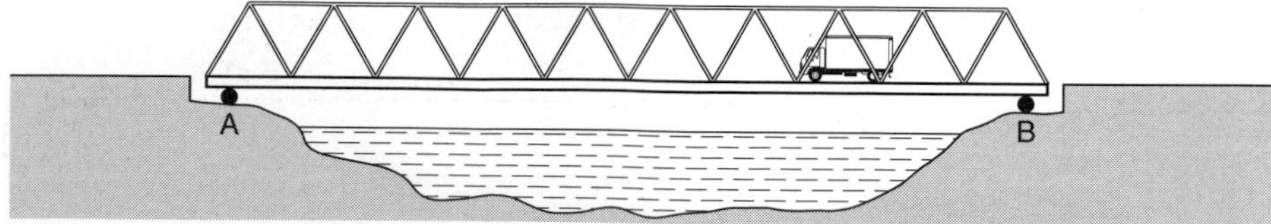

2.20 A baby-buggy and baby together weigh 140 N. The diagram shows the position of the centre of gravity G of the baby and buggy.

(a) In order to lift the front wheels up a step whilst moving forward the pusher exerts a downward force *F* on the handle. Calculate the minimum value of *F*.

(b) What is the corresponding upward force the pusher needs to exert to lift the back wheels of the pram up a step whilst moving backwards?

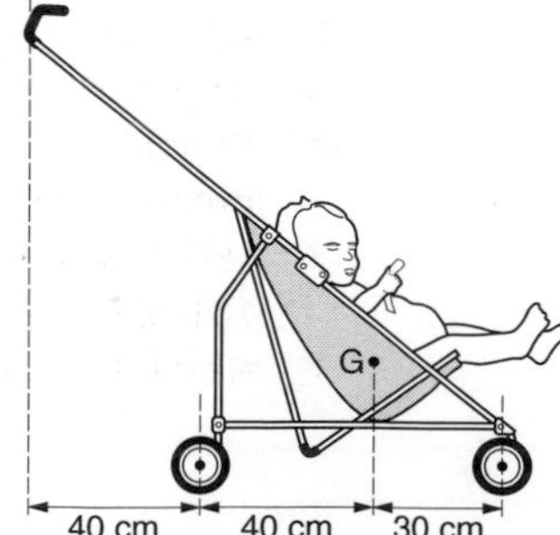

2.21 The diagram shows a simplified model of the forearm alongside a centimetre scale.

(a) Estimate how far the centre of gravity of the forearm is from the elbow joint.

(b) The forearm has a total mass of 1.7 kg. What is the moment, about the elbow joint, of the weight of the forearm?

(c) Estimate the distance of the biceps muscle from the elbow joint.

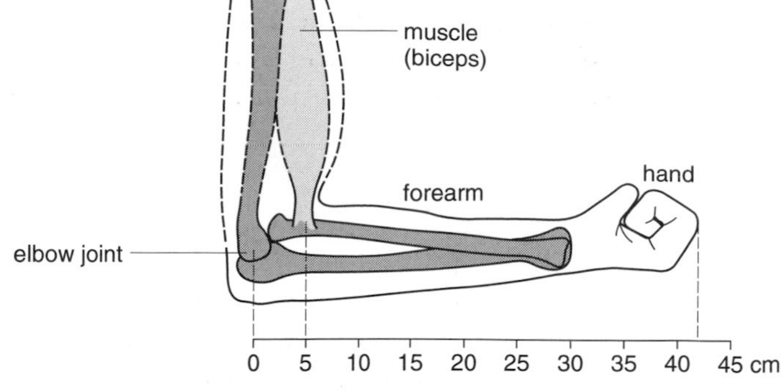

(d) Calculate the tension in the biceps muscle necessary to hold the forearm horizontal.

(e) How big is the vertical contact force between the bones at the elbow joint?

2.22 In order to find the mass m of a heavy clampstand, it is supported in equilibrium as shown in the diagram. G is its centre of gravity.

(a) Draw a free-body force diagram for the clampstand.

(b) Calculate m if, when x is 42.0 cm, the reading on the spring balance is 4.6 N. Take a to be 6.5cm.

(c) Plot a graph showing the relationship between F, the reading on the balance, and x as the balance is moved along the clampstand rod.

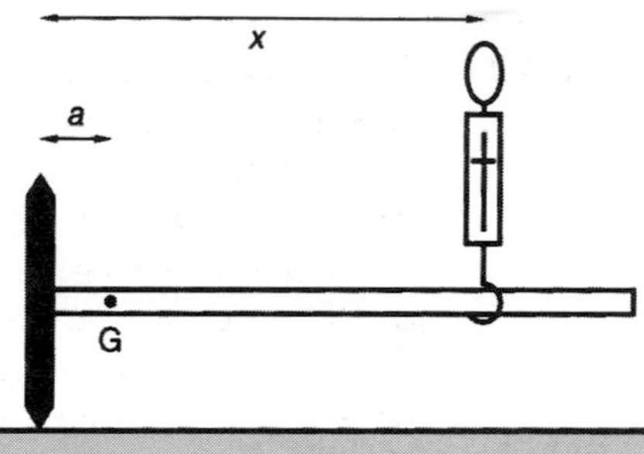

2.23 **(a)** Describe how you would use the apparatus in the diagram to locate the position of the centre of gravity, relative to his feet, of a person with his arms at his sides.

(b) Estimate the likely errors in the measurements you would take and hence the uncertainty which could be achieved in the experiment.

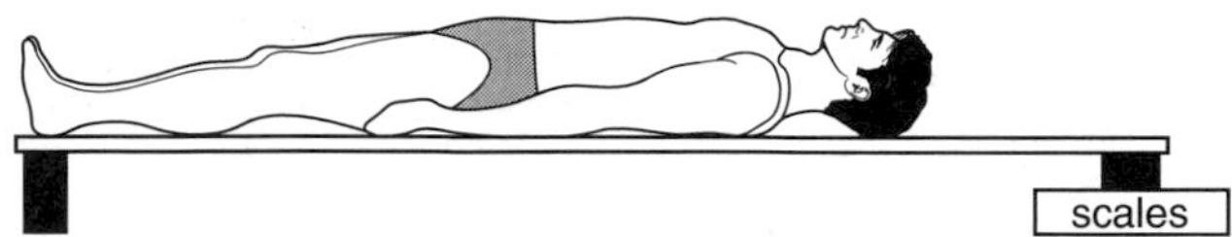

2.24 A ladder of mass 15.0 kg leans with its upper end against a frictionless wall as shown on the next page.

(a) Describe the forces N and F shown on the free-body force diagram of the ladder.

(b) By taking moments about the bottom of the ladder, calculate M, the normal contact push of the wall on the top of the ladder.

(c) Write down the sizes of the forces N and F.

(d) Calculate **(i)** the total push of the ground on the bottom of the ladder **(ii)** the angle which this force makes with the vertical.

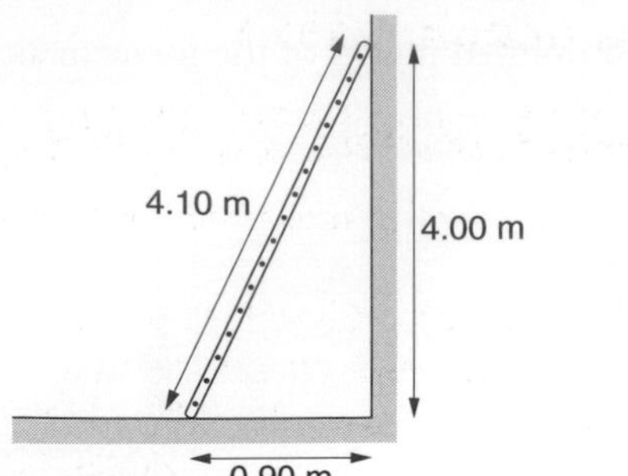

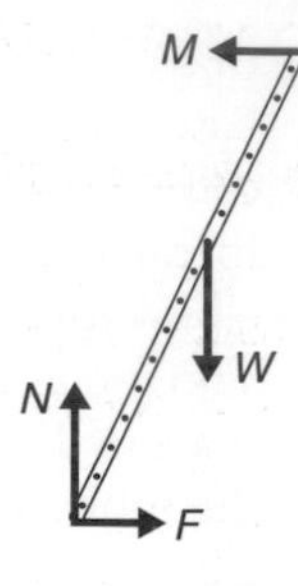

2.25 A uniform heavy window, freely hinged along its top edge, is swung open to 35° from the vertical by pulling horizontally on a handle at its bottom edge.

(a) Draw a free-body force diagram for the window seen sideways on.

(b) If the weight of the window is 80 N, calculate the pull on its bottom edge.

(c) What is the size of the total pull of the hinge on the top of the window?

2.26 A uniform horizontal beam of weight 200 N and length 2.5 m is freely hinged at one end to a wall. A rope is attached to the free end. The other end of the rope is attached to the wall at a point vertically above the beam so that the rope makes an angle of 30° with the vertical.

(a) Draw a free-body force diagram for the beam.

(b) Calculate the tension in the rope.

(c) Calculate the horizontal and vertical resolved parts X and Y of the push of the wall on the beam.

(d) Show on your diagram the direction of the resultant of X and Y.

2.27* Two men are carrying a large stone slab of side $2l$ up steps which are inclined at an angle θ to the horizontal. A free-body force diagram for the box is shown, the centre of gravity of the box being at G in the middle of the uniform slab. P and Q are the upward pull the front man and rear man exert on the slab.

(a) Show that, when $\theta = 30°$, $Q = 3P = 0.75W$.

(b) Sketch a graph to show how Q varies with θ as θ rises from 0 to 45°.

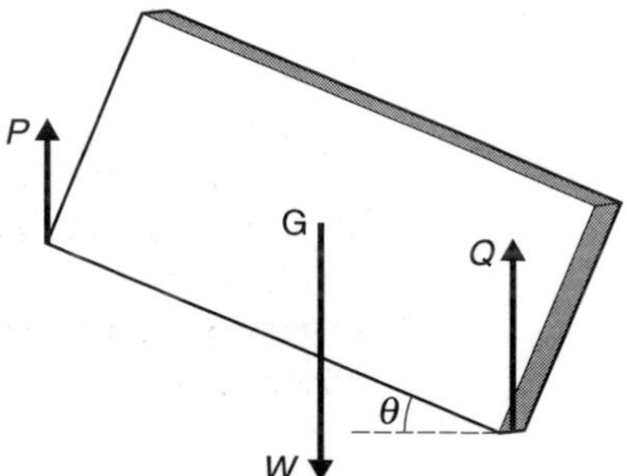

2.3 Forces causing acceleration

In this section you will need to

- draw a free-body diagram for the accelerated body when solving problems involving Newton's second law
- use the equation, Newton's second law, for accelerated motion

$$ma = \text{ sum of forces resolved parallel to } a$$

- describe experiments to demonstrate the validity of Newton's second law
- understand that all animals and vehicles accelerate forwards by pushing the ground, some water or the air backwards.

2.28 The diagram shows five bodies **(a)** – **(e)** together with the mass of each. Calculate the acceleration of each body.

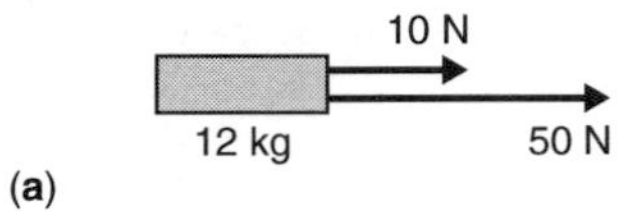

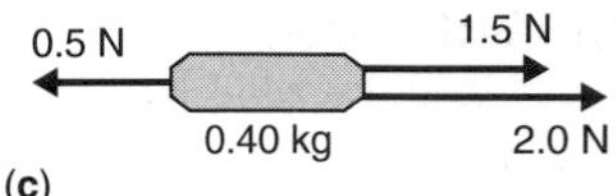

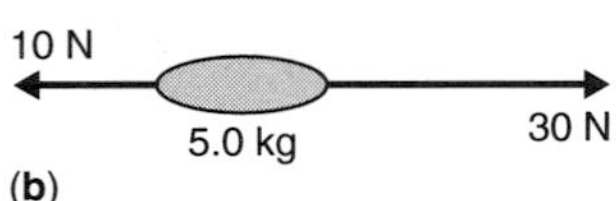

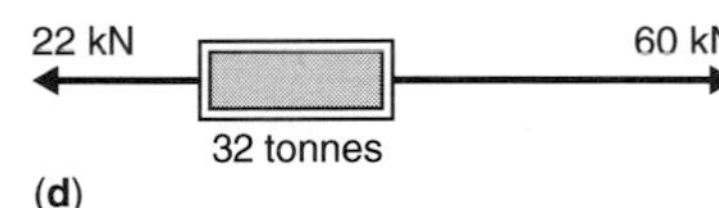

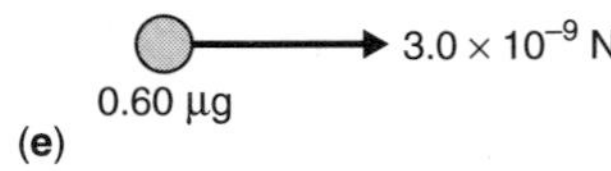

2.29 Calculate the extra force, in size and direction, which would need to be added to each of the situations above in order to produce an acceleration of 2.0 m s^{-2} to the right for every object.

2.30 Two women push a car of mass 800 kg to get it started. Each pushes with a force of 300 N and the resistance forces are equivalent to an opposing force of 160 N. What is the acceleration of the car?

2.31 A person is unlikely to be killed in a car crash if, held by a seatbelt, he or she accelerates at –250 m s^{-2} or less.

(a) What is the pull exerted by a seatbelt on **(i)** a man of mass 84 kg **(ii)** a child of mass 32 kg, at this maximum safe acceleration?

(b) Express your answers to **(a)** as a multiple of the person's weight.

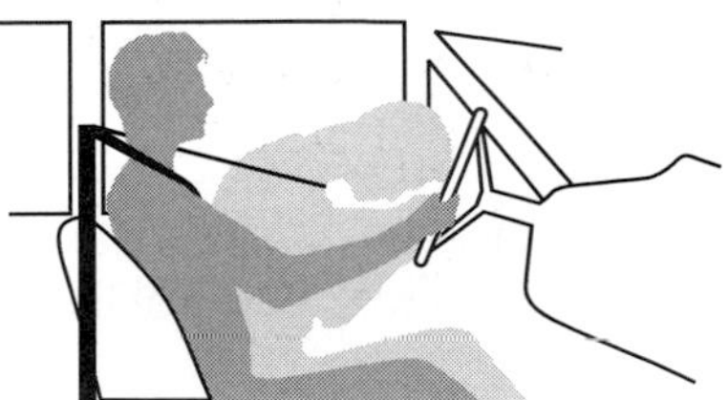

2.32 A tractor pulls a log of mass 2000 kg. When the tractor is pulling with a force of 1300 N the acceleration of the log is 0.050 m s^{-2}. What resistance force does the ground exert on the log?

2.33 Superman slams head-on into a train speeding along at 30 m s^{-1}, bringing it to rest in an amazing 0.010 s and saving Lois Lane, who was tied to the tracks ahead of the train.

(a) What is the acceleration of the train?

(b) If the train's mass is 200 tonnes (2.0×10^5 kg), what is the push which Superman exerts on the train?

2.34 In one 10 minute interval during the *Apollo 11* flight to the Moon the spacecraft's speed decreased from 5374 m s^{-1} to 5102 m s^{-1}(with the rocket motors not in use). The mass of the space craft was 4.4×10^4 kg. Calculate the average force exerted on the spacecraft during this time.

2.35 A free-body force diagram for a rear wheel drive car is shown in the diagram.

(a) Write a phrase describing each of the forces S and T.

(b) How are S and T related when the car is moving **(i)** at a constant velocity v **(ii)** with constant acceleration a?

(c) Describe what happens to the forces S and T when the driver applies the brakes and slows the car down.

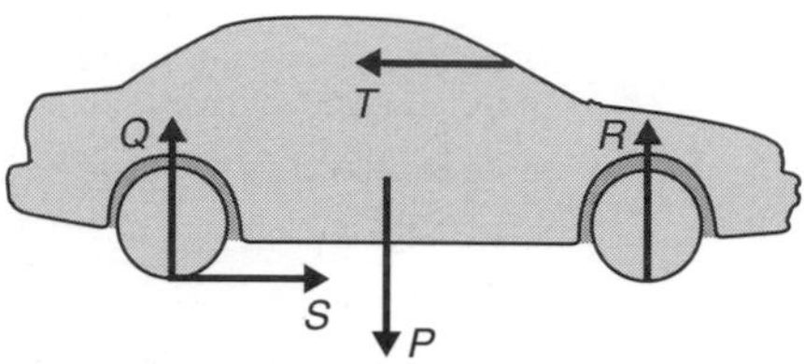

2.36 A packing case of mass 50 kg rests on a rough horizontal floor. The surface is such that the maximum frictional force which the floor can exert on the case is 0.40 times the perpendicular contact force. Describe what happens when someone pushes horizontally on the packing case with a force of **(a)** 98 N **(b)** 196 N **(c)** 294 N.

2.37 Just after the gun a sprinter of mass 65 kg is pushing against the starting block with a force of 800 N. This force acts at an angle of 65° to the horizontal.
Calculate

(a) **(i)** the resultant horizontal force acting on her
(ii) the resultant vertical force acting on her

(b) **(i)** the forward acceleration of her centre of gravity
(ii) the upward acceleration of her centre of gravity.

2.38 An articulated lorry consists of a tractor unit of mass 4.0 tonnes and a trailer of mass 26 tonnes. The lorry accelerates at 0.20 m s^{-2}.

(a) Ignoring all resistive forces calculate
(i) the forward push of the road on the driving wheels of the tractor unit
(ii) the forward pull of the tractor unit on the trailer.

(b) Draw separate free-body force diagrams for the tractor and trailer. Which pair of forces are equal because of Newton's third law?

2.39 A sprinter of mass 60 kg reaches her top speed of 12 m s^{-1} in the first 15 m of her run.

(a) Calculate her acceleration during this process.

(b) Hence calculate the horizontal force (assumed to be constant) which the ground has been exerting on her.

(c) What size horizontal force has she been exerting on the ground? Explain.

2.40 Explain how, in the laboratory, you would demonstrate that the acceleration of a body is inversely proportional to its mass for a fixed resultant force.
State how you would process and present any measurements so as to achieve the aim of the demonstration.

2.41 A javelin thrower accelerates a javelin of mass 0.80 kg from 5.5 m s^{-1} to 31.5 m s^{-1} in 0.30 s. What average pull does he exert on the javelin?

2.42 The table gives data for a skydiver during the first phase of the jump:

time from start of jump/s	0	3.0	6.0	9.0
vertical velocity/m s^{-1}	0	28	46	53
vertical acceleration/m s^{-2}	9.8	7.9	4.0	1.0

The skydiver has a mass of 85 kg.

(a) Calculate at each of these four times
  (i) the resultant force on the skydiver
  (ii) the upward push of the air on the skydiver.

(b) Draw four free-body force diagrams showing the forces acting on the skydiver at each time. Draw the force arrows (vectors) in proportion to the size of the forces. Add the velocity of the skydiver at each time to your diagrams.

(c) Explain why it is *not* sensible to say that a skydiver is a free-fall parachutist.

2.43 A lift has a mass of 1200 kg. Calculate the tension in the cable supporting the lift when the lift is

(a) ascending at a constant velocity
(b) ascending with an upward acceleration of 2.0 m s^{-2}
(c) descending with a downward acceleration of 3.0 m s^{-2}
(d) descending with an upward acceleration of 3.0 m s^{-2}.

2.44 The table gives the results of a standing-start acceleration test for a car of mass 1100 kg.

t/s	0	2	4	6	8	10	12
v/m s^{-1}	0	9	15	19.5	23	26	29

(a) Draw a graph of its speed v against time t.

(b) Estimate the resultant force acting on the car when its speed was **(i)** 15 m s^{-1} **(ii)** 25 m s^{-1}. [Hint: you will have to draw tangents to your graph.]

2.45 The diagram shows a trampolinist at the bottom of her jump where she is instantaneously stationary. Her mass is 65 kg and her upward acceleration at this instant is 85 m s^{-2}. Calculate the upward push of the trampoline on her at this instant.

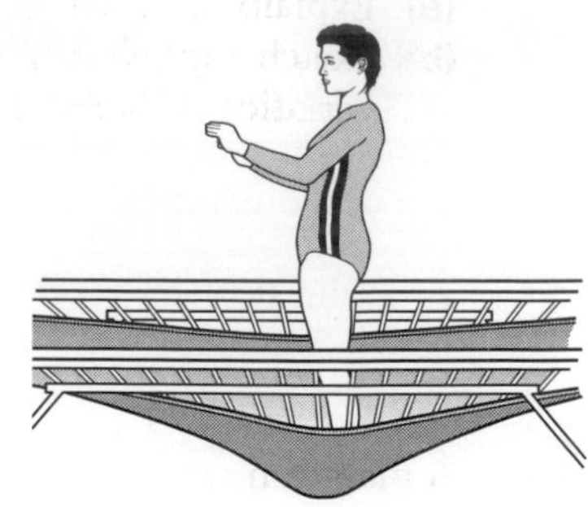

2.46 The flea in question 1.25 has a mass of 0.50 mg and its take-off acceleration is 800 m s^{-2}. Calculate the average push of the ground on the flea as it jumps.

2.47 A boy catches a cricket ball of mass 160 g which is moving at 20 m s^{-1}.
(a) Find the force which he must exert to stop it in **(i)** 0.10 s **(ii)** 0.50 s.
(b) Describe how he can vary the time in this way, and explain the advantage of lengthening the time in which the ball is stopped.
(c) Describe two other situations (as different as possible from this one) in which care is taken to lengthen the time in which a moving object is brought to rest.

2.48 The diagram shows a block on a horizontal frictionless table. A thread attached to it runs horizontally to a pulley at the edge of the table, passes over the pulley, and supports a load of mass 1.0 kg. The size of the acceleration of both the block and the load is 2.0 m s^{-2}.
(a) Draw free-body force diagrams for **(i)** the block **(ii)** the load, labelling as T the pull of the thread on each of these bodies.
(b) Use the free-body force diagram for the load to find the size of T.
(c) Now use the free-body force diagram for the block to find the mass of the block.

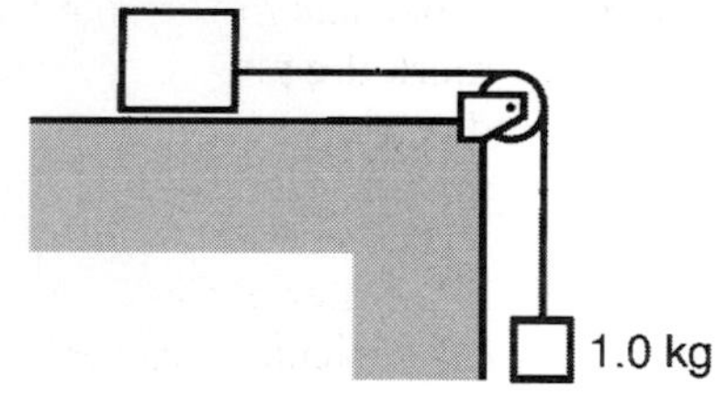

2.49 A pendulum bob hangs by a thread from the roof of a railway carriage. Describe and explain, using free-body force diagrams for the pendulum bob, what happens to the bob when the train is
(a) accelerating forwards **(b)** moving at a constant velocity **(c)** slowing down.
For **(a)** make any possible calculations if the train acceleration is 0.80 m s^{-2}.

2.50 Three boxes in contact are being pushed along a system of rollers. The two outer boxes each have a mass of 140 kg and the middle box has a mass of 80 kg. The single pushing force is 220 N. By drawing suitable free-body force diagrams, calculate
(a) the acceleration of the boxes
(b) the push of each of the outer boxes on the middle box.

2.51 In the diagram the dashed line shows the weight of a parachutist as she falls: her weight is constant. The solid line shows the size of the air resistance (drag) force on her (and her parachute, when open) as she moves downwards. She opens her parachute at time t_1.

(a) Explain the shape of the graph.
(b) Sketch a graph to show the variation of her acceleration with time.

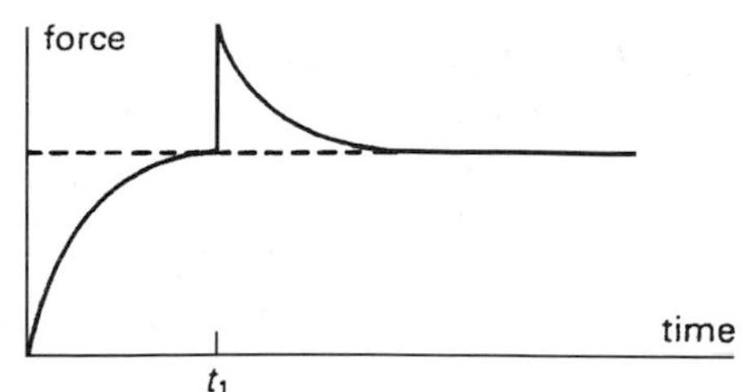

2.52 A man of mass 65 kg stands on a weighing machine in a lift which has a downwards acceleration of 3.0 m s^{-2}. What is the reading on the weighing machine? Make it clear at what stage you need to use Newton's third law and explain why it does not matter whether the lift is moving up or down.

2.4 Centripetal forces

In this section you will need to

- understand that the resultant force acting on a body moving in a circle at a constant speed is directed towards the centre of the circle
- draw free-body force diagrams and use Newton's second law in the form $mv^2/r = F_{res}$ for uniform circular motion.

2.53 A particle moves in a circular path at a constant speed. Draw a diagram of the path, mark the position of the particle at some point on the path and, for that point, show the direction of
(a) its velocity **(b)** its acceleration **(c)** the resultant force on it.

2.54 A sprinter of mass 70 kg is cornering at 9.4 m s^{-1} during a 200 m race.
(a) Calculate the centripetal force needed to corner at this speed if the radius of the bend is 38 m.
(b) What provides this centripetal force? Sketch a free-body force diagram for the sprinter seen from the front.

2.55 Refer to question 1.56 and take the mass of the ball to be 7.3 kg. Calculate the pull of the ball on the wire, assuming that the weight of the ball is negligible.

2.56 A child of mass 30 kg is playing on a swing. Her centre of mass is 3.2 m below the supports when she moves through the bottom of her swing at 6.0 m s^{-1}.
(a) Draw a free-body force diagram for the child at this moment.
(b) Calculate **(i)** her centripetal acceleration **(ii)** the push of the seat of the swing on her.

2.57 In the free-body force diagram, over the page, for a bicycle and cyclist seen from the front the push of the road on him has been resolved into two components: N the normal contact push of the ground on him and F the frictional push of the ground on him.
(a) The total mass of cyclist plus bicycle is 95 kg. Explain why $N = 930$ N.
(b) He is moving at 18 m s^{-1} in a circle of radius 50 m. Calculate F.
(c) What is **(i)** the resultant force acting on the bicycle and cyclist **(ii)** the resultant push of the ground on him?

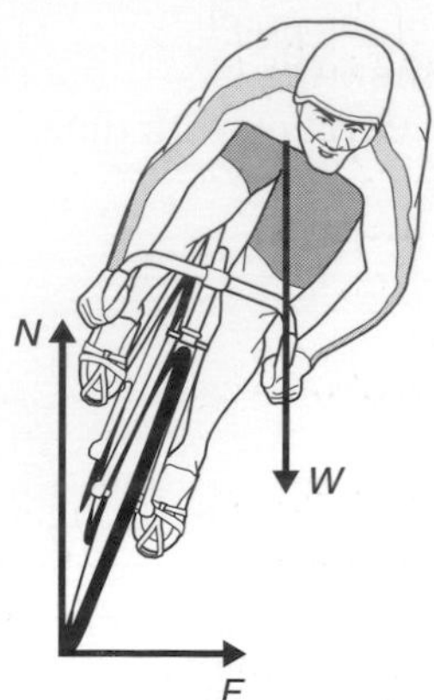

2.58 The designer of an amusement park ride wants a centripetal acceleration of 19.6 m s^{-2} at the top of a loop of radius 7.0 m.

(a) Calculate the minimum speed he must ensure the car has at the top of the loop.

(b) The free-body force diagram for the car shows the two forces acting on it at the top of the loop. The resultant centripetal force is $P + W$.

(i) Describe the two forces as the push or pull of something on the car.

(ii) Calculate W and P for a car of mass 400 kg.

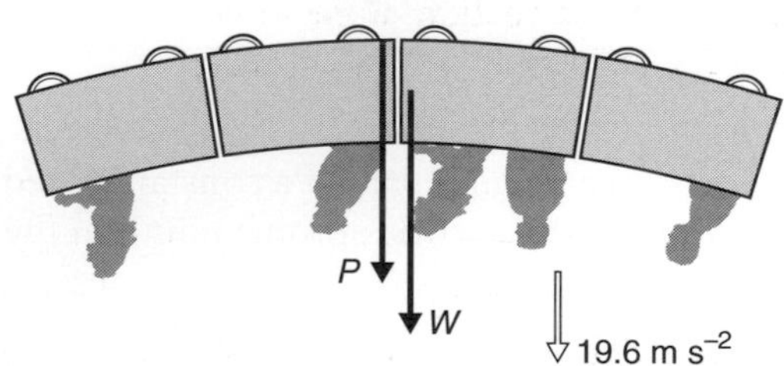

2.59 A car of mass 900 kg is driven over a hump-backed bridge at a speed of 18 m s^{-1}. The road surface of the bridge forms part of a circular arc of radius 50 m.

(a) Draw a free-body force diagram for the car when it is on top of the bridge.

(b) Calculate the push of the road on the car then.

(c) What is the greatest speed at which the car may be driven over the bridge if its wheels are not to lose contact with the road?

2.60 Draw a sketch of a man of mass 75.0 kg standing on a weighing machine (a pair of scales calibrated in newtons) at the Earth's equator where, because the Earth is rotating, he has a centripetal acceleration of 0.034 m s^{-2}.

(a) What is the pull of the Earth on him? Take g to be 9.780 N kg^{-1}.

(b) Draw a free-body diagram for the man and add the forces acting on him.

(c) Calculate the push of the weighing machine on him.

(d) What will the weighing machine record as his weight? Explain in words why this is not equal to mg.

2.61 A metal bob is whirled on the end of a string once every second as shown in the diagram. A free-body force diagram for the bob is also shown.

(a) Calculate
  (i) the radius of the circle in which it is moving
  (ii) its constant speed v as it moves round the circle
  (iii) the size of its centripetal acceleration a.

(b) The bob has a mass of 0.20 kg. Calculate
  (i) the size of the centripetal force needed to produce the acceleration
  (ii) the pull of the string on the bob T.

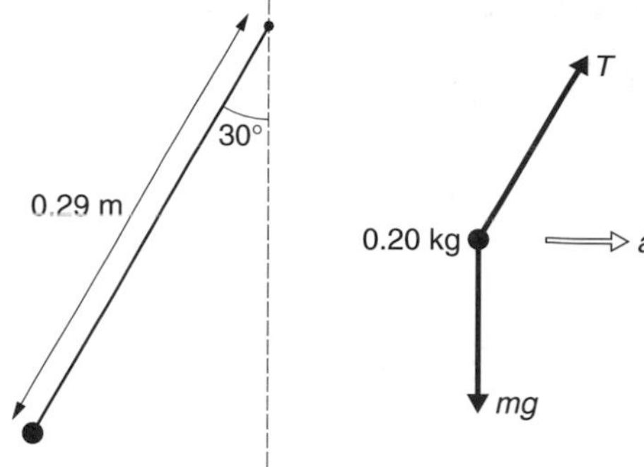

2.62* The free-body force diagram shows a racing car of mass 800 kg moving in a circle of radius 300 m on a track banked at an angle $\theta = 26°$. P and F represent the total perpendicular contact force and the total frictional force of the track on the car.

(a) At a certain speed v, the frictional force F is zero. For this speed, calculate
  (i) the size of the force P
  (ii) the value of v.

(b) Explain how the direction of the force F depends on the speed of the car for speeds greater and less than v.

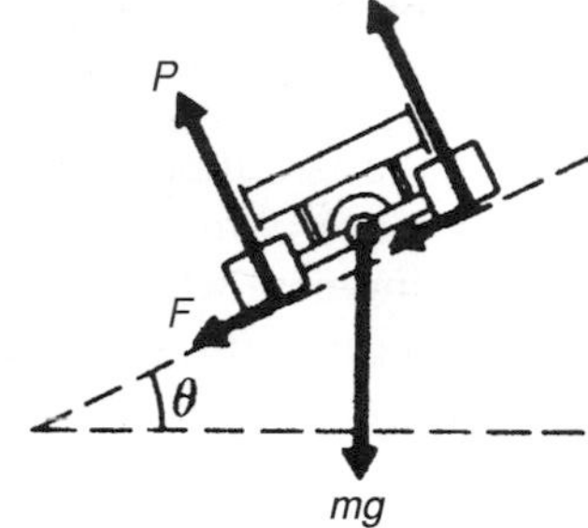

2.63* An aeroplane of mass m is moving at a constant speed v in a horizontal circle of radius r. It does this by banking at an angle θ to the horizontal. The diagram is a free-body force diagram for the plane. Show that $v = \sqrt{(rg\tan\theta)}$.

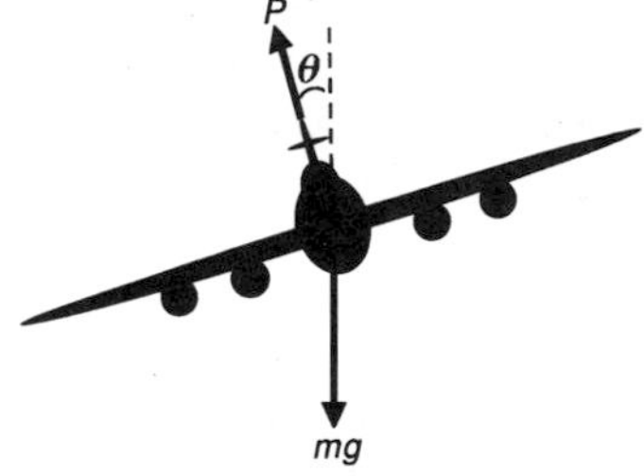

Other exercises on particles moving in circles appear in Chapter 19 on electromagnetism and in Chapter 20 on inverse square law fields.

3 Linear momentum

3.1 Conservation of linear momentum

In this section you will need to

- remember that linear momentum, defined as mass times velocity, is a vector quantity
- understand that linear momentum is conserved in collisions and explosions provided no external forces act during the interaction
- draw sketches when applying the principle of conservation of linear momentum which show what is happening immediately before and after an interaction
- describe experiments to demonstrate the validity of the principle of conservation of linear momentum.

3.1 What is the momentum of
(a) a boy of mass 50 kg running round a track at a constant speed of 3.0 m s^{-1} at the moment when he is **(i)** moving to the north **(ii)** moving to the south
(b) a car of mass 800 kg moving east at a speed of 25 m s^{-1}
(c) an oil tanker of mass 250 000 tonnes which is moving west at a speed of 20 m s^{-1}?

3.2 Explain why, when a stationary object explodes into two pieces,
(a) the pieces move off in opposite directions
(b) the piece with the smaller mass moves off with the greater speed.

3.3 The unit for momentum, kg m s^{-1}, can also be written as N s, i.e. newton times second. Explain why these units are equivalent.

3.4 Calculate the unknown velocity v in the following collisions between two trolleys seen from above. The diagrams show what is happening before and after the collision in each case.

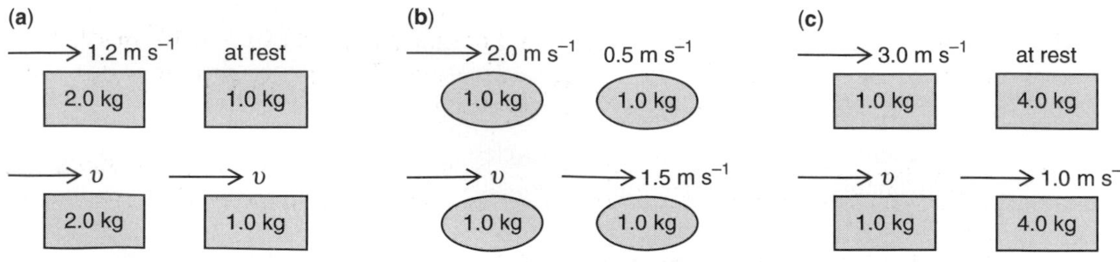

3.5 A woman of mass 60 kg steps out of a canoe of mass 40 kg onto the river bank.
(a) If she steps out at a speed of 2.0 m s^{-1}, what happens to the canoe?
(b) Can she step out without the canoe recoiling?

3.6 Two air-track gliders are at rest with a spring compressed between them. A thread tied to each prevents them moving apart. When the thread is burned, one glider moves away with a speed of 0.32 m s^{-1}, and the other moves in the opposite direction with a speed of 0.45 m s^{-1}. If the first glider has a mass of 0.40 kg, what is the mass of the other?

3.7 An army rifle fires an 11 g bullet with a muzzle speed of 880 m s^{-1}.
(a) Draw diagrams of the rifle and bullet before and after the explosion.
(b) If the rifle has a mass of 3.2 kg, with what speed does it recoil?

3.8 In the diagram two trolleys are at rest and are then 'exploded' by the release of a spring-loaded piston or plunger at the front of one of them. Each trolley has a mass of 1.0 kg, but one carries a block of unknown mass *m*. The result of the explosion is also shown. Calculate *m*.

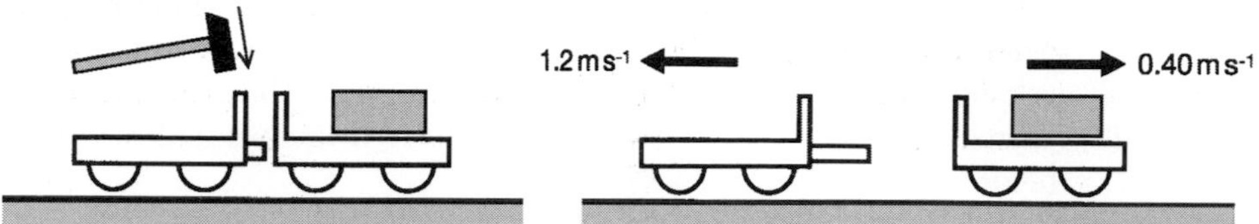

3.9 A rugby player of mass 70 kg, running south at 6.0 m s^{-1}, tackles another player whose mass is 85 kg and who is running directly towards him at a speed of 4.0 m s^{-1}. If in the tackle they cling together, what will be their common velocity immediately after the tackle?

3.10 A loaded supermarket trolley was rolled towards a stationary stack of five empty trolleys. After colliding and linking to them it was noticed that the speed of the six trolleys was half the speed of the loaded trolley.
An unloaded trolley has a mass of 8.0 kg. Calculate the mass of the goods in the loaded trolley.

3.11 A bullet of mass 10 g is fired into a block of wood of mass 200 g and becomes embedded in it. If the speed of the bullet was 500 m s^{-1}, what is the speed of the block immediately after the impact?

3.12 When a bullet hits a 'baddie' in the chest in a TV western, he sometimes slumps forward. As a physics student you know he ought to jerk backwards. Explain why.

3.13 An α-particle is emitted from a polonium nucleus at a speed of 1.8×10^7 m s^{-1}. The relative masses of the α-particle and the remaining part of the nucleus are 4.002 and 212.0. Calculate the recoil speed of the nucleus.

3.14 The graph on the next page shows how the momentum of two colliding railway trucks varies with time. Truck A has a mass of 20 tonnes (20 000 kg) and truck B has a mass of 30 tonnes.
(a) What is the *change* of momentum of **(i)** truck A **(ii)** truck B?
(b) Calculate **(i)** the initial velocities **(ii)** the final velocities of the two trucks.
(c) Calculate the *total* momentum of the two trucks at t/s = 0.4, 0.8, 1.2 and comment on your answers.

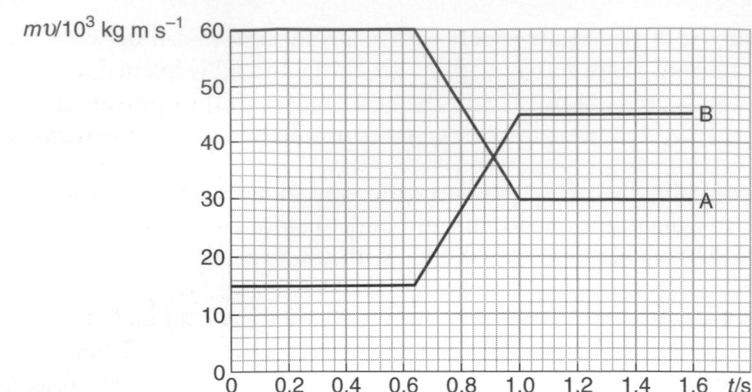

3.15 **(a)** Body A of mass 3.0 kg has a velocity of +4.0 m s^{-1} and collides head-on with a body B, which has a mass of 2.0 kg and a velocity of –2.0 m s^{-1}. After the collision the velocity of B is found to be +3.0 m s^{-1}. Find the velocity of A.

(b) Sketch a graph to show how the momentum of the two bodies varies with time before and after the collision.

3.16 A pair of skaters such as those shown in question 2.6 are together moving across the ice at a speed of 6.0 m s^{-1}. The man has a mass of 80 kg and the woman a mass of 60 kg. They push each other apart along their line of motion so that after they separate the man is moving in the same direction at 4.0 m s^{-1}. What is the woman skater's new velocity?

3.17 Explain how you would use air track gliders to demonstrate the validity of the principle of conservation of linear momentum. How would you measure the speeds of the gliders?

3.18 Two boys stand a few metres apart on a long flat boat – a punt. One throws a heavy medicine ball to the other, who catches it. Describe what happens to
(a) the speed of the punt **(b)** the position of the punt on the water
during each of the three phases – throwing/ball in air/catching – of the process.

3.19 Discuss how momentum is conserved when
(a) a train accelerates from rest **(b)** a ball falls to the ground and bounces up again.

There are some further questions on collisions and explosions in Section 4.4.

3.2 Force and rate of change of momentum

In this section you will need to

- remember that the impulse of a force is calculated as the average force times the time for which the force acts
- use the impulse–momentum equation $F\Delta t = \Delta(mv)$
- understand that Newton's third law applies at every instant of an interaction
- use Newton's second law in the form

$$\frac{\text{rate of change of}}{\text{momentum of a body}} \quad = \quad \frac{\text{resultant force}}{\text{acting on the body}}$$

3.20 What is the impulse of the forces in the following situations:

(a) a man pulling a garden roller with a horizontal force of 300 N for 10 s

(b) a rock of weight 20 N moving vertically downwards for 5.0 s

(c) a hammer hitting a nail with a vertical force of 800 N for 0.60 ms?

3.21 A tennis ball of mass 58 g is moving horizontally, at right angles to the net, with a speed of 20 m s^{-1}. A player hits it straight back so that it leaves his racket with a speed of 25 m s^{-1}. What is

(a) the size of the change of momentum of the ball

(b) the impulse of the force which the racket exerts on the ball?

3.22 A stationary snooker ball of mass 0.21 kg is struck by a cue which exerts an average horizontal force of 70 N on it. The cue is in contact with the ball for 8.0 ms. Calculate the speed of the ball after the impact.

3.23 The graph shows how a varying force acts on a body.

(a) Calculate the impulse of the force for the first 2.0 s.

(b) Calculate the impulse of the force from 2.0 s to 6.0 s.

(c) Estimate the impulse of the force from 6.0 s to 9.0 s.

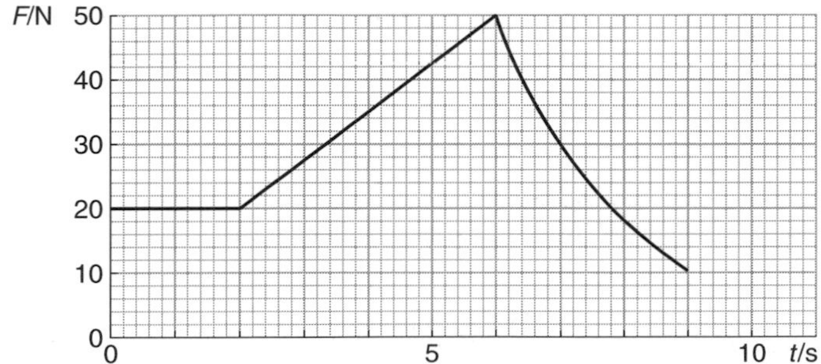

3.24 **(a)** Calculate the average force exerted by a golf club on a golf ball of mass 46 g, if the ball leaves the club at a speed of 80 m s^{-1} and the contact between the club and the ball lasts for 0.50 ms. Suggest an object which would have a weight approximately equal to this force.

(b) Sketch a graph to show how the force on the golf club might vary with time. Add scales to both axes.

3.25 The diagram shows how the push of a tennis racket on a tennis ball of mass 58 g varies with time.

(a) Estimate the change of momentum of the ball.

(b) Hence estimate the speed of the ball assuming it was at rest when it was struck.

(c) What is the maximum acceleration of the ball?

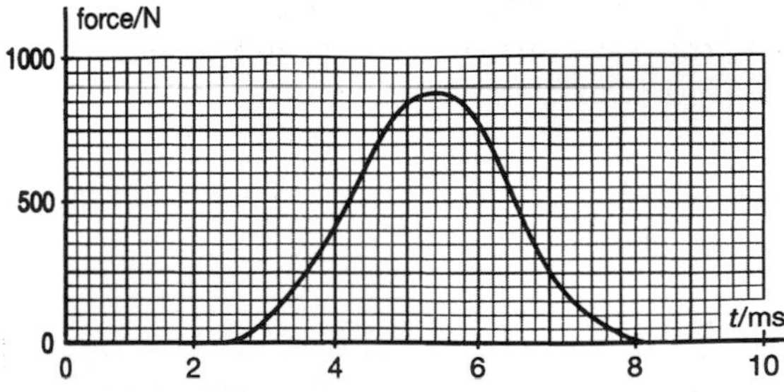

3.26 The diagrams show two strides for a sprinter during a 100 m race together with graphs showing how the *horizontal* push of the ground on the sprinter varies with time during each stride.

(a) Explain why the area under the graph represents the change of momentum of the sprinter during each stride.

(b) For stride A estimate this change of momentum and hence deduce the increase in velocity of the sprinter if he has a mass of 80 kg.

(c) Describe what is happening to the sprinter during stride B.

(d) Sketch a graph of the horizontal push of the ground on the sprinter for a stride after the end of the race during which he is slowing down.

A

F_H/N 800, 600, 400, 200, 0, −200, −400

0.1 0.2 0.3 0.4 0.5 0.6 t/s

B

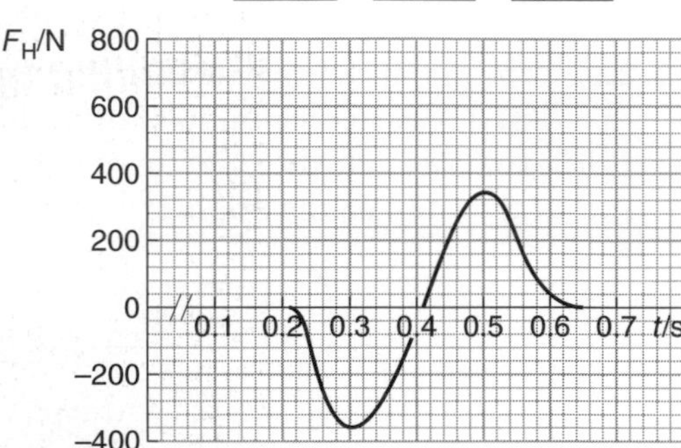

3.27 Use the impulse–momentum equation to explain why a rubber-headed hammer is not as good for driving nails into a block of wood as a hammer with a head made of metal, even if both hammers have the same mass.

3.28 See question 3.14.

(a) What is the physical significance of the slope of these momentum–time graphs?

(b) Use the graph to calculate **(i)** the force which truck A exerts on truck B (**ii**) the force which truck B exerts on truck A.

(c) Comment on your answers to **(b)**.

3.29 The diagram shows two free-body force diagrams, one for a man and one for the Earth. The man has just jumped off a wall and landed on the ground.

(a) Describe the forces P and P'.

(b) P and P' vary with time as shown. Explain the shapes of the two graphs and state how they are related.

(c) Copy the P–t graph and indicate on it the size of the force W.

P

W

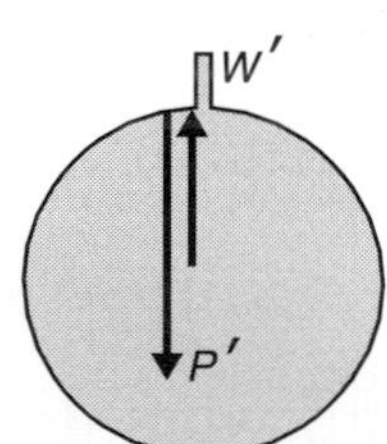

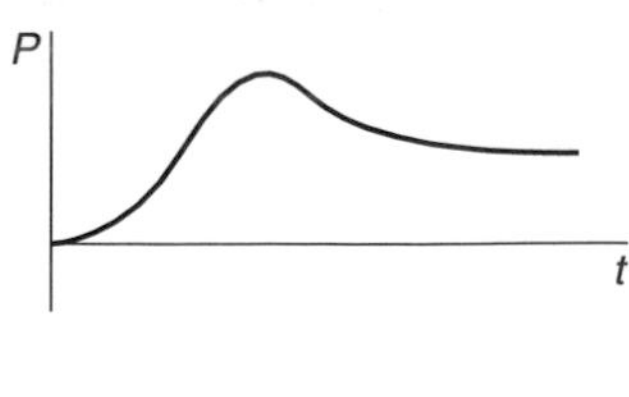

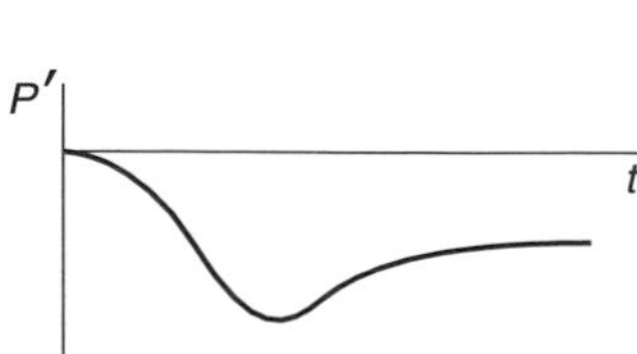

3.30 A railway wagon of mass 10 tonnes is seen to be moving at a speed of 2.0 m s^{-1} to the right. After 3.0 s it collides with another wagon of mass 40 tonnes moving at a speed of 1.0 m s^{-1} to the right. They link after the collision and move on to the right at the same velocity. (1 tonne = 1000 kg.)

(a) Calculate their common velocity after the collision.

(b) Draw a graph of momentum against time for the period $t = 0$ to $t = 7$ s if the collision took 1.0 s, i.e. from 3.0 s to 4.0 s. Assume that during the collision the momentum of each wagon changes uniformly with time.

(c) Calculate the gradient, with units, of the two graphs during the collision.

3.31 The diagram shows a car being used to test seat-belts when crashing into a solid concrete block at 20 m s^{-1}. For a car of mass 1600 kg the average stopping force is 640 kN. Calculate how long it takes for the car to come to rest in this crash.

3.32 The push F of a horizontal water jet hitting a wall can be calculated as $F = v^2 A\rho$, where v is the speed of the water, A is the cross-sectional area of the jet and ρ is the density of water. Show that the unit of the right hand side of the equation is N.

3.33 70 kg of air pass through an aircraft jet engine every second. The exhaust speed of the air is 600 m s^{-1} greater than the intake speed. Calculate

(a) the change of momentum of the air in one second

(b) the force exerted on the air to change its momentum in this way

(c) the thrust produced by four such engines.

3.34* Gas molecules, each of mass 4.8×10^{-26} kg, collide with a flat surface. The average speed of the molecules perpendicular to the surface is 550 m s^{-1} both before and after they collide with it.

(a) Calculate the change of momentum of a molecule as a result of one collision.

(b) If the force on a square millimetre of the surface is 0.10 N, how many molecules collide with that square millimetre every second?

(c) What pressure is produced by this molecular bombardment?

4 Work and energy

Data: gravitational field strength $g = 9.81$ N kg^{-1}

4.1 Work

In this section you will need to

- remember that work, energy and power are all scalar quantities
- express units, e.g. the joule, in base units
- use the equation for work done by a constant force: work done by a force = the force times the distance moved in the direction of the force
- understand that the work done by a force can, in certain circumstances, be zero or have a negative value
- understand how to calculate the work done by a variable force by using the average value of the force
- understand that doing work on a body alters its energy.

4.1 To cut a lawn, Mum has to push a lawnmower 80 m.

(a) If her average horizontal push on the mower is 100 N, how much work does she do on the lawnmower?

(b) As the mower has no kinetic energy when she has finished the task, how much work is done on the lawnmower by frictional forces?

4.2 Two tugs pull a large vessel which has lost power. The tension in each cable is 36 kN.

(a) Resolve each cable pull into forces parallel to and perpendicular to the motion.

(b) How much work is done in pulling the vessel 2.0 km by **(i)** the forces parallel to the direction of motion **(ii)** the forces perpendicular to the direction of motion?

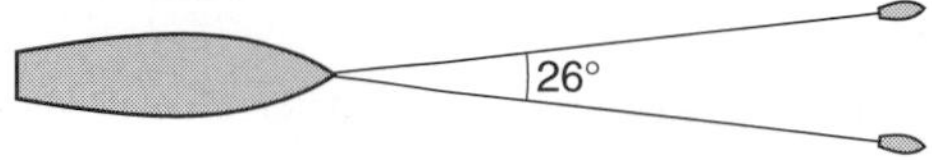

4.3 During the first 0.60 m of the lift, a heavyweight weightlifter produced an upward pull on the barbell of 3800 N.

(a) If the mass of the barbell was 240 kg, calculate

(i) the work done by the weightlifter on the barbell

(ii) the work done by the pull of the Earth on the barbell.

(b) Draw an energy flow or Sankey diagram for this lift. What can you deduce about the barbell after it has been raised 0.60 m?

4.4 A dog pulls on a lead with a force of 25 N. Calculate the work done by the dog as it moves 10 m along the pavement when

(a) the lead makes an angle of 15° to the horizontal

(b) the lead makes an angle of 40° to the horizontal.

How much work is done by the person holding the lead in each case?

4.5 **(a)** Draw a free-body force diagram for a block of mass 6.0 kg which is being pulled across a rough horizontal table by a horizontal force of 30 N. The maximum frictional force which the table can exert on the block is 20 N.

(b) What is the work done by each of the forces in a horizontal displacement of 0.80 m?

4.6 A skier is being pulled up a 20° slope by a drag lift. The pulling wire is at an angle of 15° to the slope and the forces acting on the skier are:

the pull of the drag lift wire, $P = 320$ N

the perpendicular contact push of the snow, $N = 620$ N

the frictional push of the snow, $F = 55$ N

the pull of the Earth, $W = 750$ N.

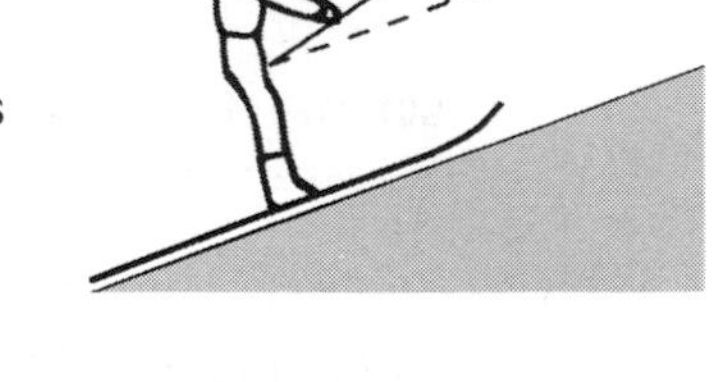

(a) Calculate the work done by each force on the skier as she is dragged 30 m up the slope.

(b) Hence show that the total work done on the skier is zero.

4.7 How much work is done by the pull of the Earth on the Moon as the Moon completes one orbit round the Earth? [Pull of Earth on Moon = 2.0×10^{20} N, radius of Moon's orbit = 3.8×10^{8} m.]

4.8 A simple pendulum consists of a thread 700 mm long and a bob of mass 60 g. The bob is pulled aside until its vertical height above its lowest position is 20 mm, and is then released.

(a) What is the size of the pull of the Earth on the bob? Draw a free-body force diagram for the bob when it is moving.

(b) Calculate the work done by each of the forces on the bob while

(i) it moves from its point of release to its lowest point

(ii) it moves from its lowest point to its extreme position on the other side.

4.9 A spring of constant stiffness 40 N m^{-1} is fixed at one end. A man pulls the other end horizontally.

(a) How much work does he do in stretching the spring **(i)** 0.20 m from its unstretched position **(ii)** 0.40 m from its unstretched position?
(b) Explain why the answer to **(i)** is not twice the answer to **(ii)**.

4.10 **(a)** Draw a sketch graph of the force *F* required to stretch a simple (Hooke's law) spring against its extension *x* for a spring on an exercise machine which stretches 0.80 m with a force of 80 N.
(b) Calculate the work done in stretching the spring
(i) from 0 to 0.20 m **(ii)** from 0.20 m to 0.40 m **(iii)** from 0.40 m to 0.60 m **(iv)** from 0.60 m to 0.80 m.

4.11* The graph shows how the Moon's gravitational pull *F* on a lunar lander varies with the distance *h* from the Moon's surface. Estimate the work done by the pull of the Moon on the lunar lander as it approaches the lunar surface from a height of 200 km. Explain how you made your estimate.

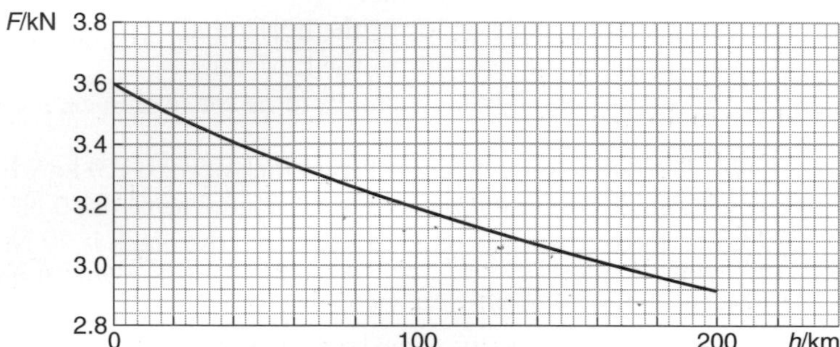

4.2 Power

In this section you will need to

- use the equations for power: $P = W/t$ and $P = Fv$ where W is the work done or energy transferred
- understand that although energy is always conserved, the effect of transferring energy is nearly always to produce some which is effectively wasted
- remember that the efficiency of energy (or power) transfer is defined as the useful energy (or power) output divided by the total energy (or power) input
- draw energy flow diagrams (Sankey diagrams) to illustrate energy transfer processes.

4.12 The data give the rate at which energy is used by a typical advanced-level student in some common activities.

Sleeping	80 W	Sitting	120 W
Standing	40 W	Walking	250 W
Running	600 W	Eating	170 W

Estimate **(a)** the energy you use in a day **(b)** the heating power of a class of 15 students.

4.13 **(a)** A man digging a trench converts energy at a rate of about 1200 W. He could not keep this up for long, but if he could, how much work would he do in a working day of 8 hours?

(b) Electrical energy is available at a cost of about 10 p per kW h (kilowatt-hour).
  (i) Show that a kilowatt-hour is equivalent to 3.6 MJ.
  (ii) Calculate the cost of the electrical energy equal to the work done by the man and discuss the result of your calculation.

4.14 Write the units of these expressions in terms of base units:
(a) work ÷ time
(b) force × speed.

4.15 Calculate the average power of
(a) Mum in question 4.1, if she cuts the lawn in 5 minutes
(b) the weightlifter in question 4.3, if he achieves the first 0.60 m of his lift in 0.32 s.

4.16 A motor drives a pulley which lifts a box of mass 5.0 kg at a steady speed of 2.0 m s^{-1}. What is the power output of the motor?

4.17 A piano of mass 300 kg is being lifted to a window 12 m above the ground using a system of pulleys and a diesel motor.
(a) If the motor has a power output of 800 W, how long will it take to raise the piano to the window?
(b) How much chemical energy is converted by the motor during the lift if its efficiency is 20%?
(c) The diagram is an unlabelled Sankey diagram for this process. Copy the diagram and label it to describe the lifting of the piano.

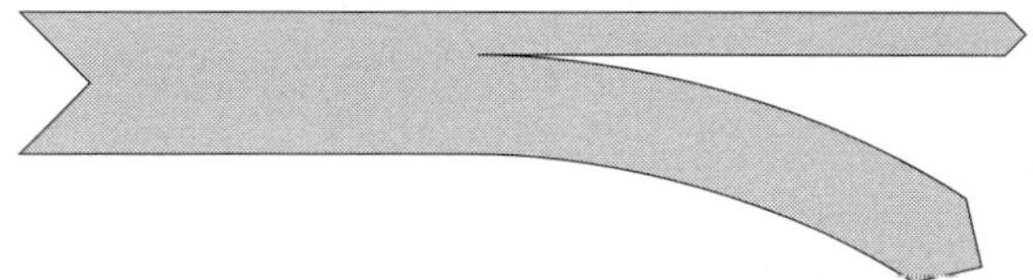

4.18 A man does 20 press-ups (i.e. lying on his front, he straightens his arms to lift his shoulders from the ground while keeping his body straight) in 50 s. Estimate the power output of the muscles he is using.

4.19 A hang glider and pilot have a combined mass of 92 kg. In flight, the glider moves forward at 20 m s^{-1} and sinks, i.e. moves vertically downwards, at 1.2 m s^{-1}.
(a) Calculate the work done by the pull of the Earth on the hang glider in one minute.
(b) Explain what happens to this energy.
(c) What rate of working would be needed to maintain level flight?

4.20 The world reserves of geothermal energy are estimated to be about 4.5×10^{25} J. Of these reserves only 2% is hot enough for use in the generation of electricity and a maximum of a quarter of this is expected to be recoverable using existing methods of extraction. The conversion efficiency would be as low as 3% .
The present consumption of energy in the world is about 2×10^{20} J per year. Discuss whether geothermal energy could make a significant contribution to this consumption.

4.21 The power of the electric motor of a locomotive unit pulling a train at a constant speed of 50 m s^{-1} is 2.5 MW. What is the total resistance force on the train?

4.22 The diagram, a Sankey diagram, shows the power transfers in a car moving at a steady speed of 18 m s^{-1} along a level road.

(a) What percentage of the energy available from the petrol is transferred **(i)** to internal energy in the engine **(ii)** to internal energy overall?

(b) Calculate the effective frictional force opposing the motion of the car produced by **(i)** the air **(ii)** the wheels.

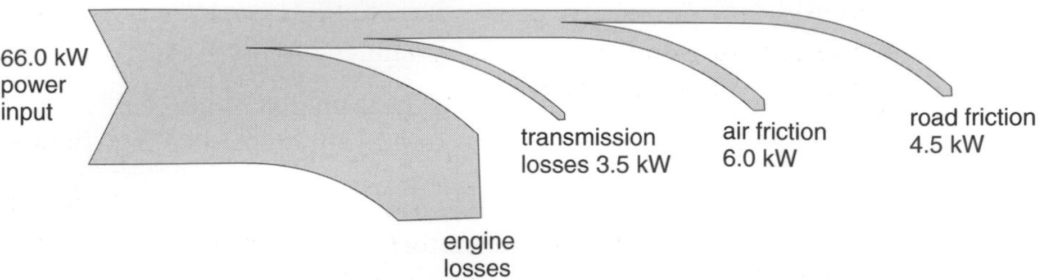

4.23 The power output P of a windmill can be expressed as

$$P = kA\rho v^3$$

where A is the area swept out by the windmill blades (sails), ρ is the density of air, v is the wind speed and k is a dimensionless constant.

(a) Show that the units on both sides of this expression are the same.

(b) Sketch a graph to show how the power increases with wind speed as v rises from zero to 15 m s^{-1}.

4.24* A student looks up a formula for the power P in a sea wave. He finds

$$P/l = kh\rho gv$$

where k is a dimensionless constant, ρ is the density of sea water, h is the amplitude of the wave (its height), g the Earth's gravitational field strength, v the wave speed and l the length of the wave being considered.

Show that the student must have copied the equation down wrongly and guess what the correct equation, which differs only in that one of the terms on the right must be squared, would look like.

4.25 The radiation received from the Sun at the Earth's surface in Great Britain is about 600 W m^{-2} averaged over 8 hours in the absence of cloud.

(a) What area of solar panel would be needed to replace a power station of 2.0 GW output, if the solar panels used could convert solar radiation to electrical energy at an efficiency of 20%?

(b) What percentage is this area of the total area of Great Britain (which is about 3×10^{11} m^2)?

(c) If the total power station capacity is about 140 GW, what percentage of the surface of Great Britain would be covered by solar panels if all the power stations were replaced?

4.3 Kinetic energy and gravitational potential energy

In this section you will need to

- use the kinetic energy (k.e) of a body calculated as $\frac{1}{2}mv^2$
- use the work–energy equation $Fs = \Delta(\frac{1}{2}mv^2)$
- understand that changes in gravitational potential energy (g.p.e.) close to the Earth's surface are calculated as mgh where h is the vertical displacement
- use $\Delta(\frac{1}{2}mv^2) = mgh$ in the simplified form $v^2 = 2gh$ for bodies released from rest in the Earth's gravitational field.

4.26 **(a)** The lift in the World Trade Centre in New York rises 400 m from ground level to the observatory on top of the building. What is the gain in g.p.e. of a passenger in the lift who has a mass of 80 kg?
(b) The speed of the lift is about 6.5 m s^{-1}. Calculate his k.e. while rising up the building.

4.27 Estimate the kinetic energy of
(a) a tennis ball which has just been served [Tennis balls have a mass of 58 g.]
(b) a rifle bullet [The speed will be less than the speed of sound.]
(c) a world class sprinter
(d) a family car on a motorway. [70 m.p.h. is a little more than 30 m s^{-1}.]

4.28 A hockey ball is held 2.0 m above the ground and has 3.2 J of g.p.e. It is then dropped.
(a) How much k.e. does it have just before it hits the ground?
(b) It bounces off the ground and is found to have 0.5 J of k.e. as it leaves the ground. How much internal energy has been transferred in the bounce?
(c) When it was momentarily at rest on the ground it had no k.e. and no g.p.e., yet a moment later it had 0.5 J of k.e. Where did this energy come from?
(d) How much g.p.e. will it have when it is at the top of its first bounce?

4.29 The energy for a 'grandfather' clock is stored in a heavy cylinder. The cylinder, of mass 4.8 kg, gradually transfers g.p.e. to keep the pendulum of the clock swinging as the cylinder descends 1.2 m in seven days. Calculate the power transfer during the descent.

4.30 The gradient of a road is 1 in 6, i.e. it rises vertically 1.0 m for every 6.0 m along the road. During a mountain stage of the Tour de France a cyclist travels 1200 m along the road at an average speed of 6.5 m s^{-1}. The mass of the cyclist and bicycle is 78 kg.
(a) Calculate the gain of g.p.e. of the cyclist and bicycle.
(b) What is the average output power of the cyclist?

4.31 To push-start a car on a winter's morning (its battery is flat), two people each push with a force of 300 N. After pushing for 15 m the car's engine starts. If at that moment the car's kinetic energy is 7500 J, calculate
(a) the work done by the push of the people on the car
(b) the work done by frictional forces on the car.
Draw an energy flow or Sankey diagram for this process.

4.32 The figure shows experiments to compare the power outputs of different sets of muscles in the body. Suppose that the bodyweight of both the boy and the girl is 600 N. Her step-up distance is 25 cm and his pull-up lifts his centre of mass 35 cm. She does 24 step-ups in 50 s and could go on easily; he does 8 pull-ups in 30 s and is exhausted. Compare the power outputs of **(a)** the girl and **(b)** the boy in these exercises.

4.33 **(a)** A car of mass 800 kg reaches a speed of 20 m s^{-1} in a distance of 100 m, as a result of the pistons in the engine doing internal work. What constant external force would have been needed to do the same amount of work?

(b) If the motor is then switched off, and frictional forces equivalent to a constant horizontal force of 200 N act on it, in what distance will it come to rest?

4.34 A car which has a mass of 1200 kg is moving at a speed of 18 m s^{-1} (i.e. about 40 m.p.h.). On a dry day the maximum braking force is 8500 N.

(a) In what distance can it stop?

(b) On a wet day the braking force is halved. In what distance can it then stop?

(c) If the car is travelling at 27 m s^{-1} (i.e. about 60 m.p.h.) on a wet day, in what distance can it stop?

4.35 A climber of mass 70 kg falls vertically off a cliff. She is attached to a rope which allows her to fall freely for 20 m. Then it becomes taut, but stretches, bringing her to rest in a further 4.0 m. An energy flow diagram for this accident would show g.p.e. being converted to k.e. while she is falling freely and, while she is being slowed down, the k.e., plus some more g.p.e., being converted to internal energy in the rope. Assume there is no elastic energy stored in the rope at the end of her fall.

(a) Draw an energy flow or Sankey diagram for this fall, attaching numerical values for changes in k.e. and g.p.e.

(b) Calculate the added internal energy in the rope, and the average force exerted on the climber by the rope.

4.36 The figure shows an energy flow diagram. Suggest a process to which it might refer, giving numerical information where possible.
[Remember $g = 9.81$ N kg^{-1}.]

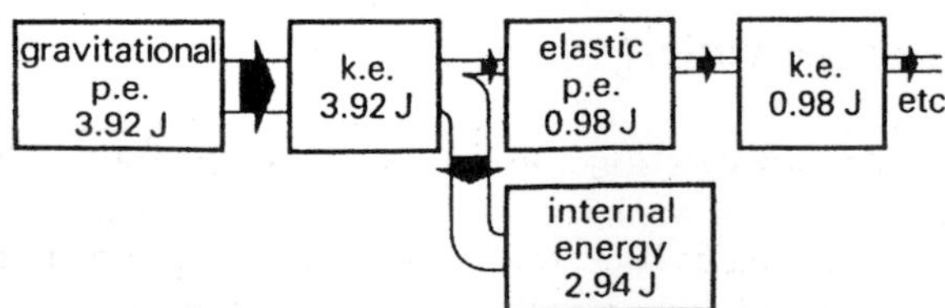

4.37 A cricket ball, of mass 0.16 kg, is thrown from the boundary to the wicket keeper. The thrower gives it 50 J of kinetic energy and it reaches a maximum height of 12 m above his arm. Calculate

(a) the speed at which the ball was thrown

(b) the k.e. of the ball at its maximum height

(c) the speed of the ball at its maximum height.

4.38 During a human heart beat, 20 g of blood are pushed into the main arteries. This blood is accelerated from a speed of 0.20 m s^{-1} to 0.34 m s^{-1}. For a heart pulsing at 70 beats per minute, calculate the average power of the heart pump.

4.39 In order to demonstrate the principle of conservation of energy, a teacher attaches a trolley of mass 1.0 kg to a 50 g mass with a piece of string. Using a pulley attached to the edge of the laboratory bench, as shown in the diagram, he allows the 50 g mass to accelerate the trolley along the bench and measures both the gain of k.e. of the trolley and the loss of g.p.e. of the 50 g mass.

(a) Explain how you would measure the two energies.

It is found that the loss of g.p.e. is 10% greater than the gain of k.e. measured.

(b) Suggest why this is so and draw an energy flow or Sankey diagram for the experiment.

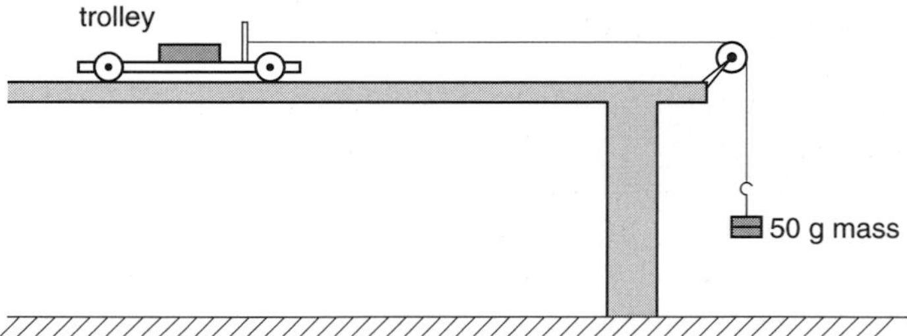

4.40 An object of mass 300 g falls from rest for 2.5 m in a vacuum.

(a) Calculate its final speed.

When it fell in air its final speed was 6.8 m s^{-1}.

(b) How much work was then done on it by air resistance?

(c) Calculate the average resistive force.

4.41 A baby drops a glass from a high chair onto a vinyl covered floor. The glass will probably break if it hits the floor at a speed of more than 3 m s^{-1}. Suppose the glass has a mass of 160 g and the chair's tray is 1.0 m above the ground. Calculate

(a) the loss of g.p.e. of the glass in falling to the ground

(b) the k.e. of the glass as it hits the ground.

(c) Hence decide whether the glass will break and discuss whether it matters what the mass of the glass is.

4.42 A trampolinist falls vertically from a height of 3.0 m; on the rebound she rises to a height of 3.5 m.

(a) Discuss the energy changes and draw an energy flow or Sankey diagram for the process.

(b) Calculate her speeds **(i)** as she reaches the trampoline mat on the way down **(ii)** as she leaves the trampoline mat on the way up.

4.43 Describe in detail how you would measure the percentage of elastic energy stored in a catapult which is transferred to the k.e. of a stone when the catapult is fired.

4.44 A high-jumper of mass 80 kg reaches the end of his run-up with 2100 J of k.e. At take-off he drives off the ground, adding 800 J to his k.e. Stating any assumptions you make, estimate

(a) the speed of the high-jumper at the end of his run-up

(b) the height of the bar which he could clear if his k.e. at the top of his jump is 1700 J.

4.45 A skydiver of mass 70 kg reaches a speed of 45 m s^{-1} after falling 150 m. By finding the loss of g.p.e. and the gain of k.e. determine the work done on the skydiver by the push of the air and hence find the average vertical push of the air on him during the first 150 m of his fall.

4.46 A supermarket trolley of total mass 22 kg is pushed up a ramp with a force of 14 N. The ramp is 5.5 m long and rises 0.35 m. There are no frictional losses.

(a) Calculate **(i)** the work done by the shopper **(ii)** the gain in g.p.e. of the trolley.

(b) If the speed of the trolley was 1.0 m s^{-1} at the bottom of the ramp, what is its speed at the top?

4.47 During the night, when power to the grid is not required, 250 MW of electrical energy are used to pump water from a reservoir up to a lake 300 m higher. The efficiency of the pumping system is 95%.

(a) Calculate the mass of water which can be pumped to the lake in two hours.

(b) What volume of water is this? Express this volume as an increase in depth of a lake which measures 300 m by 1200 m.

4.48 The diagram shows part of a roller-coaster ride at an adventure park. The carriages are pulled from A to B at a steady speed by an electric motor of power output 52 kW. At B they have effectively no kinetic energy and they then run freely down to C. The carriage and passengers have a mass of 3400 kg.

(a) How long do the carriages take to rise from A to B?

(b) Calculate the speed of the carriages at C.

(c) The actual speed at C is found to be 33 m s^{-1}. If the track from B to C is 95 m long, calculate the average resistive force acting on this part of the ride.

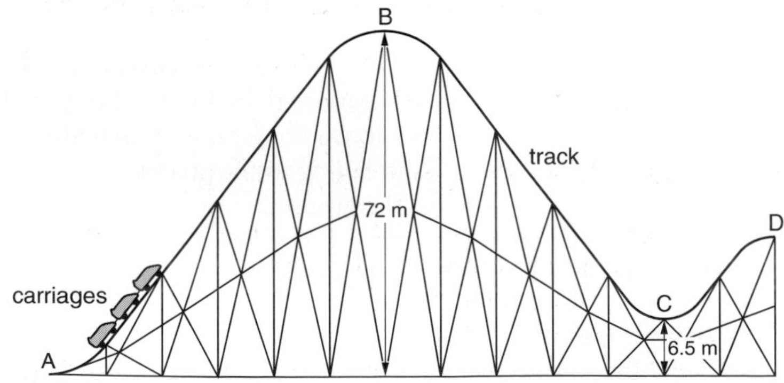

4.49 In this question, use $g = 10$ N kg^{-1}. A lump of weight 4.0 N is held stationary at the bottom of a vertical spring of stiffness 20 N m^{-1} to which it is attached. Initially the spring is taut but unstretched. The lump is then released. After it has fallen 0.20 m, find the changes in

(a) the g.p.e. of the lump
(b) the elastic potential energy of the spring [Remember to use average force.]
(c) the k.e. of the lump.
The lump passes through this position. When it has fallen a total distance of 0.40 m, find the change (from the initial values)
(d) in the g.p.e of the lump
(e) in the elastic potential energy of the spring.
What can you deduce from these figures?

4.50* The diagram shows a projectile photographed every $\frac{1}{30}$ s. The horizontal lines are 152 mm apart.
(a) By measuring the distances between the centres of adjacent images determine the speed of the ball at the mid-point between each pair of images and the height of each mid-point above the bottom edge of the photograph. [You should try to estimate the distances to the nearest 0.1 mm in the photograph before scaling up.]
(b) The mass of the ball is 0.10 kg. Calculate the k.e. and g.p.e. of the ball at each mid-point. Take the bottom of the photograph as the zero for g.p.e.
(c) Draw graphs of k.e. and g.p.e. against time, and add a graph of k.e. + g.p.e. against time – all three graphs on the same axes.

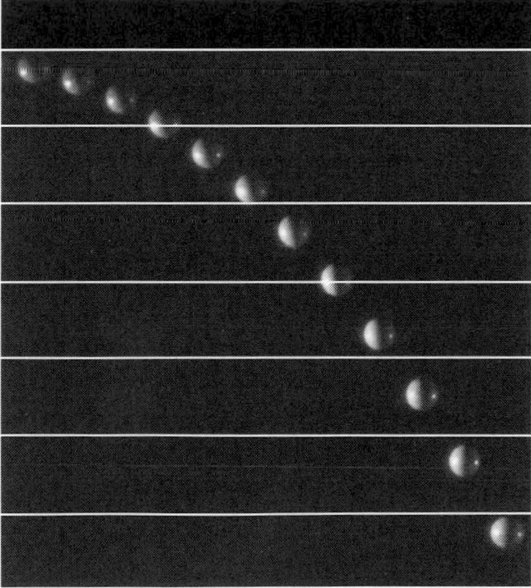

4.4 Collisions and explosions

In this section you will need to

- use the principles of conservation of momentum and mechanical energy to analyse collisions and explosions in one dimension
- understand that in an elastic collision no kinetic energy (k.e.) is transferred to internal energy
- understand that a totally inelastic collision is one in which the colliding bodies stick together.

There are a number of questions on momentum conservation in collisions and explosions in Section 3.1.

4.51 A railway wagon of mass 15 tonnes (15 000 kg) which has a velocity of 5.0 m s^{-1} north collides with a wagon of mass 10 tonnes which has a velocity of 2.0 m s^{-1} south. They couple together on impact.

(a) Find their velocity after the collision.

(b) Calculate the k.e. transformed to other forms in the collision.

4.52 Refer to question 3.8.

(a) How much energy does the spring release in this explosion? [Look up the value of *m* if necessary.]

(b) If the piston moves 5.0 cm before the two trolleys separate, calculate the average force between them.

4.53 Refer to question 3.16.
How much energy is transferred to k.e. as the skaters push each other apart?

4.54 A trolley of mass 2.0 kg moving to the right at 0.3 m s^{-1} collides with a trolley of mass 4.0 kg which is at rest. The 4.0 kg trolley moves off to the right at 0.7 m s^{-1}.

(a) Show that the 2.0 kg trolley rebounds and moves off to the left at 1.1 m s^{-1}.

(b) How much energy was added to the kinetic energy of the trolleys as a result of the collision?

4.55 A bullet of mass 16 g is fired into a block of wood of mass 4.0 kg which is supported by vertical threads. After the bullet has become embedded in the block, the block swings and rises (vertically) 50 mm.

(a) Draw an energy flow or Sankey diagram for this event.

(b) Calculate the speed of the block and bullet immediately after the collision.

(c) What was the speed of the bullet before the collision?

4.56 A 950 kg car is stopped on an icy road at the bottom of a hill which rises 1 m for every 12 m along the road. The car is hit from behind by a truck of mass 7600 kg which is moving at 50 km h^{-1}. The collision is totally inelastic. Calculate the distance up the hill which the combined wreckage slides.

4.57 A Newton's cradle consists of five steel balls suspended by threads so that the balls lie in a horizontal line, almost touching one another. When one ball is pulled back and released so that it strikes head-on the row of the remaining four balls, the ball stops on impact, and the ball at the far end moves off with the velocity which the first ball had, the others now being stationary. Clearly both linear momentum and energy are conserved. Suppose now that two balls are pulled back together and then released. Show that it is not possible for one ball to move off from the other end of the row without contradicting one of the conservation principles.

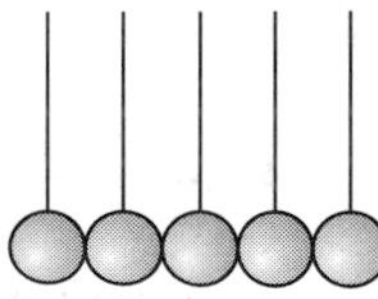

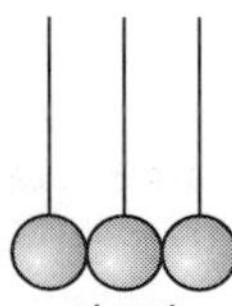

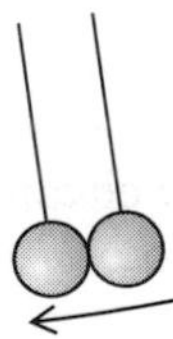

4.58* A particle, travelling along a straight line with velocity v, explodes into two equal parts. The explosion causes the k.e. of the system to be doubled. Assuming that the forces of the explosion act along the initial line of travel, determine the velocities of the two parts of the body after the explosion.

5 Electric currents and electrical energy

Data: gravitational field strength $g = 9.81 \text{ N kg}^{-1}$

electronic charge $e = 1.60 \times 10^{-19}$ C

number of conduction electrons per m^3 for copper, $n_{Cu} = 1.0 \times 10^{29} \text{ m}^{-3}$

5.1 Electric current

In this section you will need to

- remember that the current is the rate of flow of electric charge and use the equation $Q = It$
- remember that the total current leaving a junction is equal to the total current entering it
- remember that an ammeter can be used to measure current.

5.1 A door bell is powered by a battery and is to be operated by a push-switch at the front door and also by a push-switch at the back door. Draw the circuit.

5.2 **(a)** Is the current in the circuit shown in the diagram clockwise or anti-clockwise?
(b) Do the electrons in this circuit flow clockwise or anti-clockwise?
(c) Your answers to **(a)** and **(b)** should be different. Explain how this difference arises, and say whether you think it matters.

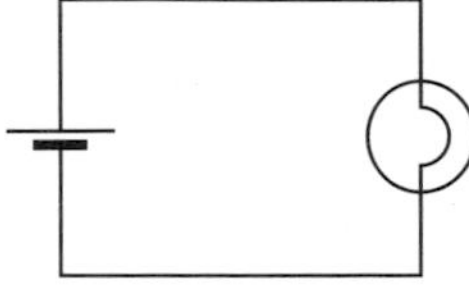

5.3 Suppose you are supplied with a battery, three identical bulbs A, B and C and only two switches. The bulbs are lit properly when each is connected separately to the battery. Draw a diagram of a circuit which would enable you to have no bulbs lit or bulb A lit or bulbs B and C lit or all three bulbs lit. You are allowed to press both switches at the same time, but the bulbs must all be properly lit.

5.4 In the circuit shown the bulbs are identical. The current at P is 0.20 A. What are the currents at **(a)** Q **(b)** R **(c)** S **(d)** T?

5.5 In the circuit shown the three bulbs are identical and reach full brightness when the current is 0.20 A. Bulb B is observed to be at full brightness. Which (if any) of the other bulbs will be at full brightness, and what is the current in C and D?

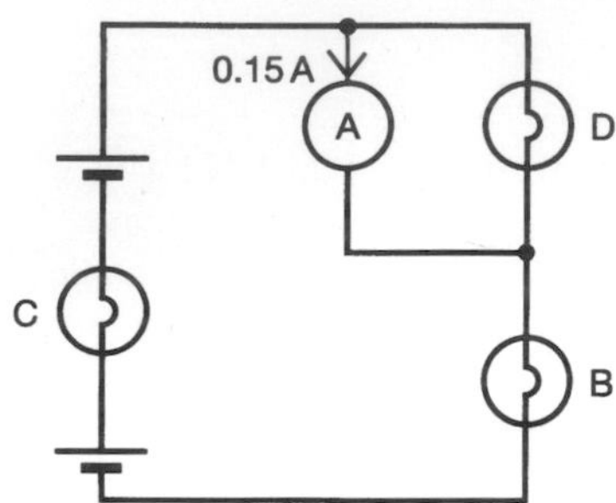

5.6 What is the average electric current in a wire when a charge of 150 C passes in 30 s?

5.7 The current in a small torch bulb is 0.20 A.
(a) What is the total electric charge which passes a point in the circuit in 12 minutes?
(b) How many electrons pass this point in this time? [Use data.]

5.8 How long will it take 1.0×10^{18} electrons to enter the filament of a torch bulb which carries a current of 0.15 A? [Use data.]

5.9 In a television tube the picture is formed by streams of electrons hitting the screen. The current is 20 mA. How much electric charge hits the screen in 30 minutes?

5.10 In a camera flash lamp a charge of 5.0 C passes through the lamp in 10 ms. What is the average current?

5.11 In a lightning flash a typical amount of charge which reaches the Earth is 10 C. If the flash lasts for 0.50 ms, what is the average current?

5.12 A car battery sends a current of 5.0 A through each of two headlamps and a current of 0.50 A through each of two sidelamps.
(a) Draw a circuit diagram for the battery and the lamps.
(b) In 20 minutes how much charge passes through **(i)** each headlamp **(ii)** each side-lamp **(iii)** the battery?

5.13 The graphs show how the current through an ammeter varies with time in three different situations. Calculate the amount of electric charge passing through the ammeter in **(a)** 15 s **(b)** 10 s **(c)** 6.0 s.

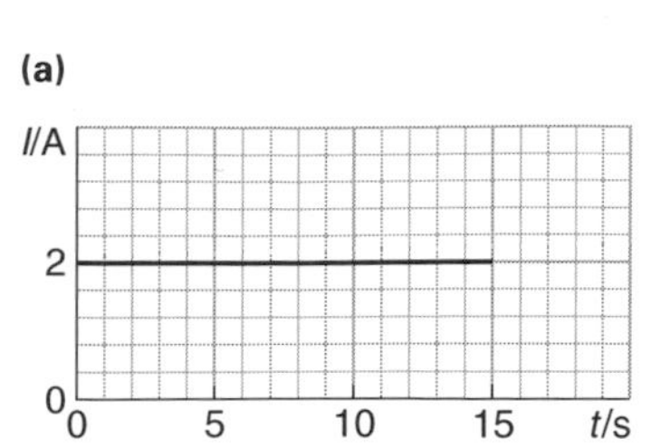

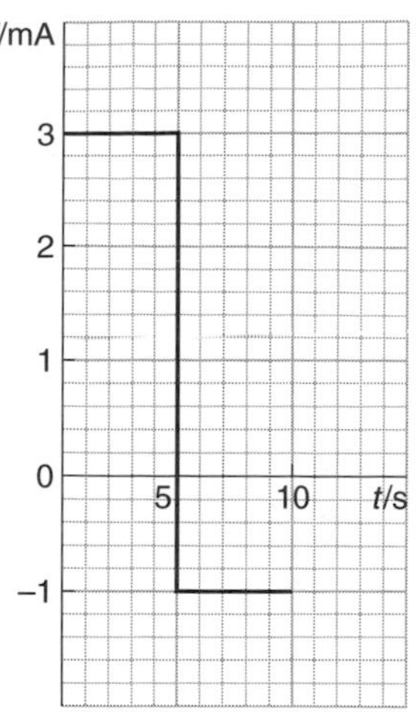

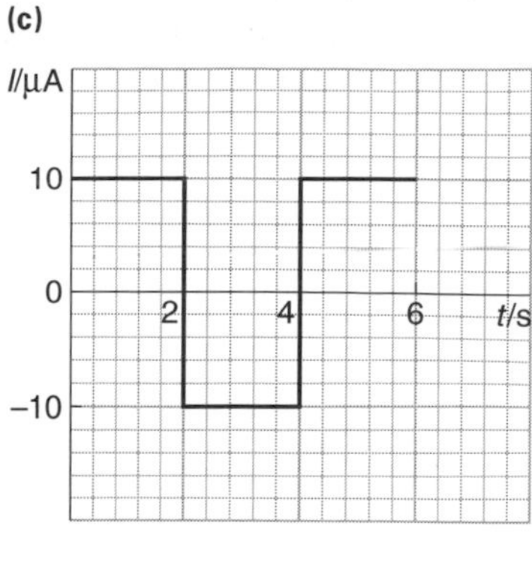

5.14 A new electric cell was joined in series with a bulb and an ammeter. The initial current was 0.30 A. At subsequent intervals of 1 hour the readings on the ammeter were: 0.27 A, 0.27 A, 0.26 A, 0.25 A, 0.23 A, 0.19 A, 0.09 A, 0.03 A and at 9 hours the ammeter reading had become negligibly small.

(a) Plot a graph of current against time.
(b) What does the area under the graph represent?
(c) How much electric charge passes through the circuit when a current of 0.10 A passes for 1 hour?
(d) What is the total electric charge which passes through the cell in the 9 hours?

5.2 Currents in solids, liquids and gases

In this section you will need to

- use the equation $I = Aqnv$ for the current in a conductor
- remember that there is a difference between the drift speed of the charge carriers and the speed of transmission of the electric field
- understand that the current in liquids and gases is caused by the movement of electrically charged particles such as ions and electrons.

5.15 The equation $I = Aqnv$ shows how the current I in a conductor depends on the quantities A, q, n and v, where these symbols have their usual meanings. Write down the units of each of these four quantities, multiply the units together (showing your working), and leave your answer in its simplest form.

5.16 Some copper fuse wire has a diameter of 0.22 mm and is designed to carry currents of up to 5.0 A. What is the mean drift speed of the electrons in the fuse wire when it carries a current of 5.0 A? [Use data.]

5.17 Fuse wire which is labelled '15 A' will melt when it carries a current of 1.5 times that current. Its diameter is 0.51 mm. What is the maximum drift speed of the electrons in this wire? [Use data.]

5.18 A copper wire joins a car battery to one of the tail lamps and carries a current of 1.8 A. The wire has a cross-sectional area of 1.0 mm^2 and is 6.0 m long. Calculate how long it takes an electron to travel along this length of wire. [Use data.]

5.19 'I'm sure the answer to the last question must be wrong. It can't possibly take 15 hours for the tail lamps to come on!' What would you say to this?

5.20 Two copper wires of diameter 2.0 mm and 1.0 mm are joined end to end.

(a) What is the ratio of the average drift speeds of the electrons in the two wires when a steady current passes through them?

(b) In which wire are the electrons moving faster?

5.21 A shallow trough has the shape shown in the diagram and contains a liquid with positive and negative ions. Electrodes are fixed to the ends of the trough as shown, and connected to a battery.

(a) What is the direction of the current in the liquid?

(b) In which direction do **(i)** the positive ions move **(ii)** the negative ions move?

(c) If the speed of the positive ions is 1.2×10^{-7} m s^{-1} between A and B, what is their speed when they are moving between C and D?

(d) Draw a graph of current against distance from A, to show how the average speed of the positive ions varies along the line ABCD.

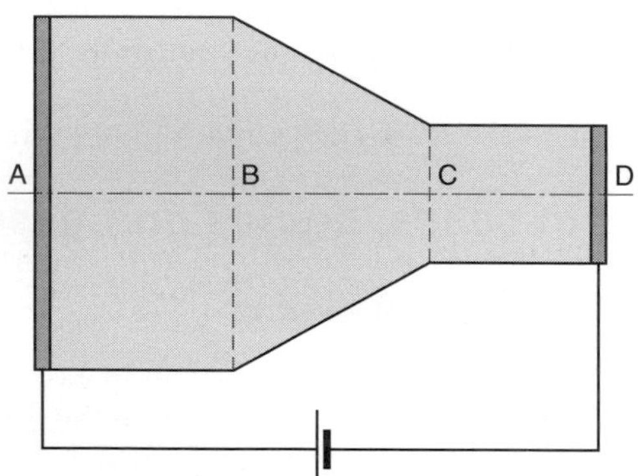

5.22 A piece of n-type germanium is 2.0 mm wide and 0.25 mm thick. At a certain temperature the number of conduction electrons per cubic metre is 6.0×10^{20} m^{-3}. What is the average drift speed of the electrons when a current of 1.5 mA flows? [Use data.]

5.23 In an electroplating process to deposit silver on the surface of another metal an electric current of 10.0 A passes for 30 minutes through an electrolytic tank. A thickness of 0.10 mm of silver is deposited on an area of 192 cm^2.

(a) How much charge reaches the silvered surface?

(b) The density of silver is 1.05×10^4 kg m^{-3}. What is the mass of silver deposited?

(c) The mass of a silver ion is 1.79×10^{-25} kg. How many ions are deposited?

(d) What is the charge on each silver ion?

5.3 Electric cells and e.m.f.

In this section you will need to

- understand that the e.m.f. $\mathcal{E}$ of a cell is defined by the equation $\mathcal{E} = W/q$ where W is the energy transferred to electrical energy from some other form of energy, and q is the charge passing through the cell
- understand that for cells in series the total e.m.f. is the sum of the separate e.m.f.s
- understand that for cells of e.m.f. $\mathcal{E}$ in parallel the total e.m.f. is $\mathcal{E}$ (but that the cells can now supply a larger current)
- remember that a voltmeter can be used to measure e.m.f.
- understand what is meant by the capacity of a battery
- draw a circuit diagram to show how a secondary cell may be recharged.

5.24 How much chemical energy is transferred to electrical energy when
(a) a charge of 10 C flows through a cell of e.m.f. 1.5 V
(b) a charge of 30 C flows through a cell of e.m.f. 1.5 V
(c) a charge of 10 C flows through two cells, each of e.m.f. 1.5 V, connected in series?

5.25 A cell has an e.m.f. of 1.50 V.
(a) How much energy is transferred when an electric charge of 100 C flows through the cell?
(b) Which form of energy is there less of afterwards, and which form of energy is there more of?

5.26 The diagram shows a circuit with four 1.5 V cells (A, B, C and D) in it. How much energy is gained when a charge of 3.0 C flows through
(a) A
(b) A and B
(c) A, B, C and D?

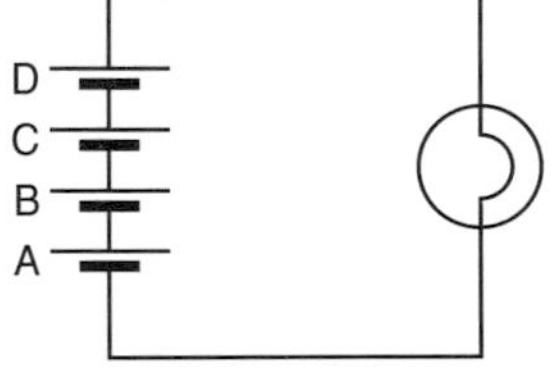

5.27 Five identical cells each provide an e.m.f. of 1.5 V. What does a voltmeter read when connected between A and B when the cells are arranged as shown in the diagram?

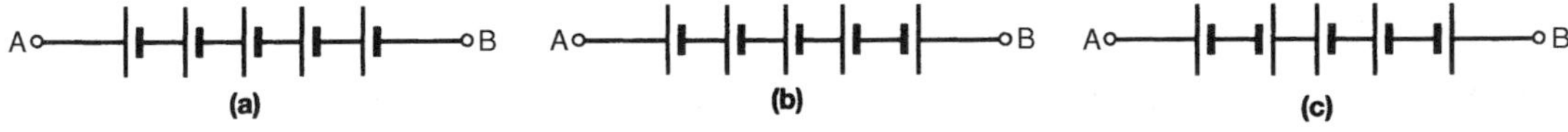

5.28 A radio cassette recorder might need six 1.5 V cells to make it work. The diagram shows what one cell would look like. Draw a diagram to show how they would be arranged if
(a) they were all connected side by side
(b) they were in two side-by-side 'columns' each of which consisted of three connected cells.

\+ −

In each diagram show the points between which you could connect a voltmeter so that it would read 4.5 V.

5.29 **(a)** How many cells each of e.m.f. 1.2 V are needed to provide a 6.0 V battery for a calculator?
(b) In fitting these cells the user puts one of the cells in the wrong way round by mistake. What is the e.m.f. of the battery now?

5.30 The top of a 12 V car battery, which consists of six 2 V cells, might look like the diagram. The six cells are connected by thick metal strips, as shown. The battery terminals are labelled '+' and '–'.
(a) Copy the diagram and mark the terminals of each of the cells + or –.
(b) What would a voltmeter read if it was connected between
(i) A and B
(ii) B and C
(iii) D and E
(iv) E and F?

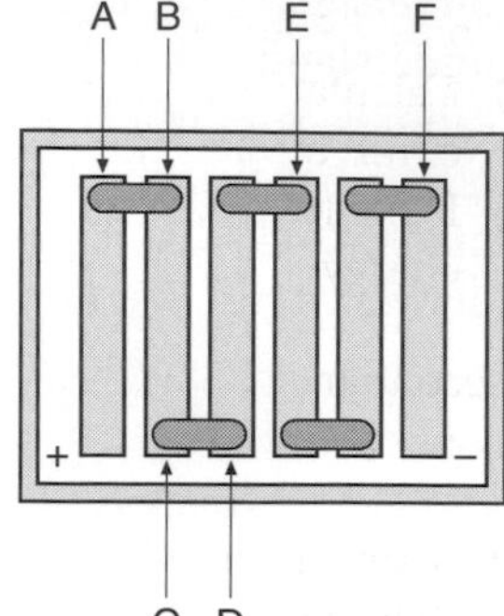

5.31 Two cells of e.m.f. 2.0 V are connected **(i)** in series and **(ii)** in parallel to a resistor of resistance 10 Ω. The cells may be assumed to have negligible resistance.
(a) What is the total e.m.f. of the cells, and the current in the resistor **(i)** when connected in series **(ii)** when connected in parallel?
(b) What are the answers to these questions if one of the cells is replaced by a rechargeable nicad cell of e.m.f. 1.2 V in series with a 1.0 Ω resistor?

5.32 A cell has an e.m.f. of 1.2 V.
(a) How much energy is transferred to a charge of 10 C which passes through it?
(b) A second cell has an e.m.f. of 1.5 V. How much energy is transferred to a charge of 10 C which passes through it?
(c) What is the total energy transferred to the charge of 10 C when it passes through the two cells in succession?

5.33 Prove that the combined e.m.f. of two cells of e.m.f. $\mathcal{E}_1$ and $\mathcal{E}_2$ is $\mathcal{E}_1 + \mathcal{E}_2$ when they are connected in series.

5.34 Explain why the combined e.m.f. of two cells each of e.m.f. $\mathcal{E}$ is still $\mathcal{E}$ when they are connected in parallel.

5.35 A battery of e.m.f. 6.0 V passes a current of 0.30 A through a torch bulb for 5 minutes. How much energy is transferred from the cell?

5.36 A lead-acid cell of e.m.f. 2.0 V can drive a current of 0.50 A round a circuit for 10 hours.
(a) How much chemical energy is transferred to electrical energy in this time?
(b) How long would you expect the same cell to maintain a current of 0.20 A?

5.37 A cell is said to have an e.m.f. of 1.5 V and a 'capacity' of 10 A h, e.g. it can pass a current of 1.0 A for 10 hours.
(a) How much charge would pass?
(b) How much chemical energy is stored in the cell?

5.38 The e.m.f. of a small cell is 1.5 V and it can pass a current of 20 mA for 9 hours before its chemical energy is completely transferred.

(a) How much chemical energy does the cell have?

(b) How much energy would be transferred if the same total charge were passed by three such cells connected in **(i)** series **(ii)** parallel?

5.39 A battery manufacturer makes three different D-size 1.5 V cells. Their capacities are 5.2 A h, 7.9 A h and 16 A h.

(a) Calculate the energy stored in each of these cells, assuming that the cells provide an e.m.f. of 1.5 V throughout their lives.

(b) If the total masses of the cells are 79 g, 100 g and 131 g, respectively, calculate the energy per unit mass for each of the cells.

5.40 A secondary cell, such as a lead-acid battery, can be recharged if a source of higher e.m.f. is connected to it.

(a) Why would it be more correct to say that the battery can be 're-energised'?

(b) A simple circuit for recharging a battery is shown in the diagram. The resistor is there to limit the current that flows to a safe value. Write down the type of energy transfer that takes place in

(i) A, the battery of higher e.m.f.

(ii) the resistor

(iii) B, the battery of lower e.m.f.

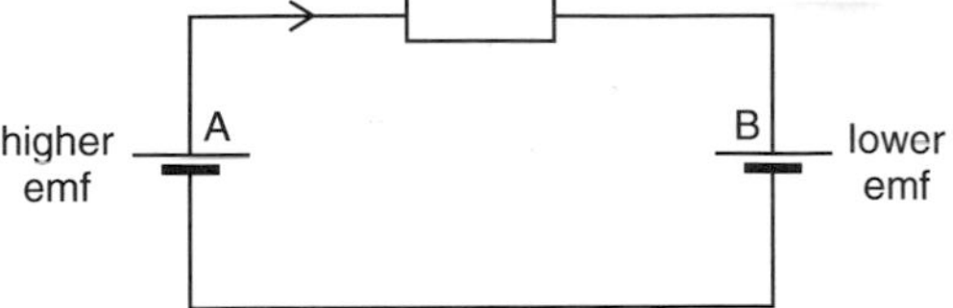

5.4 Electrical energy and potential difference

In this section you will need to

- understand the meaning of potential difference V and use the equation $W = VQ$
- understand that when two components are connected in series, the total p.d. is equal to the sum of the separate p.ds
- understand that when two components are connected in parallel, the p.d. between the ends of each of them is the same
- understand that we can assign values of potential to points in a circuit if we decide to fix the value of the potential at some point (usually calling the cell's negative terminal zero).

5.41 When you connect a voltmeter between two points in a circuit it tells you how much energy is transferred per coulomb of charge passing between those points. You can assume that in the circuit shown on the following page, the connecting wire and the ammeters have negligible resistance. Between which of the following pairs of points would a voltmeter read zero (or nearly zero): A and B, B and C, C and D, D and E, E and F, F and G, G and H, H and A? Explain your answers.

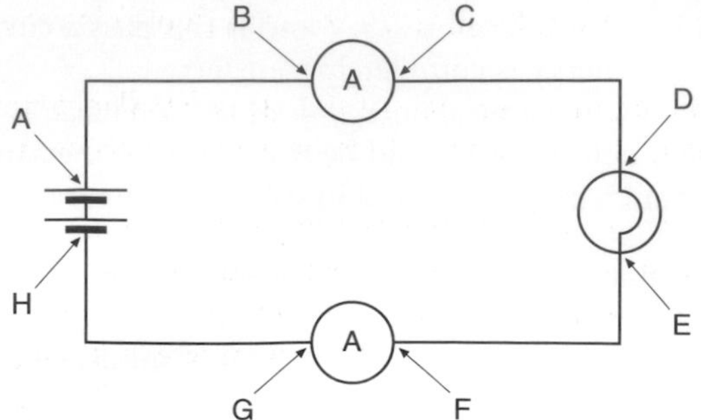

5.42 When 1.0 C passes through each of the lamps in the circuit 4.0 J of electric p.e. is transferred to internal energy in the lamp. What is

(a) the p.d. across each bulb

(b) the charge which passes through the battery

(c) the energy supplied by the battery

(d) the p.d. across the battery

If it took 20 s for this to happen, calculate

(e) the current in each bulb

(f) the current in the battery

(g) the rate at which energy is supplied by the battery.

5.43 A teacher wants her students to understand what potential difference means, so she connects up the circuit shown in the diagram. Before the switch is closed, the joulemeter reads 66 100 J. She closes the switch and while the lamp is lit the ammeter reads 2.0 A. After 5 minutes she opens the switch and notes that the joulemeter reading is now 73 300 J.

(a) How much charge passed through the lamp in this experiment?

(b) How much energy was transferred?

(c) What was the potential difference across the bulb?

(d) How would you use this experiment to explain to someone what potential difference means?

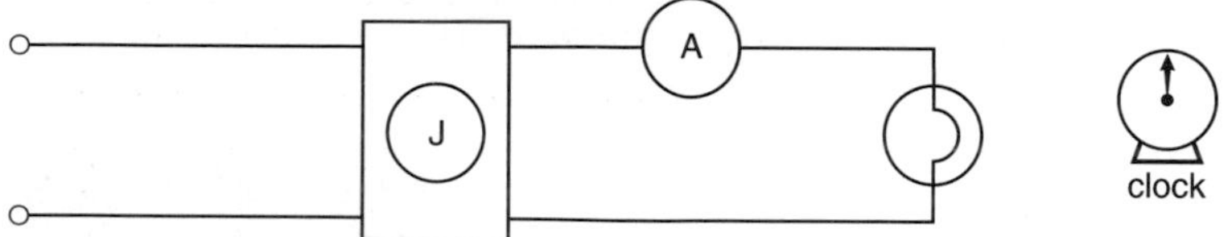

5.44 In the circuit shown in the figure a voltmeter reads 1.46 V when connected across the cell and 0.67 V when connected across the bulb. The wires have negligible resistance.

(a) What will the voltmeter read when connected between

(i) A and B

(ii) C and D

(iii) E and F?

(b) What will the voltmeter read when connected across the resistor?

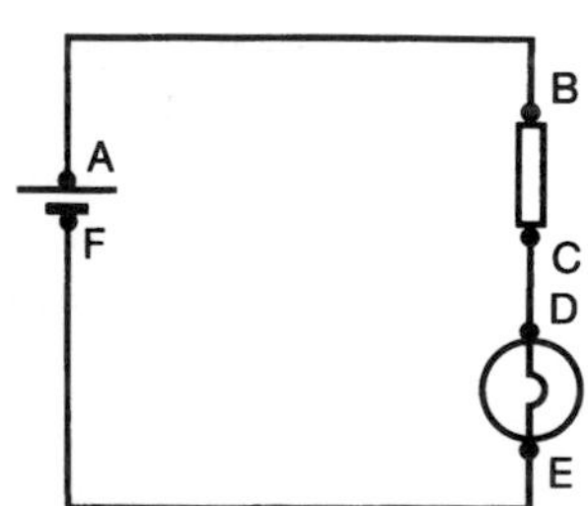

5.45 The point F in the circuit (see previous figure) is earthed, i.e. its potential is taken to be zero.
(a) What are the potentials of the other labelled points?
(b) Where are the points which have potentials of **(i)** 1.30 V **(ii)** 0.70 V **(iii)** 0.64 V?

5.46 The diagram shows a battery connected to a length AB of 'resistance' wire (i.e. wire which has much more resistance than the other wires or the battery). The length of the wire is 0.50 m. The battery maintains a p.d. of 6.0 V between A and B. If B is earthed, how far from B is the point on the wire which has a potential of **(a)** 1.8 V **(b)** 5.4 V?

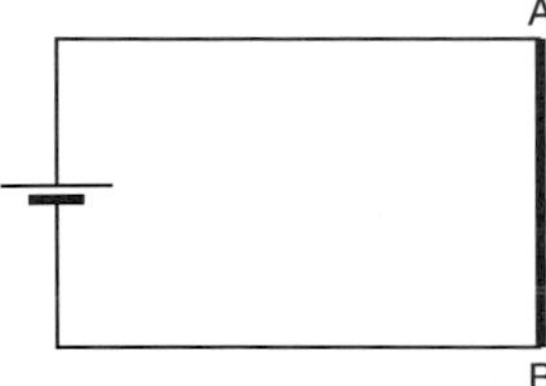

5.47 In this circuit the battery p.d. is 6.0 V and the p.d.s across A and B are measured to be 2.0 V and 4.0 V. What are the potentials of X, Y and Z if
(a) Z is earthed
(b) X is earthed
(c) Y is earthed

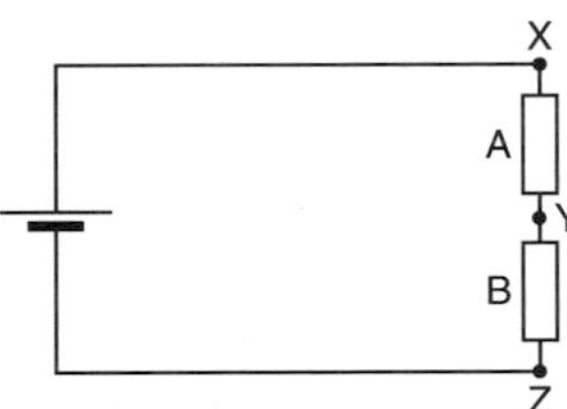

5.48 Draw circuit diagrams to show four lamps connected to a 12 V battery so that
(a) each has a p.d. of 12 V across it
(b) each has a p.d. of 6.0 V across it
(c) each has a p.d. of 3.0 V across it.

5.49 The circuit shows a bulb, a resistor and a motor connected in parallel across a battery. A voltmeter connected across the battery reads 2.89 V. What does it read when connected across
(a) the bulb
(b) the resistor
(c) the motor.

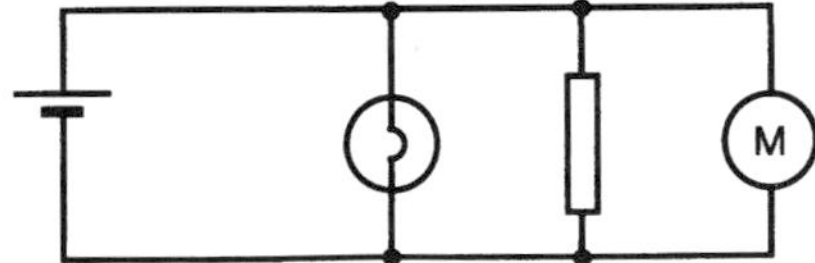

5.50 In the circuit the p.d. across the battery is 5.86 V, and the p.d. across the bulb is 2.45 V. If the wires have negligible resistance, what is the p.d. across
(a) resistor R_1
(b) resistor R_2?

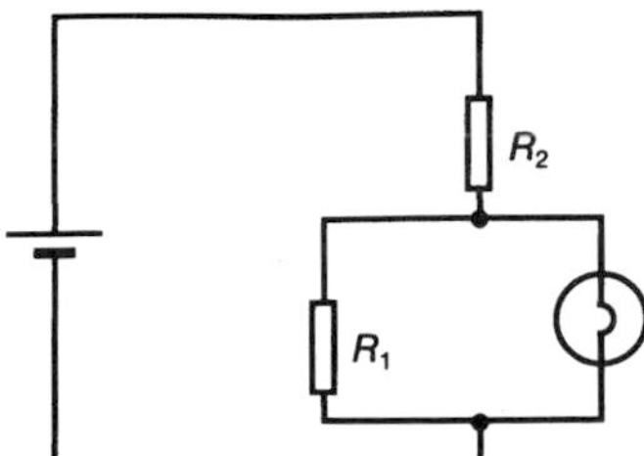

5.51 Five resistors are connected to a cell as shown. The p.d.s across R_1 and R_2 are 0.60 V and 0.75 V respectively, and the other three resistors are identical. The potential at point A is zero.

(a) What are the potentials at
- **(i)** point G
- **(ii)** point E?

(b) State the p.d. between
- **(i)** B and E
- **(ii)** B and F, in each case saying which point has the higher potential.

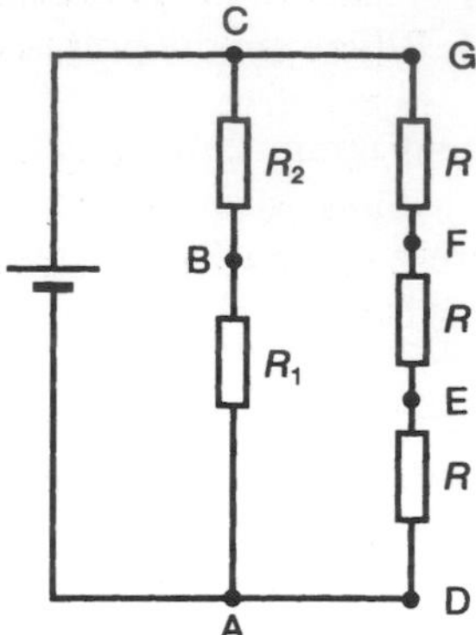

5.52 Two identical cells of e.m.f. 1.50 V are connected to two identical resistors in the circuits shown. In each case point C is earthed.

(a) What are the potentials at A, B and D in circuit **(i)** and in circuit **(ii)**?

(b) In each case, what would be the effect of connecting a wire between C and D?

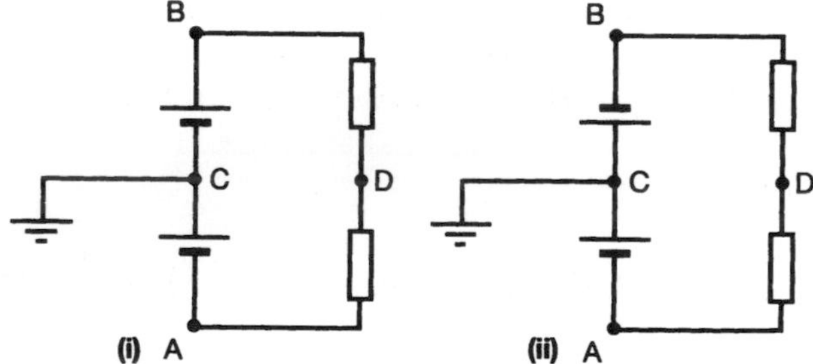

5.53 A uniform sheet of conducting paper is cut into a rectangular shape as shown in part **(a)** of the diagram. The more heavily shaded areas show where strips of conducting paint have been painted on to the paper. The paint is thick enough for its resistance to be negligible. A battery is connected to the lines of paint and maintains a p.d. of 6.0 V between A and C. C is earthed. Copy the diagram and draw on it a line which joins all the points which have a potential of 4.0 V, and another line which joins all the points which have a potential of 2.0 V.

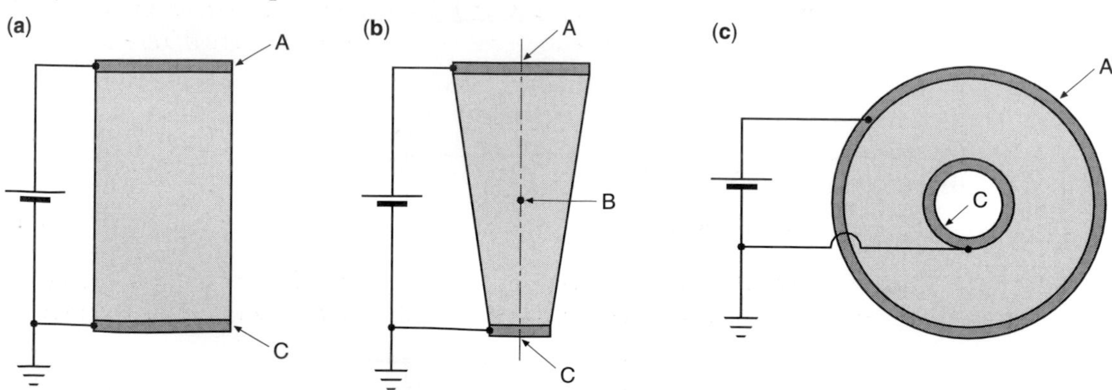

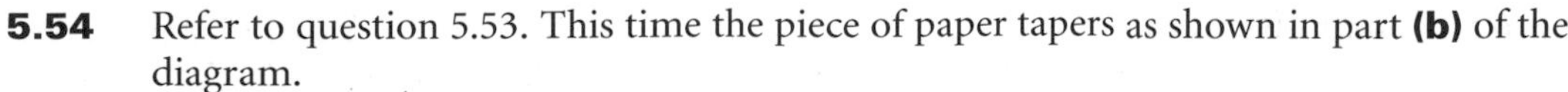

5.54 Refer to question 5.53. This time the piece of paper tapers as shown in part **(b)** of the diagram.

(a) Copy this part of the diagram.

(b) Explain whether the potential at B, which is the mid-point of AC, is more or less than 3.0 V.

(c) Mark on your diagram the approximate positions of the lines joining points which have a potential of 4.0 V, and the points which have a potential of 2.0 V.

(d) Hence sketch a rough graph of how the potential varies with distance along the line ABC.

5.55 Refer to question 5.53. This time, in part **(c)** of the diagram, the piece of paper has a circular shape. One strip of conducting paint runs round the edge of the circle. The other forms a smaller circle near the centre of the shape.

(a) Copy this part of the diagram.

(b) Mark on your diagram the approximate positions of the lines joining points which have a potential of 4.0 V, and the points which have a potential of 2.0 V.

5.56* **(a)** It is estimated that the average electric charge carried in a lightning flash is 5 C. If the p.d. between the cloud and the ground is about 800 MV, approximately how much energy is transferred in the flash?

(b) In a typical thunderstorm lightning flashes strike the ground at intervals of about 3 minutes. Over the whole surface of the Earth the total current carried in this way between the atmosphere and the ground averages 1800 A. Estimate the average number of thunderstorms taking place at any instant over the whole Earth.

5.5 Electrical power

In this section you will need to

- use the equations $W = Pt$ and $P = W/t$
- use the equations $P = \mathcal{E}I$ and $P = VI$ to calculate the power of a cell or the power of another circuit component.

5.57 A lead-acid battery of e.m.f. 12 V is supplying a current of 10 A to some car headlamps. What is the rate of transfer of chemical energy to electrical energy?

5.58 The power rating of a portable tape recorder is 2.4 W and the current it needs is 0.4 A. How many 1.5 V cells would be needed, and would they be connected in series or in parallel?

5.59 A hand-held vacuum cleaner uses 4 nicad cells, each of which has an e.m.f. of 1.2 V, in series. If the current drawn is 20 A, what is the power of the vacuum cleaner?

5.60 An electric toaster is labelled 800 W. How much energy is transferred in it in 3 minutes?

5.61 The current in a small immersion heater is 3.8 A and the p.d. across its terminals is 11.9 V. How much electrical energy is transferred to internal energy in 20 minutes?

5.62 An electric light bulb is labelled 100 W and is designed to be used with a p.d. of 240 V. What current flows in it, and how much energy is transferred in 1 hour?

5.63 Two lamp bulbs are labelled '240 V 100 W' and '240 V 60 W'. What do these markings mean, and what is the current in each of them when it is working normally?

5.64 The energy of a single flash of light from a stroboscopic lamp is 0.60 J. The p.d. across the bulb is 240 V.

(a) How much charge passes through the lamp during the flash?
(b) If the flash lasts for 10 μs, what is the average current?
(c) What is the average power?

5.65 The circuit shown in the diagram contains four lamps. The powers of the lamps are shown. The supply provides a p.d. of 240 V. Calculate

(a) the current at A **(b)** the current at B **(c)** the p.d. across the 60 W lamps
(d) the current at C **(e)** the current drawn from the supply
(f) the total power provided by the supply.

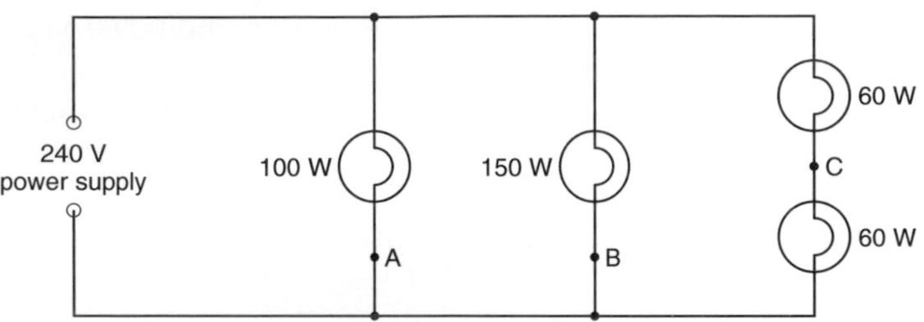

5.66 A car has two headlamps, two side-lamps and two tail-lamps. The electrical circuit for these lamps is shown in the diagram. The earthing signs to the bottom show that the metal frame of the car acts as the return wire to complete the circuit.

(a) Copy the diagram and label the switches and lamps.
(b) Explain why the headlamps cannot be on unless the side-lamps and tail-lamps are on.
(c) What will happen if, with the headlamps on, one of the side-lamps fails?
(d) If each side- and tail-lamp carries a current of 0.50 A, what is the power transfer from the battery with only the side- and tail-lamps on?
(e) The headlamps are known to have a power of 48 W each. What total current passes through the battery when all the lamps are on?
(f) A car also has a heater for the rear window. Add this to your circuit diagram.

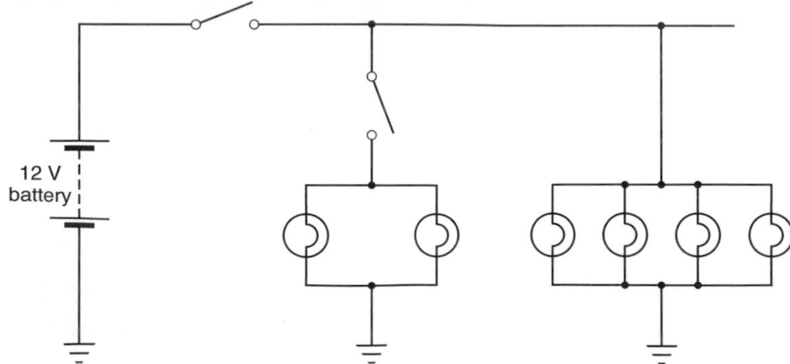

5.67 A power station generates electrical energy at a p.d. of 25 kV and an average rate of 500 MW.

(a) What is the current in the cables leaving the power station?

(b) How much energy is generated in one day?

(c) If the efficiency of the power station is 40%, how much internal energy is delivered to the surroundings in one day?

6 Electrical resistance

Data: electronic charge $e = 1.60 \times 10^{-19}$ C

material	copper	aluminium	steel	nichrome
resistivity/10^{-8} Ω m	1.7	3.2	14	130

6.1 Resistance

In this section you will need to

- use the equation $R = V/I$
- understand that Ohm's law states that the resistance of a metallic conductor is constant if physical conditions are constant
- describe how to check on whether Ohm's law is obeyed for a particular circuit component
- describe how to plot characteristics (i.e. graphs of current I against p.d. V) for circuit components
- draw the characteristics for a lamp filament, a thermistor and a silicon diode
- use the equations $R_{ser} = R_1 + R_2$ and $\frac{1}{R_{par}} = \frac{1}{R_1} + \frac{1}{R_2}$
- use the equations $P = I^2R$ and $P = V^2/R$ to calculate the power transfer in a resistor
- explain why the resistance of two resistors in parallel is always less than the resistance of the smaller of the two.

6.1 There is a current of 0.20 A in a wire when the potential difference between its ends is 5.0 V. What is its resistance?

6.2 The opposite faces of a sheet of polythene are covered with metal foil. When the potential difference between the two layers of foil is 12 V, the current through the polythene is 1.4×10^{-10} A. What is the resistance of the polythene?

6.3 What potential difference must be applied to a resistor of resistance of 10 MΩ to drive a current of 5.0 μA through it?

6.4 The current I in a resistor was measured for various values of the applied p.d. V and the values shown below were obtained:

V/V	50	100	200	300	400
I/mA	0.60	1.15	2.20	3.15	4.04

(a) Plot a graph of I against V.
(b) Was the resistor made from metal or from carbon?
(c) What was the resistance of the resistor at an applied p.d. of 200 V?
(d) What was the resistance of the resistor at an applied p.d. of 400 V?
(e) Estimate the value of the resistance for a very small applied p.d.

6.5 A lamp bulb is connected to the mains supply by cable, each wire of which has a resistance of 0.025 Ω per metre. If the length of the cable between the supply and the lamp is 8.0 m, what is
(a) the resistance of each wire
(b) the p.d. between the ends of each wire when the current in it is 0.60 A
(c) the power in the cable then?

6.6 The heating element of one kind of toaster consists of a ribbon of nichrome wire which is wound round an insulating support. A girl finds that it is broken and decides to mend it by twisting the two broken ends together (thus making the ribbon shorter). Explain whether the power of the toaster will be larger or smaller than before.

6.7 The diagram shows the kind of rheostat, or variable resistor, often found in laboratories. The slider makes contact with the wire beneath it. The wires which are wound round the 'former' (i.e. the supporting cylinder) are insulated from each other. Suppose the slider stays in the position shown. Which terminals would you connect to a circuit if you wanted **(a)** the least resistance **(b)** the most resistance **(c)** a resistance between the values in **(a)** and **(b)**?

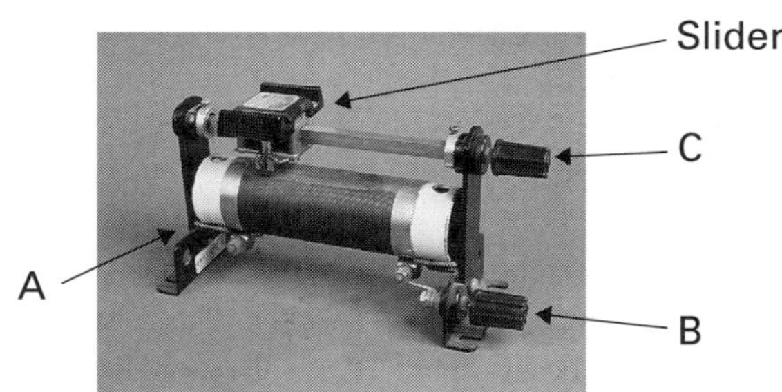

6.8 Refer to the previous diagram, and suppose that the total resistance of the resistance wire is 100 Ω. Complete the table to show the value of resistance between the different terminals as the slider is moved away from C:

position of slider	resistance between A and C	resistance between B and C	resistance between A and B
slider at end, next to C			
slider moved one-quarter of the length of the bar			
slider moved one-half of the length of the bar			
slider moved to the end of the bar (next to A)			

6.9 A rheostat consists of resistance wire uniformly wound on a former of length 300 mm. The resistance of the wire is 100 Ω. Initially the slider is at the centre of the rheostat (so that its resistance is 50 Ω), and the rheostat is connected in series with a battery which provides a constant p.d. of 6.0 V across the rheostat.

(a) What is the current in the circuit initially?

(b) The slider is now moved 30 mm so as to reduce the resistance of the rheostat. What is now the resistance of the rheostat, and the current in the circuit?

(c) The slider is now moved 30 mm on three more occasions, each time in the direction which reduces the resistance. What are the new resistances, and the currents?

(d) Comment on the suitability of the rheostat for adjusting the current in the circuit.

6.10 When the current through a resistor is 2.0 A its power is 10 W. Assuming that the resistance is unchanged, what does the power become when the current is increased to 6.0 A?

6.11 Two resistors R_1 and R_2, of constant resistance 10 Ω and 15 Ω, respectively, are each in turn connected to a power supply which provides a constant p.d. of 6.0 V. Which resistor provides the greater power, and what is that power?

6.12 Two resistors, of resistance 3.3 Ω and 4.7 Ω are connected first in series and then in parallel to a power supply which provides a constant p.d. of 6.0 V. Which resistor has the greater power, and what is that power, when the resistors are

(a) in series **(b)** in parallel?

6.13 A car headlamp bulb has a power of 60 W when it is connected to a potential difference of 12 V.

(a) What is the resistance of the filament?

(b) If its resistance remained constant, what would be the power if a p.d. of 6.0 V were connected across it?

(c) Its resistance will not remain constant: how will it change, and will the power of the bulb be larger or smaller than the power calculated in **(b)**?

6.14 The resistance of some wire is 14 ohms per metre. What length is needed to provide a power of 20 W when a p.d. of 12 V is available?

6.15 A set of Christmas tree lights consists of 20 lamps in series, and is designed to be connected to a 240 V supply. Each bulb is rated at 1.2 W.

(a) Draw a circuit diagram showing the lamps connected to the supply. What is the main disadvantage of this way of connecting the lamps? Can you think of any advantage?

(b) Calculate **(i)** the p.d. across each bulb **(ii)** the current in each lamp **(iii)** the current drawn from the supply.

(c) When one lamp fails the others do not go out. How might this result be achieved? Assuming that your solution is correct, what would happen if several lamps failed?

6.16 Resistances of 10 Ω and 15 Ω are joined **(a)** in series **(b)** in parallel. What is the total resistance in each case?

6.17 Calculate the combined resistance of each of the arrangements of resistors shown in the figure.

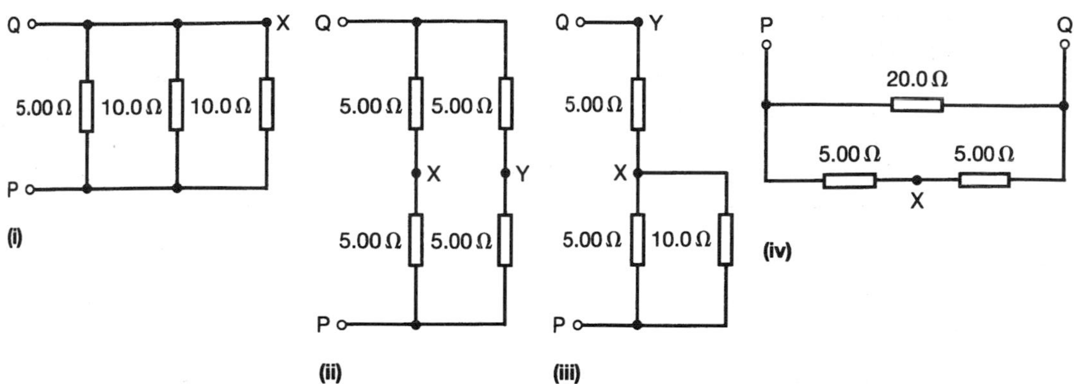

6.18 The diagram shows fifteen 100 Ω resistors mounted in a small package. The pins are labelled 1 to 16. What is the resistance between these pins (the '&' sign means that the pins are connected together)?

(a) 1 & 2 and 16 **(b)** 1 & 2 and 15 **(c)** 1 & 2 and 15 & 16
(d) 1 & 2 & 3 & 4 and 16 **(e)** 1 & 2 & 3 & 4 and 15 **(f)** 1 & 2 & 3 & 4 and 15 & 16

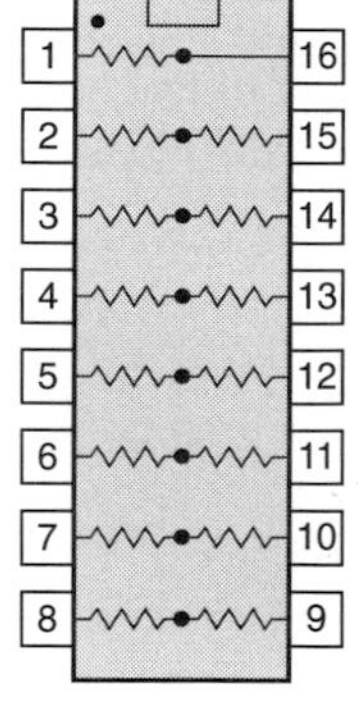

6.19 The manufacturer sets a limit of 0.125 W as the power which may be transferred in each resistor from question 6.18. A circuit designer wishes to use a resistor of resistance 100 Ω but needs to pass a current of 60 mA.

(a) Show that a current of 60 mA which enters at pin 1 and leaves at pin 16 will transfer energy at a rate of 0.36 Ω, which is too great.

(b) Explain how to use the set of resistors so as to obtain a resistance of 100 Ω and allow a current of 60 mA to pass without overheating.

6.20 What is the current in each of the resistors shown in the diagram?

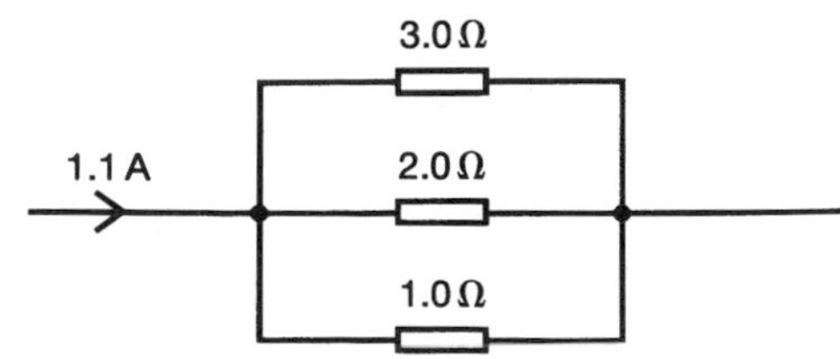

6.21 Each of the arrangements of resistors in question 6.17 is joined to a power supply, so that P is earthed and Q is at a potential of 6.0 V.
(a) What is the potential at X in each case?
(b) In **(ii)** and **(iii)** what are the potential differences between X and Y?

6.22 A connecting lead of length 1.0 m used in a laboratory consists of 55 strands of wire. Each strand has a resistance of 2.3 Ω. What is the resistance of the complete wire?

6.23 What is the total resistance when **(a)** two **(b)** ten 10 Ω resistors are connected in parallel?

6.24 In the two arrangements **(a)** and **(b)** of resistors shown in the figure, is the combined resistance about 100 Ω, between 1 Ω and 100 Ω, or less than 1 Ω? What general rule could you state for calculating a rough value of the resistance of resistors connected in parallel?

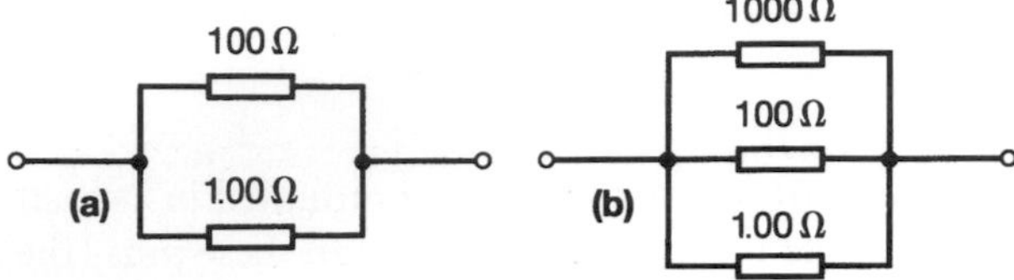

6.25 Calculate the combined resistance of each of the arrangements of resistors in the figure for question 6.24.

6.26 In the first circuit in the diagram a power supply maintains a p.d. V across two resistors R_1 and R_2 in parallel. The currents which flow in the circuit are I, I_1 and I_2. In the second circuit the two resistors have been replaced by a single resistor which draws the same current I from the power supply.
(a) What is the relationship between I, I_1 and I_2?
(b) Write down equations connecting
(i) V, I_1 and R_1
(ii) V, I_2 and R_2
(iii) V, I and R.
(c) Hence show that $\frac{1}{R} = \frac{1}{R_1} + \frac{1}{R_2}$

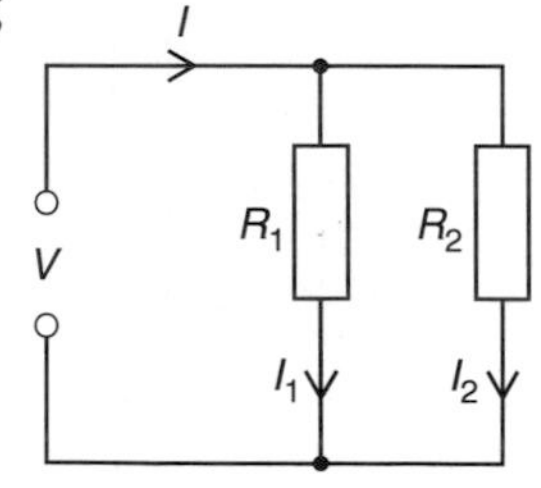

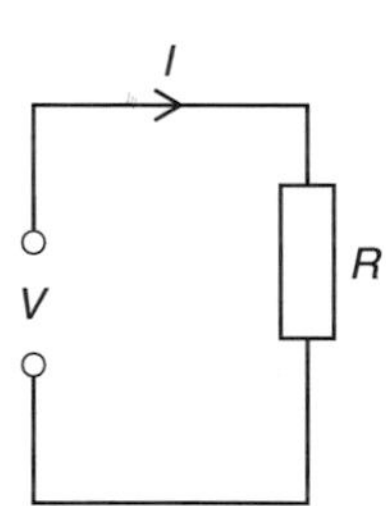

6.27 You are given a box in which there are three 10 Ω resistors connected in series, as shown in the diagram. Connections can be made to the ends of any resistor, but they cannot be removed from the box. You are asked to obtain a resistance of **(a)** 30 Ω **(b)** 5 Ω **(c)** 15 Ω. Make three copies of the diagram and add any connections you would need to do this. Label X and Y, the points between which the network has the required resistance.

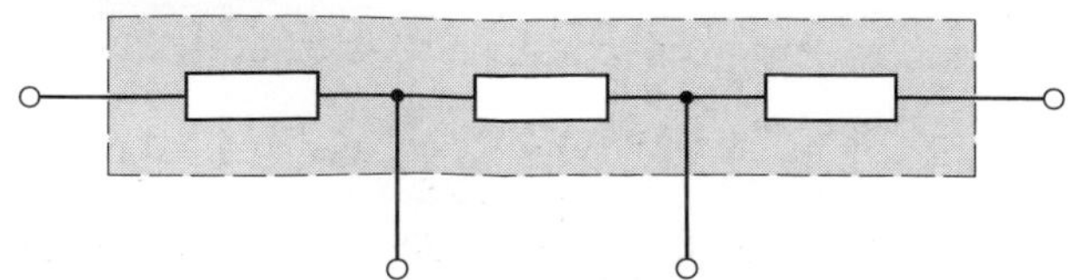

6.28 What is the smallest number of resistors you need to provide a resistance of
(a) 5 Ω, given a supply of 3 Ω resistors
(b) 7 Ω, given a supply of 4 Ω resistors?
In each case draw a diagram to show how you would connect them.

6.29 The diagram shows a resistor used as a shunt for an ammeter. The resistor consists of a metal film mounted on a plastic base. It is difficult to make the film of exactly the required thickness, so the resistance is adjusted by cutting through the film at points like A, B etc. Explain whether this will make the resistance larger or smaller.

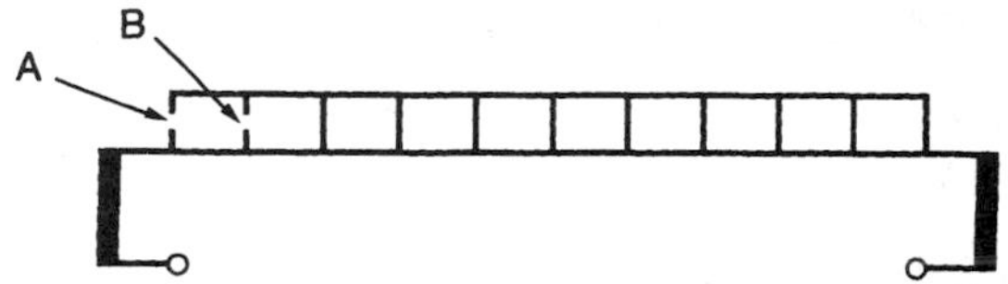

6.30 A manufacturer makes a series of 0.25 W resistors. That means that the maximum power transfer in one of these resistors is 0.25 W. Calculate the maximum p.d. which may be applied to one of these resistors which has a resistance of **(a)** 22 Ω **(b)** 220 Ω **(c)** 2.2 kΩ.

6.31 The ring on an electric stove contains two heating elements. They are controlled by a switch on which the three settings are Off, Low, Medium and High. The circuit for the switches and heating elements (R_1 and R_2) is shown in the diagram. What must be the positions of the two-way switches in order to obtain
(a) Off
(b) Low
(c) Medium
(d) High?

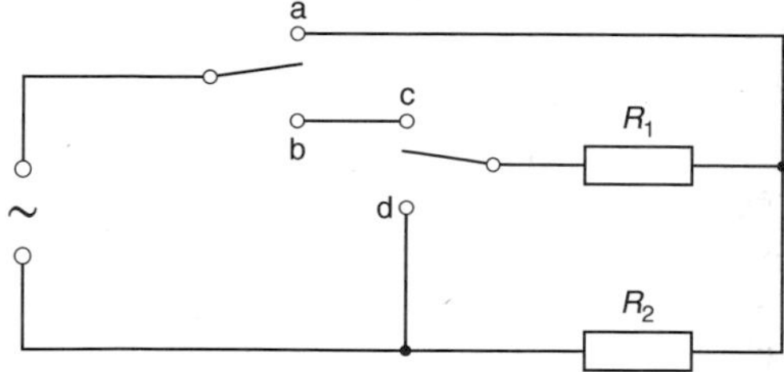

6.32 A small clothes iron (for use when travelling) is bought in Britain and has a power of 600 W when connected to a p.d. of 240 V. If it is taken to the USA, and used on a mains power supply whose p.d. is 110 V, what is its power?

6.33 The maximum current in a particular light-emitting diode (l.e.d.) is 30 mA. The maximum power which can be transferred in this type of l.e.d. is 100 mW.
(a) Show that the l.e.d. cannot be connected by itself to a 5.0 V power supply.
(b) Which of the following resistors would you place in series with the l.e.d. to limit the current to a safe value: 33 Ω, 47 Ω, 68 Ω, 82 Ω?

6.34 **(a)** What is the resistance, at full brightness, of **(i)** a 240 V 60 W bulb **(ii)** a 12 V 60 W bulb?
(b) How many 12 V 60 W bulbs must be connected in series across a 240 V supply if each is to work at full brightness?
(c) What will happen if two 240 V 60 W bulbs are connected in series across a 240 V supply?
(d) What will happen if a 240 V 60 W bulb and a 12 V 60 W bulb are connected in series across a 240 V supply?

6.35 Why is it not possible to calculate the current in the bulbs in part **(c)** of the previous question?

6.36 The graph in the diagram shows the result of an experiment to study the conducting properties of a thin nichrome wire. [Nichrome is a metal alloy often used to make resistors.]

(a) Use the graph to find the resistance *R* of the wire for *I*/mA = 10, 20, 30, 40, 50, 60, 70, 80.

(b) Plot a graph of *R* against *V*.

(c) Describe how *R* varies as *V* increases **(i)** from 0 to 1.0 V **(ii)** above 1.0 V.

(d) What do you think is happening to the nichrome wire above 1.0 V?

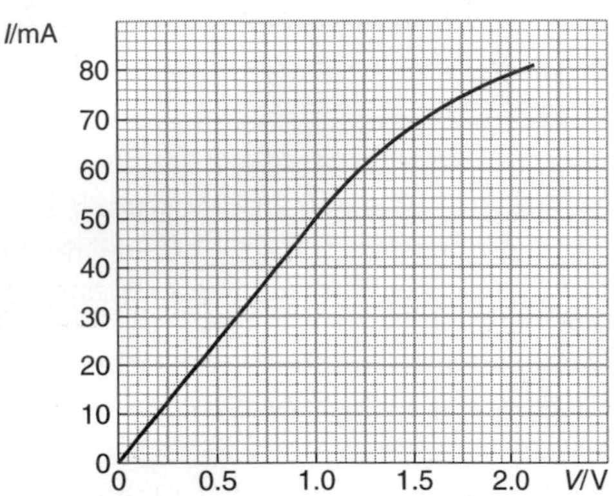

6.37 A tungsten filament bulb is connected to a variable power supply, and the current measured as the p.d. is varied. The table shows the readings obtained:

V/V	0	2.0	4.0	6.0	8.0	10	12
I/mA	0	36	57	72	84	94	102

(a) Plot a graph to show how current varies with p.d. for this bulb.

(b) On the same axes plot the graph which would be obtained if the same p.d.s were applied to a resistor of resistance 100 Ω.

(c) Use your graphs to find the current which would flow if a p.d. of 6.0 V were applied to the bulb and the resistor in parallel.

(d) The bulb and the resistor are now connected in series, and a p.d. of 12 V is applied to them. Explain why the current which flows is 67 mA.

(e) With the bulb and the resistor again connected in series, what p.d. is needed to send a current of 50 mA through them?

(f) The bulb and the resistor are now connected in parallel again. Explain why the applied p.d. must be 10 V if the current through the resistor is to be 100 mA.

6.38 A variable power supply is connected to two tungsten filament lamps A and B separately and the current is measured for different values of p.d. The results are shown in the table:

p.d./V	0	2.0	4.0	6.0	8.0	10	12
I/A for lamp A	0	1.80	2.84	3.66	4.28	4.72	5.00
I/A for lamp B	0	0.77	4.45	2.04	2.57	2.86	3.00

(a) On the same axes plot graphs of current (on the y-axis) against p.d. (on the x-axis) for both lamps.

(b) Describe how the current varies with p.d. for each lamp. What happens to the resistance of each lamp as the p.d. increases? How do you account for this change of resistance?

The bulbs are connected in parallel across a 12 V supply.

(c) Explain which bulb is brighter.

(d) What is the combined resistance of the two bulbs?

The two bulbs are now connected in series to a variable power supply.

(e) What p.d. is needed to pass a current of 2.5 A through the bulbs?

(f) The p.d. is steadily increased beyond this value. Which bulb burns out first?

6.39 A tungsten filament bulb, and a carbon filament bulb, are connected (separately) to a variable power supply, and the current measured as the p.d. is varied. The table shows the readings obtained:

V/V	0	50	100	150	200	250
tungsten I/A	0	0.11	0.16	0.20	0.23	0.26
carbon I/A	0	0.07	0.16	0.26	0.37	0.51

(a) Plot graphs to show how current varies with p.d. for each of these filaments.

(b) What is the power of each lamp when the p.d. across it is 240 V?

(c) When the bulbs are connected in parallel **(i)** what is the current drawn from the supply when the p.d. is set at 250 V **(ii)** what p.d. is needed to draw a current of 0.50 A from the supply?

(d) When the bulbs are connected in series **(i)** what is the current drawn from the supply when the p.d. is set at 250 V **(ii)** what p.d. is needed to draw a current of 0.25 A from the supply?

6.40 The diagram shows a graph of p.d. across a silicon diode against the current in it.

(a) What is the resistance of the diode when the p.d. across it is

(i) 0.25 V

(ii) 0.64 V

(iii) 0.74 V?

(b) Estimate the resistance of the diode when the p.d. across it is

(i) 0 V

(ii) 10 V.

(c) Draw a graph of resistance against p.d. for values of V between 0.55 V and 0.75 V.

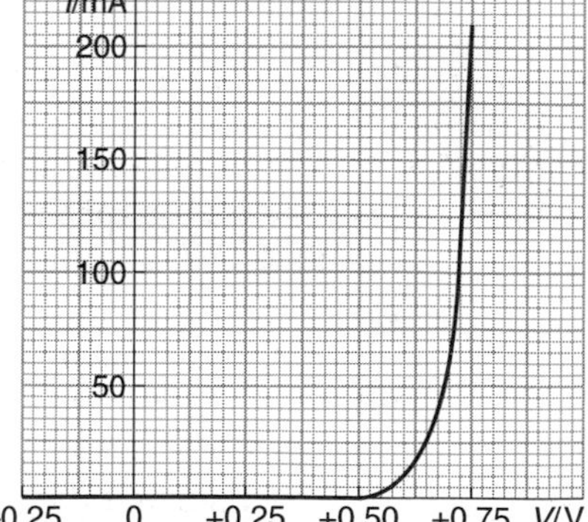

6.41 A silicon diode is connected to two resistors as shown in the figure overleaf.

(a) First assume that the silicon diode has zero resistance in the forward direction, and infinite resistance in the reverse direction. The p.d. between A and B is varied from −3.0 V to +3.0 V. Draw a graph to show how the current between A and B varies with p.d.

(b) Now assume that the diode behaves as a 'real' diode, like the diode in question 6.40. Sketch (no detailed calculations required) on the same axes how the current between A and B varies with p.d.

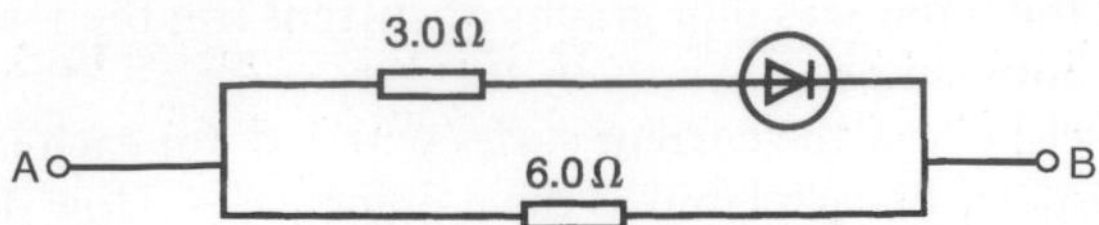

6.42 All the bulbs in the circuit in the diagram are identical. The battery has negligible internal resistance.

(a) Explain which will be brightest.

(b) The p.d. across C becomes 2.0 V when one of the bulbs fails. Explain which bulb failed.

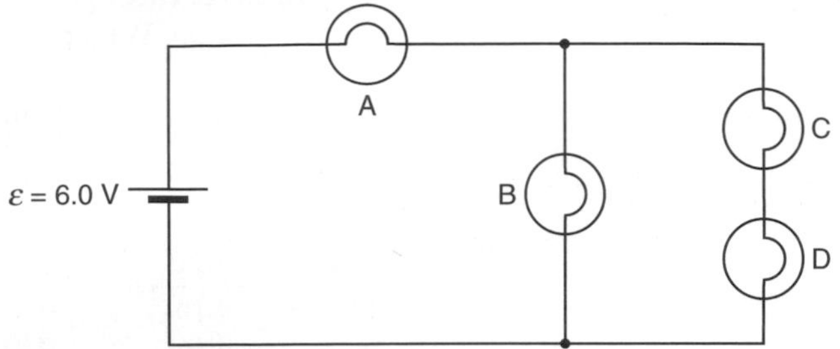

6.2 Resistivity

In this section you will need to

- use the equation $R = \rho l/A$ which defines resistivity ρ
- remember that the resistivity of metals increases with temperature but the resistivity of semiconductors decreases with temperature
- remember that the resistivity of all metals and alloys falls to zero at their transition temperature, and they are then superconducting.

6.43 Use the equation $R = \rho l/A$ to show that the unit of ρ is the Ω m.

6.44 Each of the copper wires in a three-core power cable has a cross-sectional area of 0.50 mm^2. What is the resistance of a 10 m length of this wire? [Use data.]

6.45 The filament of a 240 V 60 W lamp is made from 600 mm of tungsten wire. At its working temperature the resistivity of tungsten is 7.0×10^{-7} Ω m. What is the diameter of the wire?

6.46 Consider two pieces of copper wire. The first one has a length of 200 mm and a diameter of 0.50 mm. The second one has a length of 100 mm and a diameter of 0.25 mm. Do they have the same resistance? If not, which has the larger resistance?

6.47 Conducting putty is a material which is similar to Plasticine but it is an electrical conductor. A student rolls some of the putty into a cylindrical shape which is 60 mm long and has a diameter of 20 mm. He then rolls it into a new cylindrical shape which has a diameter of 10 mm.

(a) What is the new length?

(b) What is the ratio (new resistance)/(old resistance)?

6.48 A wire has a resistance of 6.0 Ω. It is then doubled back on itself. What is now the resistance between the ends of the doubled wire?

6.49 A power cable (for the grid system) consists of six aluminium wires enclosing a central steel wire. The purpose of the steel wire is to give the cable strength: the current in it may be assumed to be negligible. If each of the aluminium wires has a diameter of 4.0 mm, calculate

(a) the cross-sectional area of each wire, in mm^2

(b) the resistance of 1.0 km of one of the aluminium wires

(c) the resistance per km of the whole cable.

[Use data.]

6.50 The table gives the values of resistance per metre of manganin wire of different diameters. Manganin is a metal alloy used for making resistors.

diameter/mm	resistance per metre/$\Omega\ m^{-1}$
1.63	0.204
0.914	0.645
0.559	1.73
0.376	3.82
0.315	5.45
0.234	9.90

(a) Draw a graph of resistance R against area A.

(b) Draw a second graph to show that R is inversely proportional to A.

6.51 The heating element of an electric toaster consists of a ribbon of nichrome which is 1.0 mm wide and 0.050 mm thick. What length of ribbon is needed to provide a power of 800 W when the element is connected to a p.d. of 240 V? [Use data.]

6.52 **(a)** What is the resistance of a copper wire of length 1.2 m and diameter 0.50 mm? [Use data.]

(b) A power supply is connected to a 12 V 36 W lamp using two of these wires. If the lamp is working normally what is the current in each wire?

(c) What is the p.d. across each wire?

(d) What is the p.d. supplied by the power supply?

6.53 A car, which has a 'flat' battery, can be started if jump leads are connected from the battery to the battery in another car whose battery has enough energy. The current used by a starter motor may be as high as 800 A. Suppose the jump leads must be at least 1.5 m long, and that 20 strands of copper wire are used in each lead to make it flexible. The e.m.f. of each battery is 12 V and the potential difference across each of the wires themselves must be less than 1.0 V.

(a) Calculate the diameter of each of the 20 strands of wire. [Use data.]

(b) What is the rate of transfer of electrical energy to internal energy in each wire?

6.54 A slice of silicon which measures 30 mm by 30 mm and which is 0.50 mm thick, as shown in the diagram overleaf, has conducting strips fitted to two opposite edges AB and CD.

(a) If the resistivity of silicon is $4.0 \times 10^3\ \Omega$ m, calculate the resistance of the sheet measured between AB and CD.

(b) What would be the resistance of a similar sheet, measuring 15 mm by 15 mm, of the same thickness?

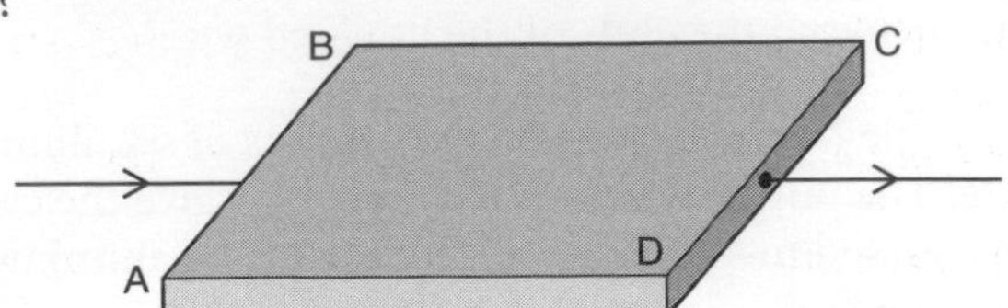

6.55 The resistance of the heating element of an electric fire is measured at 0°C and found to be 50 Ω. When it is connected to a 240 V supply the power delivered is 1.0 kW. If the resistivity of the metal increases by 0.000 17 of its resistance at 0°C for each degree rise in temperature, calculate the temperature of the heating element when it is working normally.

6.56 The resistance of the tungsten filament of an electric lamp is found to be 27 Ω at 0°C (= 273 K). If the resistivity of tungsten increases by 0.0056 of its resistivity at 0°C, for each degree rise in temperature, calculate

(a) the resistance of the filament at its working temperature of 2400 K

(b) the ratio

$$\frac{\text{power of lamp when first switched on}}{\text{power of lamp at working temperature}}$$

6.57 Explain why the power of a filament lamp falls during the first few milliseconds after it is switched on.

6.58* A copper wire and an iron wire have the same length and diameter. The resistivity of iron is about 8 times that of copper. They are connected to a variable power supply **(a)** in parallel **(b)** in series.
The p.d. provided by the power supply is gradually increased from zero. Explain which wire glows first in each situation.

6.3 Internal resistance

In this section you will need to

- understand what is meant by the internal resistance r of a cell
- use the equations $\mathcal{E} = I(R + r)$ and $\mathcal{E} = V + Ir$ to analyse circuits in which the internal resistance r of the cell has to he taken into account
- describe how to measure the internal resistance of a cell
- explain that a cell delivers maximum power when the resistance of the external circuit is equal to the cell's internal resistance.

6.59 A bulb is connected to a battery, of e.m.f. 1.50 V, which has some internal resistance. What can you say about the p.d. across the battery and the p.d. across the bulb?

6.60 A bulb of resistance 14.0 Ω is connected to a dry cell of e.m.f. 1.50 V and internal resistance 0.80 Ω. Calculate

(a) the current in the circuit

(b) the p.d. across the bulb

(c) the p.d. across the cell.

6.61 A voltmeter of resistance 100.0 Ω is used in an attempt to measure the e.m.f. of a cell.
(a) If the cell has an e.m.f. of 1.5 V and its internal resistance is 0.80 Ω, what reading does the voltmeter give for the e.m.f.?
(b) What would be the reading if the voltmeter had a resistance of 10 000 Ω?

6.62 Two cells, each of e.m.f. 1.50 V and internal resistance 0.50 Ω, are connected **(a)** in series **(b)** in parallel.
In each case what is the combined e.m.f. and internal resistance?

6.63 A student sets up the circuit shown in the diagram. He expects the ammeter to read 0.25 A but finds that it reads 0.20 A.
(a) Explain **(i)** his expected reading **(ii)** why the actual reading is lower.
(b) Calculate the internal resistance of the cell, assuming that the ammeter and the leads have negligible resistance.
(c) Calculate the rate at which
(i) the cell is transferring chemical energy to electrical potential energy
(ii) the resistor is transferring electrical potential energy to internal energy
(iii) the cell is transferring electrical potential energy to internal energy.

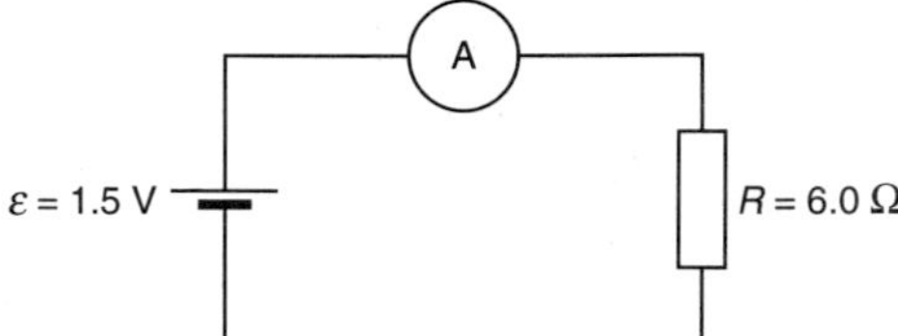

6.64 A battery has an e.m.f. of 3.0 V and is connected to a bulb. The current in the bulb is 0.30 A and the potential difference between its ends is 2.8 V.
(a) In 5 minutes how much chemical energy is transferred to electrical energy in the battery?
(b) In the same time how much electrical energy is transferred to internal energy in the bulb?
(c) How do you account for the fact that these two answers are not the same?

6.65 The diagram shows a circuit containing a cell of e.m.f. 6.0 V and two resistors of resistance 15 Ω and 10 Ω. In parts **(a)** and **(b)** assume that the cell has negligible internal resistance.
(a) What is the p.d. across the 10 Ω resistor?
(b) A second 15 Ω resistor is now connected between the points X and Y. What is now the p.d. across the 10 Ω resistor?
(c) What would your answers to **(a)** and **(b)** become if the cell had an internal resistance of 1.0 Ω?

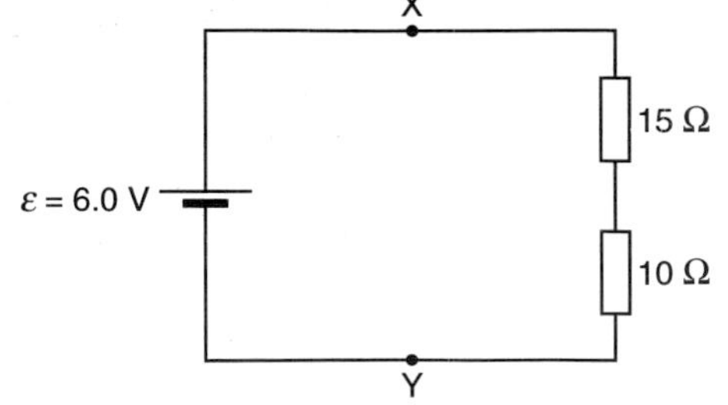

6.66 A battery is connected to an electric motor which is used to raise a load. An ammeter shows that the current in the motor is 2.6 A when the motor raises a load of 3.2 kg at a steady speed of 0.43 m s^{-1}. Assuming that there are no transfers of energy into internal energy, what is the e.m.f. of the battery?
Explain whether the actual e.m.f. of the battery would be greater or less than this calculated value if transfers of energy into internal energy cannot be neglected.

6.67 A cell of e.m.f. 1.55 V and internal resistance 0.500 Ω is connected to a resistor of resistance 22.0 Ω. What is
(a) the current in the circuit **(b)** the p.d. across the resistor **(c)** the p.d. across the cell
(d) the rate of transfer of chemical energy to electrical energy in the cell
(e) the power transfer in the resistor?
Your answers to **(d)** and **(e)** should not be the same: explain why.

6.68 A small torch bulb is marked '2.5 V 0.3 A'.
(a) What is its resistance at its normal working temperature?
(b) What resistance would you put in series with it to run it from a 6 V battery?
(c) When the torch bulb is run directly from a battery of e.m.f. 3.0 V, the correct p.d. of 2.5 V is produced across it. What is the internal resistance of the battery?

6.69 A car battery has an e.m.f. of 12.0 V and an internal resistance of 5.0 mΩ. What is the p.d. across its terminals when it supplies
(a) a current of 0.50 A to each of two side-lamps and two tail-lamps
(b) a current of 0.50 A to each of these four lamps and a current of 5.0 A to each of two headlamps
(c) a current of 800 A to a starter motor?

6.70 A car driver switches on the car's side-lamps and headlamps before starting the engine. Explain why the headlamps dim when he operates the starter switch.

6.71 A car battery has an e.m.f. of 12 V, negligible internal resistance and a capacity of 70 A h. It is connected, through a series resistor of 1.2 Ω, to a battery charger, also of negligible internal resistance, which provides a p.d. of 16.5 V.
(a) What is the current in the circuit?
(b) If the battery initially has no energy stored in it, how long will it take to fill it with chemical energy?
(c) How much energy is converted in the resistor in this time?
(d) Express this as a percentage of the total energy supplied by the battery charger.

6.72 A cell of e.m.f. $\mathcal{E}$ and internal resistance r is connected to a resistor of resistance R, and there is then a current I in the circuit and a p.d. V across the resistor (and the cell). The power of the cell is $\mathcal{E}I$, the power of the resistor is VI: these measure rates of conversion of energy from one form to another form.
(a) For the rate of conversion $\mathcal{E}I$, what are the two forms?
(b) For the rate of conversion VI, what are the two forms?
(c) Explain why $\mathcal{E}I > VI$ in this situation. Using the symbols given in the question, write down an expression for X in the equation $\mathcal{E}I = VI + X$.

6.73 When you buy a resistor from a manufacturer you are told not only its resistance but also the maximum power which can be transferred in it without it overheating. A laboratory technician is designing a circuit which contains a 6 V battery and wants to limit the current by putting a 100 Ω resistor in the circuit. He has available only 100 Ω resistors which have a power rating of 0.25 W.
(a) Check that a p.d. of 6.0 V connected across a 100 Ω resistor will develop a power of more than 0.25 W.
(b) What is the least number of identical resistors which he can use in order to provide the same total resistance without exceeding the power limit for each?
(c) Draw a circuit to show how they would be connected to the battery.

6.74 A laboratory power supply is designed to provide p.d.s of up to 6.0 kV; it is provided with an internal resistance of 50 MΩ. Suppose the power supply is set at 6.0 kV.

(a) What will be the current if the power supply is connected to a resistor of resistance 100 MΩ?

(b) What will be the current if the terminals are short-circuited?

(c) What will be the current if an ammeter of resistance 5.0 Ω is connected directly across the power supply's terminals, with no other resistance in the circuit?

6.75 High-voltage power supplies in schools and colleges can provide up to 6.0 kV. This is much higher than the mains voltage, yet they are considered safe to use.

(a) Describe what is done to make them 'safe'.

(b) Under what circumstances might they not be safe?

(c) 'I don't see the point of providing such a big voltage if you're going to stop it providing a big current.' Does it matter that the current they can provide is very small?

6.76 The battery manufacturer in question 5.39 made a D-size cell with a capacity of 16 A h and a mass of 131 g. He makes other cells using the same materials. The C, AA and AAA size cells have capacities of 7.2 A h, 2.2 A h and 1.0 A h, and their masses are 64 g, 22 g and 11 g, respectively. Making the same assumptions as in question 5.39, calculate the energy stored per unit mass for these four cells, and comment on your answers.

6.77 A student uses the circuit shown in the diagram. R is a resistance box which gives known values of resistance. For various values of its resistance R the current is measured, and the table gives the results:

R/Ω	8.0	6.0	4.0	2.0	1.2	1.0	0.80	0.60
I/mA	167	214	300	500	682	750	833	938

(a) Calculate the power of the resistor for each value of R.

(b) Predict the power of the resistor when R is zero and when R is infinitely large.

(c) Plot a graph of power against resistance.

(d) For what value of R is the power a maximum?

(e) The internal resistance of the cell is known to be 1.0 Ω. Comment on the shape of the graph.

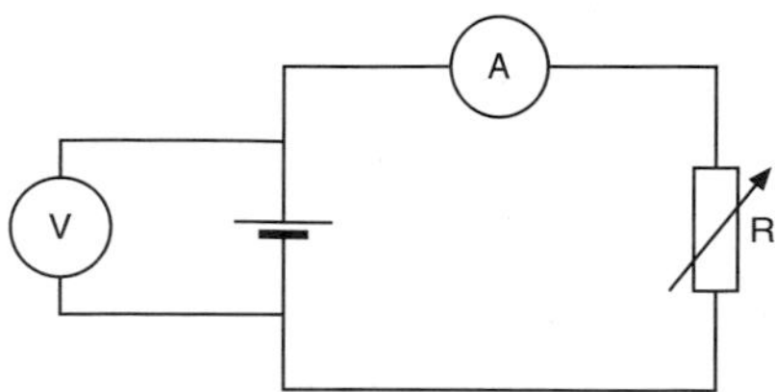

6.78* A student wants to measure the internal resistance of a cell and uses the circuit shown in the diagram for the previous question. He obtains the following readings:

V/V	1.43	1.41	1.39	1.33	1.20
I/mA	143	176	231	333	600

Plot a graph of V against I and deduce **(a)** the internal resistance **(b)** the e.m.f. of the cell.

7 Circuits and meters

7.1 Circuit calculations

In this section you will need to use what you have learnt in Chapters 5 and 6.

7.1 A circuit contains a battery, a bulb, a 15 Ω resistor, and an ammeter in series. The ammeter reads 0.23 A. A voltmeter, connected across the battery terminals, reads 5.7 V: when connected across the bulb it reads 2.4 V.
(a) What is the p.d. across the bulb?
(b) What is the resistance of the bulb?

7.2 A catalogue states that when a particular light-emitting diode is used with a 5.0 V supply a 270 Ω resistor must be connected in series with it to limit the current to 10 mA. Calculate
(a) the p.d. across the resistor
(b) the resistance of the l.e.d. in these conditions.

7.3 A power supply which provides a constant p.d. of 6.0 V is connected in series with a resistor of constant resistance 100 Ω and a thermistor. The resistance of the thermistor is 380 Ω at 25°C but falls to 28 Ω at 100°C. Calculate the p.d. between the ends of the resistor at **(a)** 25°C **(b)** 100°C.

7.4 The figure shows a circuit in which a battery provides a p.d. of 2.9 V. The p.d. across the 22 Ω resistor is found to be 1.2 V.
(a) What is the p.d. across
  (i) bulb A
  (ii) bulb B?
(b) What is the current in the resistor?
(c) If the bulbs have the same resistance, how many times greater is the current in B than the current in A (leave your answer as a fraction)?
(d) What is the current in the battery?

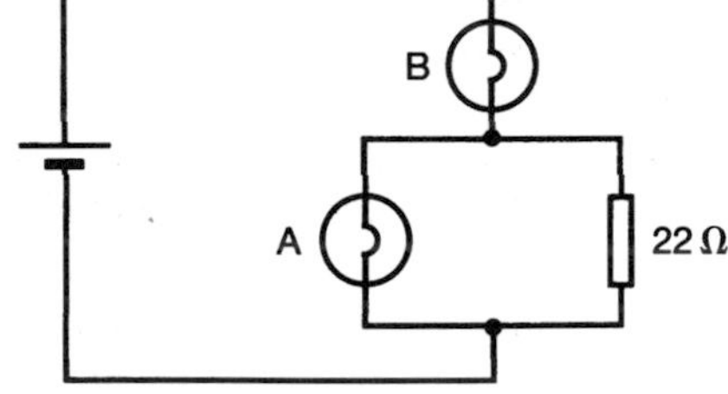

7.5 In the circuit in the previous figure the fixed resistor is replaced by a rheostat of maximum resistance 22 Ω. Explain what happens to the brightness of each bulb as the resistance of the rheostat is reduced from 22 Ω to zero.

7.6 Three resistors, of resistance 4.7 Ω, 10 Ω and 15 Ω, respectively, are connected to a battery as shown in the figure. A voltmeter connected across the battery reads 5.7 V.

(a) Calculate the resistance of the circuit between B and C.
(b) Calculate the p.d. between
  (i) A and B
  (ii) B and C.
(c) Calculate the current through
  (i) the 4.7 Ω resistor
  (ii) the 10 Ω resistor
  (iii) the 15 Ω resistor.

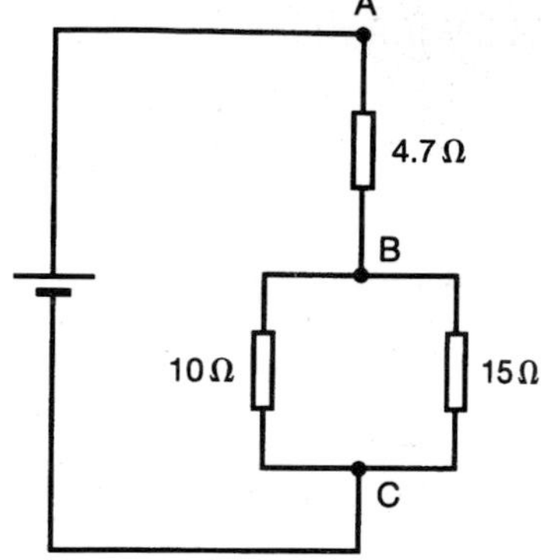

7.7 A bulb which is labelled '2.5 V 0.2 A' is connected in series with a rheostat of maximum resistance 50 Ω and a battery. Assume that the battery has a constant p.d. of 3.0 V between its terminals.
(a) Calculate the resistance of the bulb when it is working normally.
(b) Assume that the resistance of the bulb is constant, and calculate the minimum current which may be obtained by adjusting the rheostat.
(c) The resistance of the bulb is not constant: is the actual minimum current larger or smaller than your answer to **(b)**?

7.8 A power supply which provides a constant p.d. of 12 V is connected across two resistors R_1 and R_2 in series. Calculate the p.d. across each when
(a) the resistances of R_1 and R_2 are both 120 Ω
(b) the resistances of R_1 and R_2 are both 470 Ω
(c) the resistance of R_1 is 120 Ω and the resistance of R_2 is 470 Ω
(d) the resistance of R_1 is 680 Ω and the resistance of R_2 is 470 Ω
(e) the resistance of R_1 is 1.5 kΩ and the resistance of R_2 is 470 Ω.

7.9 If the negative terminal of the power supply in the previous question is connected to R_1 and earthed, what is the potential of the point between the resistors for each of the four cases?

7.10 Refer to question 7.8, part **(a)** A third resistor R_3 of resistance 120 Ω is available.
(a) When R_3 is connected in parallel with R_1, what is the p.d. across
  (i) R_1 **(ii)** R_2?
(b) When R_3 is connected in parallel with R_2, what is the p.d. across
  (i) R_1 **(ii)** R_2?

7.11 The diagram shows two identical resistors connected to the same cell in two different ways. In which case, **(a)** or **(b)**, is the total power greater, and how many times greater is it? Assume that the cell has no internal resistance, and that the resistors have constant resistance.

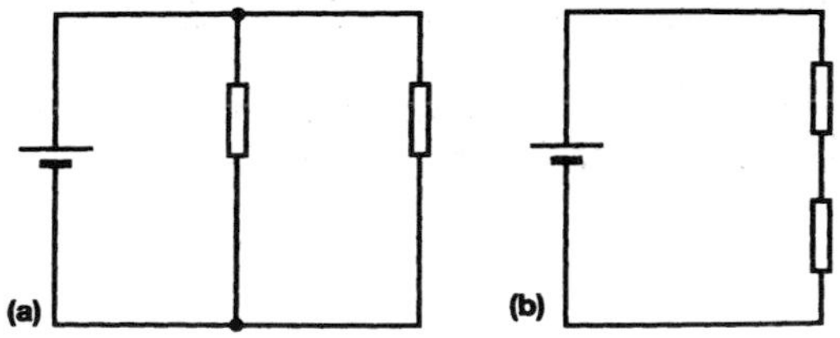

(a)

7.12 When connected to a 240 V mains supply the current in an electric lamp is 0.25 A. The current in an electric heating element connected in parallel to the same mains supply has a current in it of 5.0 A.

(a) Calculate the resistances of the devices when connected in this way (i.e., at their normal working temperatures).

(b) Assuming, in this part of the question, that the resistance of the devices remains constant, what will be the current in them when they are connected in series to the same supply, and what will be their powers?

(c) What will be the appearance of the two devices in these conditions?

(d) In practice the resistances will not have remained the same. Will they be greater or smaller when the devices are connected in series? Which device will have the greater change of resistance?

7.13 A mains lamp labelled '240V 60W' and a car headlamp bulb labelled '12 V 60 W' are connected in series to a 240 V supply. What will happen? [Hint: calculate the resistances of the lamp filaments when the lamps are in normal use.]

7.14 A mains lamp labelled '240 V 60 W' and a torch bulb labelled '1.25 V 250 mA' are connected in series to a 240 V supply. What will happen?

7.15 Your electric fan heater is not working and you need to draw a circuit diagram to help you find out what is wrong. All you know is that there are three simple on-off switches. One switches on the fan without heating, and the other two provide different powers. You assume that the circuit is arranged so that the heating cannot be switched on if the fan is not on, and that there are two separate heating elements.

(a) Draw a possible circuit diagram.

(b) The maximum power of the heater is stated to be 1.0 kW when it is connected to a 240 V supply. What is the resistance of each heater, assuming they are the same?

7.16 The diagram shows a circuit containing two cells C_1 and C_2 of e.m.f. 1.5 V, two resistors of resistance 15 Ω and two ammeters A_1 and A_2. The cells, the ammeters and the leads all have negligible resistance. What are the readings on ammeters A_1 and A_2

(a) in the circuit as drawn

(b) if a wire of negligible resistance is connected between X and Y

(c) if C_1 is reversed, with the wire still in place

(d) if the wire is removed, with C_1 still reversed?

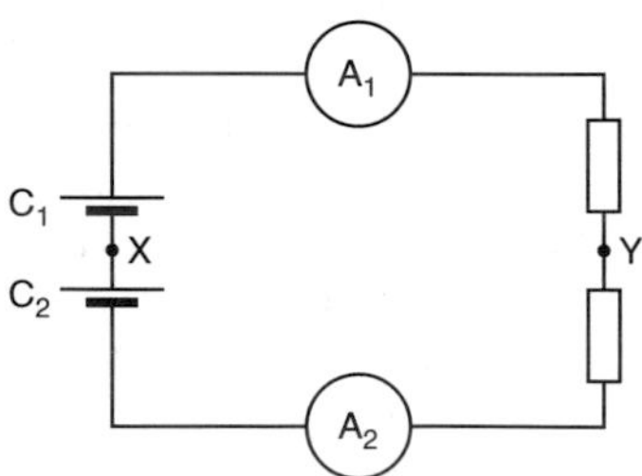

7.17 The diagram shows a cubical wire framework, made of resistance wire. The sides all have the same length, so each side has the same resistance. A current of 6 A enters at A, and leaves at H. In answering the questions remember that the circuit is symmetrical.

(a) What is the current in AB, AC, AD?

(b) What is the current in BE, BF, CF, CG, DE, DG?

(c) What is the current in EH, FH, GH?
If the potential at A is 5 V, and H is earthed, what are the potentials at
(d) B, C, D?
(e) E, F, G?

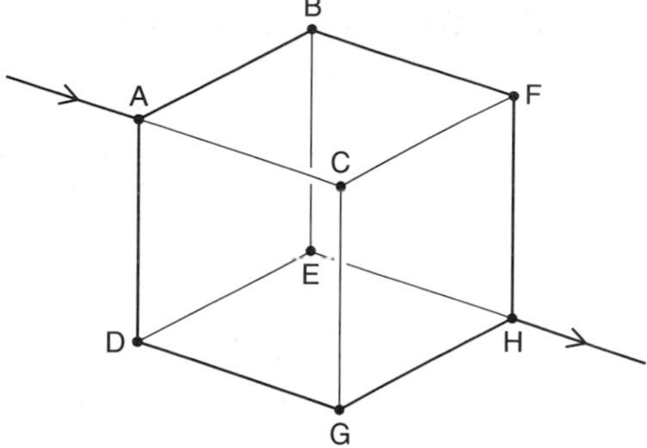

7.2 Circuits for measurement and sensing

In this section you will need to

- know how to connect an ammeter and a voltmeter in a circuit to measure current and p.d.
- describe how to measure the resistance and power of a resistor
- understand how a rheostat can be used as a current limiter, using two terminals
- understand how a rheostat can be used as a potential divider, using three terminals, and what the advantage of this is
- understand that thermistors and light-dependent resistors (l.d.r.s) are sensors which can be used to control other devices
- understand how potential dividers can be used with sensors.

7.18 The diagram shows a circuit in which a rheostat is connected to a battery. A voltmeter is also connected to the rheostat. The slider bar is moved from end A to end B of the rheostat. Explain why the voltmeter records a range of p.d.s from a value slightly less than the e.m.f. of the battery down to zero.

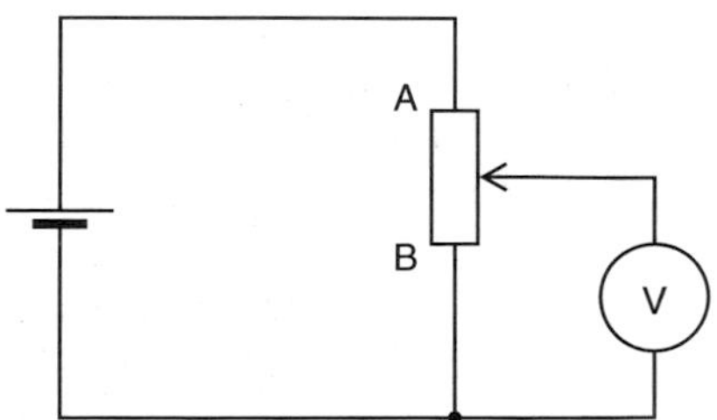

7.19 A rheostat of maximum resistance 100 Ω is connected as a potential divider, as shown in the diagram, to a power supply which provides a constant p.d. of 12 V. The slider is moved until the voltmeter (of very high resistance) reads 3.0 V.

(a) Copy the diagram and show the position of the slider.
(b) A 47 Ω resistor is now connected across the output of the potential divider, i.e. in parallel with the voltmeter. The voltmeter reading changes. Explain whether it rises or falls.
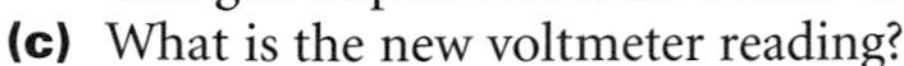
(c) What is the new voltmeter reading?
(d) Calculate the current in the power supply.
(e) What is then the current in **(i)** the 47 Ω resistor **(ii)** the part of the rheostat in parallel with the 47 Ω resistor?

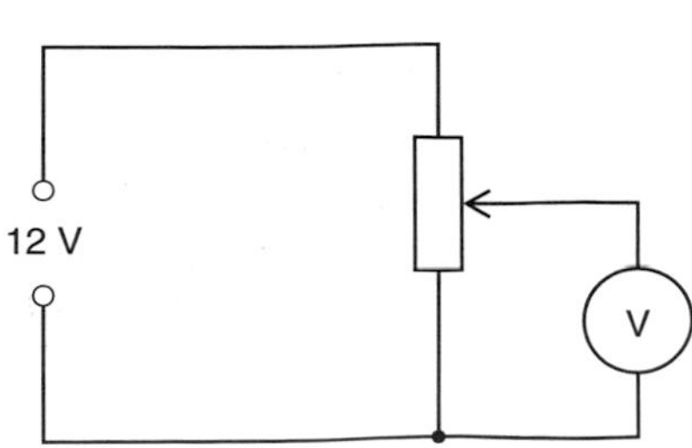

7.20 A rheostat of maximum resistance 50 Ω is connected as a potential divider to a battery which provides a constant p.d. of 3.0 V. A student adjusts the slider on the rheostat until the p.d. available, as measured by a voltmeter, is 2.5 V. He then connects a 2.5 V 0.2 A bulb in parallel with the voltmeter but finds that the bulb does not light. Explain this result.

7.21 The diagram shows two ways of connecting an ammeter and a voltmeter to measure the resistance of a bulb. Someone might say that in **(a)** the ammeter measures the total current through the bulb and the voltmeter, and in **(b)** the voltmeter measures the p.d. across the bulb and the ammeter.

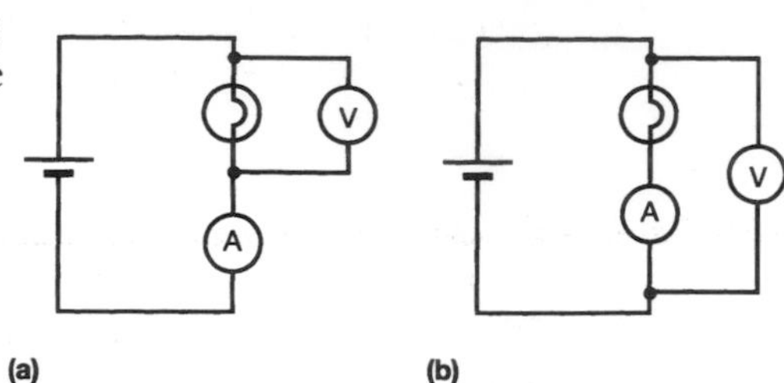

(i) Explain whether this matters.

Suppose that the bulb has a resistance of 20 Ω, the ammeter has a resistance of 0.10 Ω and the voltmeter a resistance of 100 kΩ.

(ii) Explain which method you would choose to measure the bulb's resistance. (No detailed calculations are required.)

7.22 The diagram shows a circuit containing a battery of e.m.f. 6.0 V and negligible internal resistance, and four resistors *P*, *Q*, *R* and *S*. The negative terminal of the battery is earthed, so the potential at that part of the circuit is zero.

(a) What are the potentials at X and Y if

(i) $P = Q = R = S = 10\ \Omega$?

(ii) $P = Q = 10\ \Omega$, $R = S = 20\ \Omega$?

(iii) $P = 5\ \Omega$, $Q = R = 10\ \Omega$, $S = 20\ \Omega$?

(b) What can you say about the values of *P*, *Q*, *R* and *S* if the potentials at X and Y are to be the same?

(c) What circuit component would you need to add in order to test whether the potentials at X and Y were the same?

(d) Suppose $P = 20\ \Omega$, $Q = 10\ \Omega$, *R* is a resistor of unknown size, and *S* is a variable resistor. What is *R* if $S = 23\ \Omega$?

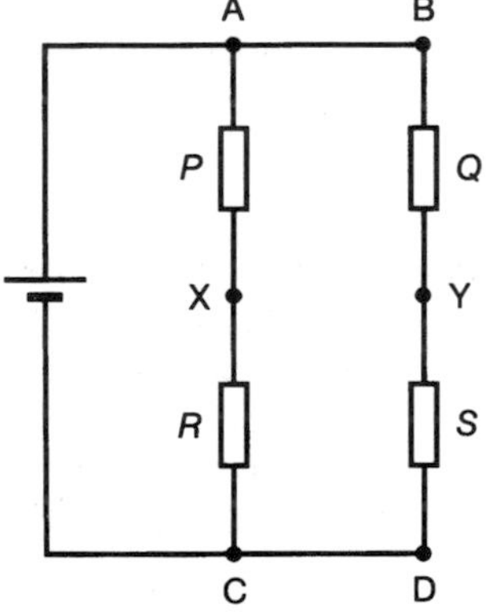

7.23 The circuit in the previous diagram was once used as a method of measuring resistance. It is called a bridge network. The bridge is said to be 'balanced' when the potentials at X and Y are the same.

(a) The resistances of *P*, *R* and *S* are 15 Ω, 33 Ω and 22 Ω. Show that the resistance of *Q* must be 10 Ω if the bridge is to be balanced.

(b) Explain why, if *P* and *R* are unchanged, and *Q* and *S* are increased to 100 Ω and 220 Ω, the bridge is still balanced.

(c) If the resistance of *Q* is now increased slightly from 100 Ω, explain in which direction the current will flow in an ammeter connected between X and Y.

7.24 A light-dependent resistor and a fixed resistor (of resistance 10 kΩ) are connected as shown in series between the terminals of a power supply which provides a constant p.d. of 5.0 V. The negative terminal of the power supply is earthed, i.e. may be taken to be 0 V. In the dark the potential of the point X is 0.21 V; when more light falls on the l.d.r. the potential of X rises to 4.2 V. Calculate the resistance of the l.d.r. in these two situations.

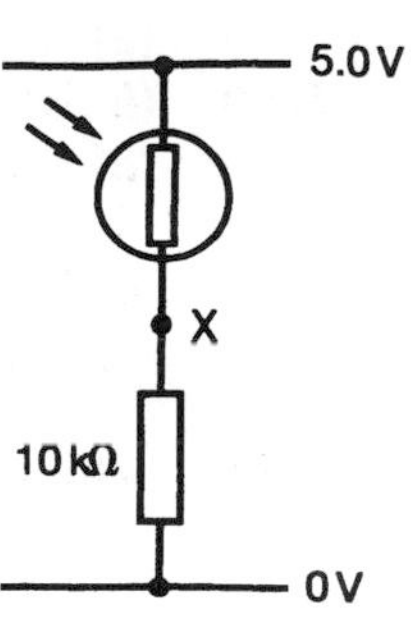

7.25 The diagram shows a circuit in which an l.d.r. is being used as a sensor. A lead at X is connected to a device called a transistor (not shown). When the potential at this point rises above about 0.7 V, the transistor 'turns on' and operates a relay which switches on a lamp.

(a) On a dull day, the resistance of the l.d.r. is 500 kΩ, and the rheostat is set to provide maximum resistance in series with the fixed resistor of 300 Ω.

(i) What is the potential at X?

(ii) Is the transistor 'on'?

(b) What happens to the resistance of the l.d.r. as the day brightens?

(c) Calculate the resistance of the l.d.r. when the transistor first turns on.

(d) Why is the rheostat included in the circuit?

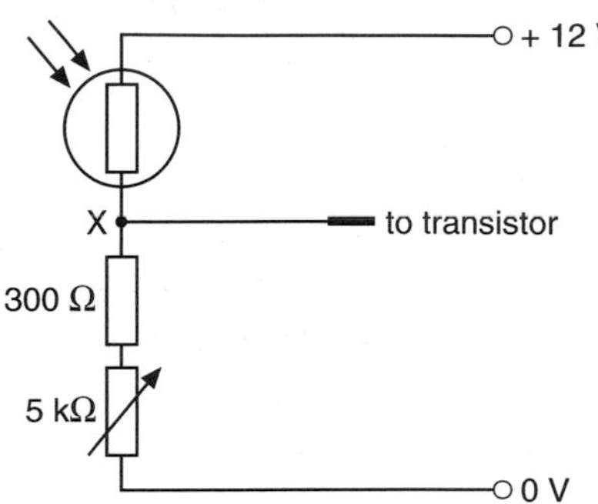

7.26 The diagram shows a circuit in which a thermistor is connected, and a graph of how its resistance varies with temperature. Draw a graph to show how the p.d. measured by the voltmeter varies with temperature.

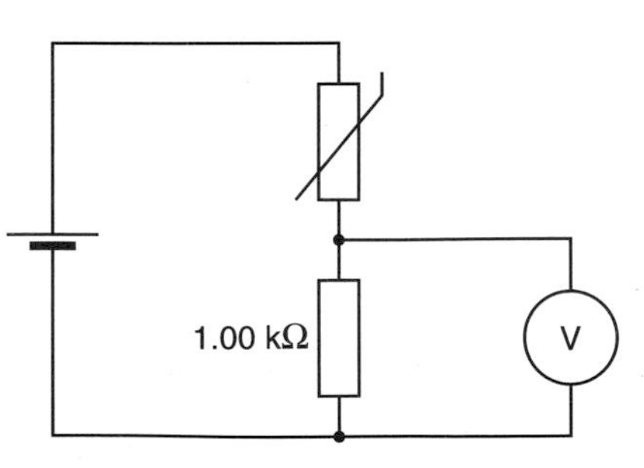

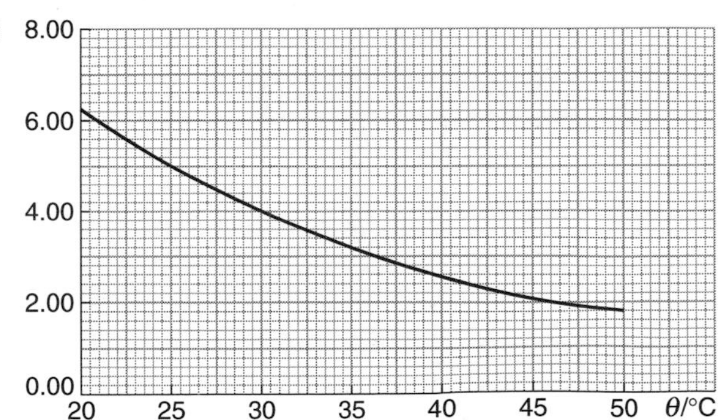

7.27* The table shows how the resistance of a thermistor varies with temperature.

temperature θ/°C	0	30	60	90	120
temperature T/K	273	303	333	363	393
R/kΩ	16.3	5.00	1.24	0.457	0.195

(a) Plot a graph of $\lg(R/\text{k}\Omega)$ (on the y-axis) against $1/(T/\text{K})$ (on the x-axis).
(b) Why is this thermistor called a *negative temperature coefficient* (NTC) thermistor?
(c) One of these thermistors is connected as shown in the circuit in the diagram. The point X is connected to a device called a transistor. When the potential at X rises above about 0.7 V, the transistor 'turns on' and operates a relay which switches on a lamp. Use your graph to find out what value the variable resistor should have if the lamp is to come on when the temperature reaches 50°C.
(d) What should be the value of the variable resistor if the lamp is to come on when the temperature reaches 80°C.
(e) What might be the purpose of the constant resistance of 15 Ω?
(f) What is the highest temperature which could be set in this circuit?

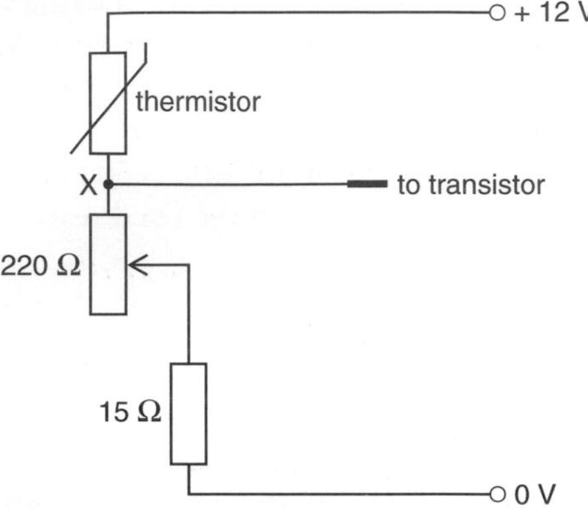

7.3 Meters and oscilloscopes

In this section you will need to

- understand why an ammeter must have a much lower resistance than the circuit in which it is placed
- understand why a voltmeter must have a much higher resistance than the component across which it is measuring the p.d.
- understand the advantages and disadvantages of light-beam meters, digital meters, data loggers and electronic meters
- describe how to use an oscilloscope to measure p.d.s and how the sensitivity (or gain) control is used
- describe how to use an oscillioscope to measure the frequency of an alternating p.d. and how the time-base control is used.

7.28 An analogue meter contains a coil which has a resistance of 100 Ω. The scale reads from 0 to 1.0 mA, as shown in the diagram.
(a) A p.d. of 50 mV is connected across the meter. What does the pointer point to on the scale?
(b) The meter is connected to a different p.d. The scale reading is 0.76 mA. What is the size of the p.d.?
(c) The scale is marked in mA, so the meter seems to be an ammeter. But it can also be used to measure p.d.s, as you have seen. Copy the diagram and adapt it so that it can be used to measure p.d.s.

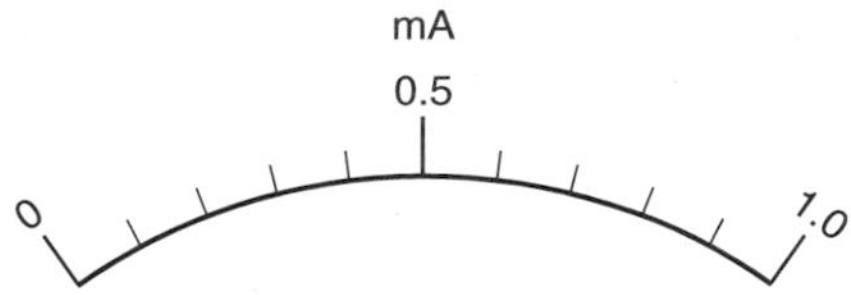

7.29 An analogue meter can measure currents of up to 100 μA when connected in a circuit. It can measure p.d.s of up to 100 mV when connected across part of a circuit. What is its resistance?

7.30 A student measures the p.d. between the terminals of a battery of negligible internal resistance and finds that it is 6.0 V. He then connects the battery to two 100 kΩ resistors as shown in the diagram, and uses a voltmeter to measure the p.d. across each of them in turn. He expects the voltmeter to record a p.d. of 3.0 V across each of them, but to his surprise the recorded p.d.s are each only 2.0 V. His teacher tells him that 'this is because the voltmeter has a resistance of only 100 kΩ'. Why does this remark explain the readings?

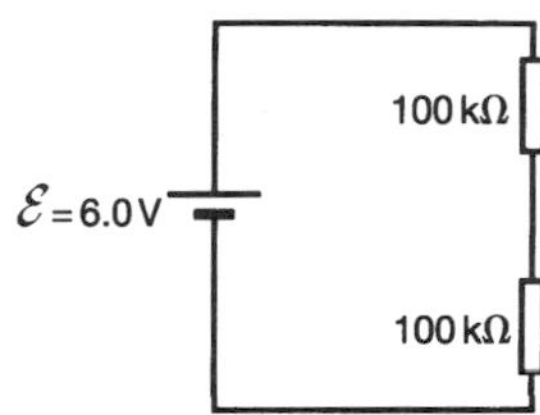

7.31 A cell of e.m.f. 1.5 V is connected in series with two resistors of resistance 47 kΩ and 68 kΩ, respectively. The p.d. across the 68 kΩ resistor is measured with an analogue meter of resistance 100 kΩ and a digital meter of resistance 100 MΩ. What readings does each meter record when

(a) the analogue meter is used by itself
(b) the digital meter is used by itself
(c) both meters are used simultaneously, both being connected in parallel across the resistor?

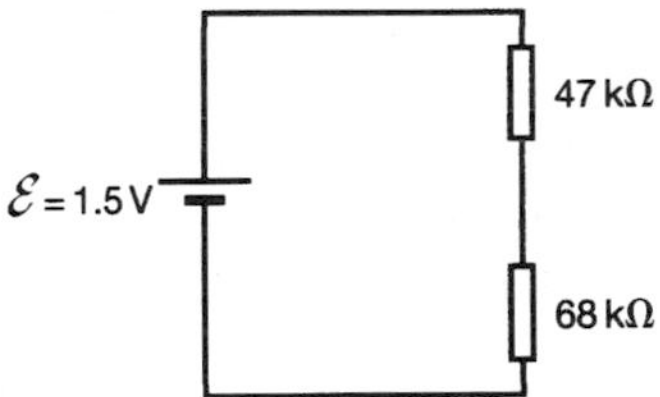

7.32 A student wants to find the resistance of a bulb which is known to be about 20 Ω. He has available a battery which has negligible internal resistance and an e.m.f. of 6.0 V, an ammeter which has a resistance of 10 Ω and a voltmeter which has a resistance of 100 kΩ. His first three attempts at connecting up a circuit, shown in the diagram, are wrong. In each case what do the meters read?

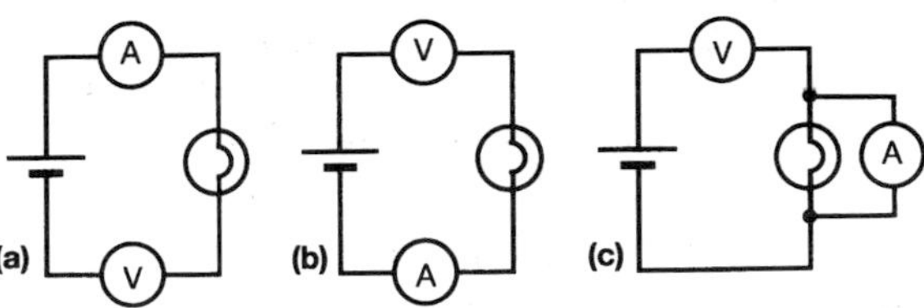

7.33 The diagram shows a circuit containing two resistors of resistance 10 Ω, connected in series with a cell whose e.m.f. is 1.5 V and internal resistance is 0.5 Ω. A voltmeter is connected across one of the resistors, as shown. The two resistors are then replaced by two 100 Ω resistors, two 1000 kΩ resistors and two 10000 kΩ resistors, and in each case the voltmeter readings are recorded, as shown:

resistance/Ω	10	100	1000	10 000
p.d./mV	732	748	746	714

(a) Account qualitatively for the variation in the voltmeter readings.
(b) Show that the measurements are consistent with the voltmeter having a resistance of 100 kΩ.

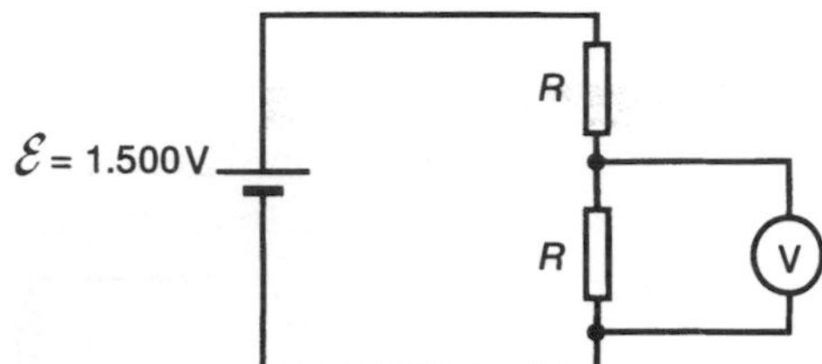

7.34 A student wishes to investigate how the current varies with time when a filament bulb is switched on. Its resistance, when operating normally, is about 100 Ω. He decides to use a data logger with a circuit which includes a 0.47 Ω resistor, as shown in the diagram.
(a) Why is it sensible to choose a resistor with such a small resistance?
(b) The graph shows the trace he obtains from the data logger. When the bulb is operating normally, what is the p.d. across the resistor?
(c) What is the current in the bulb 1.0 ms after switching on?

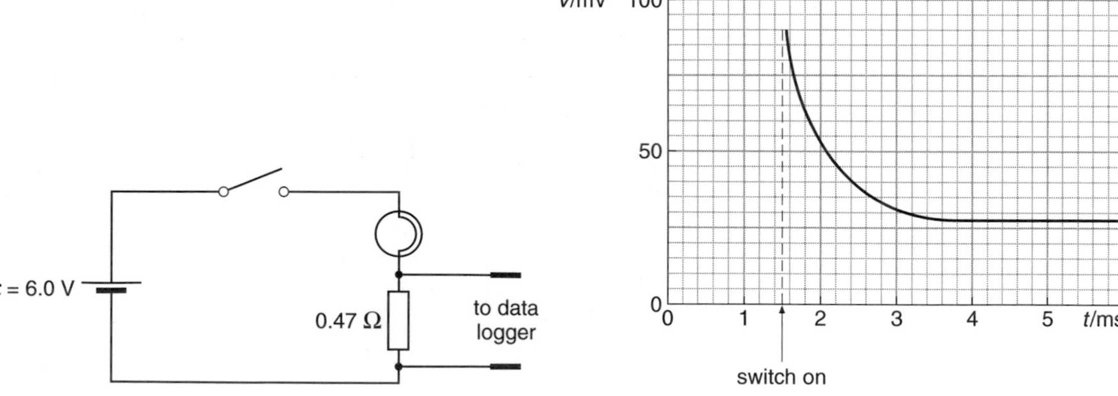

7.35 An oscilloscope has its sensitivity control set to 2.0 V div^{-1} and its time base switched off. Initially the spot is at the centre of the screen. What is the size and direction of the movement of the spot when the oscilloscope is connected to the following points in the circuit shown in the diagram (in each case the positive terminal of the oscilloscope is connected to the first-named point)? The battery has an e.m.f. of 6.0 V and negligible internal resistance, and all the resistors are identical. Indicate upward and downward movements with + and – signs respectively.

(a) A and B
(b) B and C
(c) B and D
(d) A and E
(e) E and F
(f) B and F
(g) C and F?

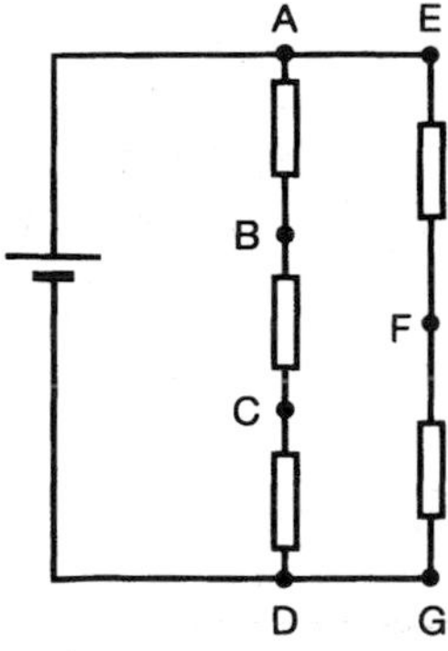

7.36 The diagram shows some traces on an oscilloscope screen. What is the frequency of the alternating p.d. if in each case the time base speed is 5.0 m s div^{-1}?

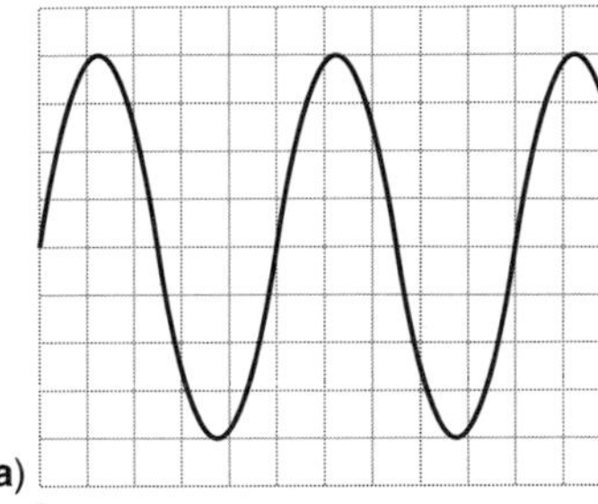
(a)

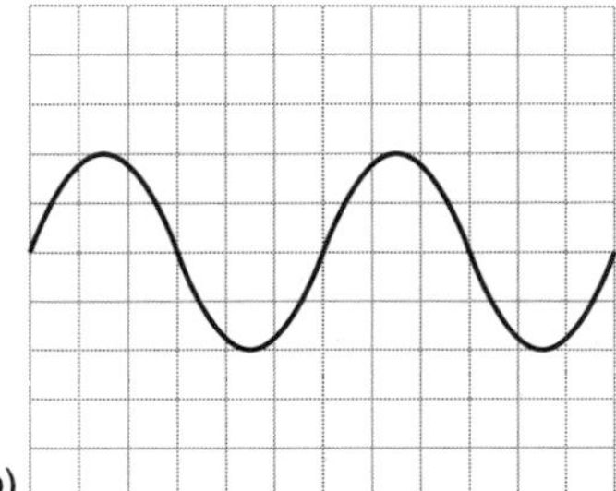
(b)

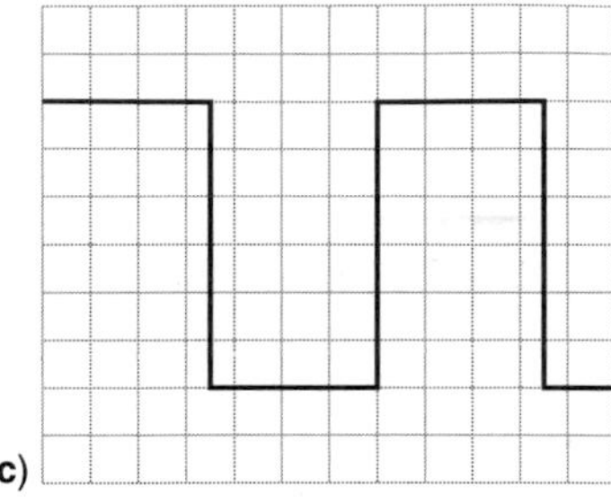
(c)

7.37 The diagram shows a trace on an oscilloscope screen. The time base is set to 10 ms div^{-1}. The sensitivity is set to 100 mV div^{-1}.

(a) Describe the variation of the p.d. across which the oscilloscope is connected.

(b) The time base is now altered to 5.0 m s div^{-1} and the sensitivity is set to 200 mV div^{-1}. Sketch the new appearance of the trace.

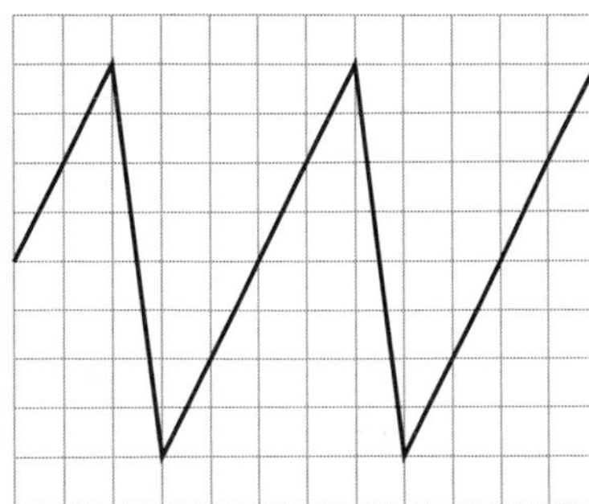

7.38 With the time base switched off, the vertical movement of the spot on an oscilloscope screen corresponds to the change in p.d. across its terminals. Knowing the sensitivity enables you to measure the p.d. The oscilloscope can also be used to measure current. The circuit diagram on the next page shows suitable connections in an experiment to measure the current in a circuit when it is varied using a 2.2 kΩ variable resistor.

(a) The fixed resistor R has a known resistance of 10.0 Ω. Why is it necessary to include this fixed, known resistance in the circuit?

(b) What is the p.d. across it when the variable resistor is set to its maximum resistance?

(c) Which sensitivity would you use on the oscilloscope: 100 mV div^{-1}, 20 mV div^{-1}, 5 mV div^{-1}, 1 mV div^{-1}?

(d) What would be your answers for **(b)** and **(c)** if the variable resistor's resistance is reduced to 50 Ω?

(e) Later, the oscilloscope spot is found to move 6.3 divisions with the sensitivity set to 10 mV div^{-1}. What was the p.d. across the 10 Ω resistor?

(f) What was the current in the circuit?

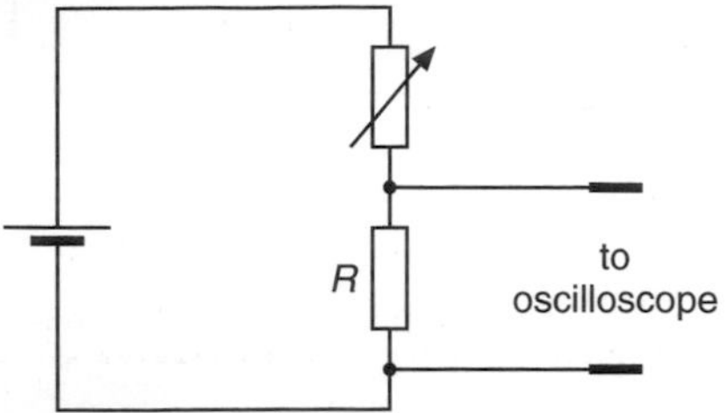

8 Density, pressure and flow

Data: $g = 9.81\ \text{N kg}^{-1}$

atmospheric pressure = 101 kPa

substance	air	water	sea water	steel	mercury
density/kg m^{-3}	1.29	1.00×10^3	1.02×10^3	7.70×10^3	13.6×10^3

8.1 Density and pressure

In this section you will need to

- use the equation $\rho = m/V$, which defines density ρ
- use the equation $p = F/A$, which defines pressure p
- understand that atmospheric pressure is caused by the layer of air which is attracted to the Earth by gravitational forces
- understand that the pressure in a fluid exerts a force at right angles to any surface with which the fluid is in contact
- use the equation $\Delta p = \rho g(\Delta h)$
- calculate pressure differences in manometers and barometers
- understand how Archimedes' principle can be used to calculate the upthrust on a body immersed in a fluid.

8.1 Copy and complete the following table:

material	mass/kg	volume/m^3	density/kg m^{-3}
aluminium	160	0.060	
lead		0.032	11×10^3
steel	60		7.7×10^3

8.2 The average radius of the Earth is 6.4×10^6 m. Its mass is 6.0×10^{24} kg. What is its average density?

8.3 A scaffolding pole has an external diameter of 40 mm and an internal diameter of 32 mm. It is made of steel. What is the mass of a 5-metre length of pole? [Use data.]

8.4 Copy and complete this table. Each object is in equilibrium on a horizontal surface.

	object	weight/N	contact area/m^2	pressure/N m^{-2}
(a)	elephant	5.5×10^4	0.14	
(b)	skier	6.9×10^2		2.5×10^3
(c)	tractor	1.5×10^4		1.2×10^4
(d)	tray	8.0	0.20	
(e)	pavement slab		0.50	8.0×10^2

8.5 Express these pressures in **(i)** Pa **(ii)** kPa, giving the numbers in standard form, to two significant figures:
(a) 1592 N m^{-2} **(b)** 234 000 N m^{-2} **(c)** 56 300 N m^{-2}.

8.6 A building brick has a mass of 2.8 kg and measures 230 mm by 110 mm by 75 mm. What pressure does it exert when stood, in turn, on each of its three faces on a horizontal surface?

8.7 Refer to question 8.3. Suppose that one of these scaffolding poles rests vertically on the ground, and the contact force is 12 kN.
(a) What pressure would the pole exert on the ground?
(b) In practice a horizontal steel plate is placed between the pole and the surface it is resting on. If the plate measures 15 cm by 15 cm, what is the new pressure on the ground? (Ignore the weight of the plate.)
(c) When the ground is particularly soft, the plate may rest on a scaffolding board. If the board measures 3.0 m by 0.25 m, what is the pressure between the board and the ground? (Ignore the weight of the board and the plate.)

8.8 Referring to the pressure which the objects produce when they are used correctly, explain the construction of **(a)** skis **(b)** drawing pins **(c)** football boots.

8.9 How far would you have to go below the surface of the sea to experience double the pressure at the surface, on a day when the atmospheric pressure was 102 kPa? [Use data.]

8.10 Explain why the equation $\Delta p = \rho g(\Delta h)$ cannot be used to calculate differences of pressure in the atmosphere when Δh is large.

8.11 If air were not easily compressible, the atmosphere would consist of a uniformly-dense (1.29 kg m^{-3}) layer of air which ended abruptly at a certain height above the Earth's surface. What height would this be on a day when the atmospheric pressure was 101 kPa?

8.12 Pressures, and pressure differences, are sometimes given in heights of a stated liquid. For example, it might be said that 'atmospheric pressure is 760 mm of mercury' or '760 mmHg'.
(a) Explain how it is possible to express a pressure in terms of a height of liquid.
(b) Express 760 mm of mercury as a pressure in kPa. [Use data.]

8.13 Airliners coming into land at an airport may have to circle round above the airport in a 'stack' while they wait for an empty runway. The usual difference in height between aircraft in the stack is 500 feet. The pilots work out their height by using an instrument to measure the atmospheric pressure. What is the difference in pressure which corresponds to this difference in height? [1 foot = 0.31 m]

8.14 A rectangular swimming pool which measures 20 m by 15 m contains water to a depth of 1.6 m. Calculate the push caused by the water on

(a) the bottom of the pool [Use data.]

(b) the larger of the two sides [remember that the pressure increases uniformly with depth]

(c) the smaller of the two sides.

8.15 Show that the units of both sides of the equation $\Delta p = \rho g(\Delta h)$ are the same.

8.16 **(a)** Calculate the force exerted by the atmosphere on the upper surface of a window in a sloping roof, if the window measures 90 cm by 60 cm, and the roof makes an angle of 40° with the horizontal, on a day when the atmospheric pressure is 103 kPa.

(b) Is it possible to calculate the force on the lower surface? If so, what is its size?

8.17 The diagram shows a U-tube *manometer* which can be used to measure differences in pressure (e.g. doctors use similar manometers to measure differences in blood pressure: see question 8.19). It contains oil of density 780 kg m^{-3} and has one side (A) connected to a gas supply. The other side (B) is open to the atmosphere. The difference in levels in the tube is 230 mm.

(a) On which side is the pressure greater?

(b) What is the pressure of the gas supply, if the atmospheric pressure is 101.2 kPa?

(c) What would have been the difference in levels if water had been used instead of the oil? [Use data.]

(d) What is the advantage of using oil in the tube, rather than water?

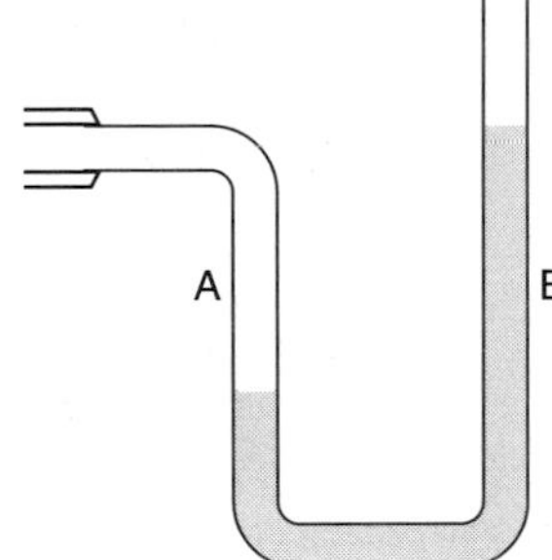

8.18 'Pressure decreases *exponentially* with height in the atmosphere, but increases *linearly* with depth in the oceans.'

(a) Explain the meaning of the terms in italics.

(b) Explain why there is this difference in behaviour.

8.19 The heart pumps blood through our arteries. As it does so, the pressure in the arteries rises and falls. The usual method of measuring the pressure of the blood in the arteries is to wrap a rubber cuff round the patient's upper arm and increase the pressure in it until the blood flow stops (as indicated by the pulse in the wrist). This gives the maximum blood pressure (the systolic pressure), which is recorded as a difference in levels (in mm) on a mercury manometer; the air in the cuff is then released until the blood begins to flow again, and the pressure (the diastolic pressure) when this happens is again recorded. (All these pressures are relative to atmospheric pressure.)

(a) If your blood pressure measurements are '120/80', what are these pressures in kPa? [Use data.]

(b) What is your mean arterial blood pressure in mmHg?

(c) Copy the diagram of a human being shown on the following page. Assume that his mean arterial blood pressure is the same as yours and on the pressure scale mark the mean arterial blood pressures at the four heights shown on the height scale. The density of blood may be assumed to be the same as that of water.

(d) What is the mean systolic pressure in this person's feet?
(e) Why is blood pressure usually measured on your upper arm?

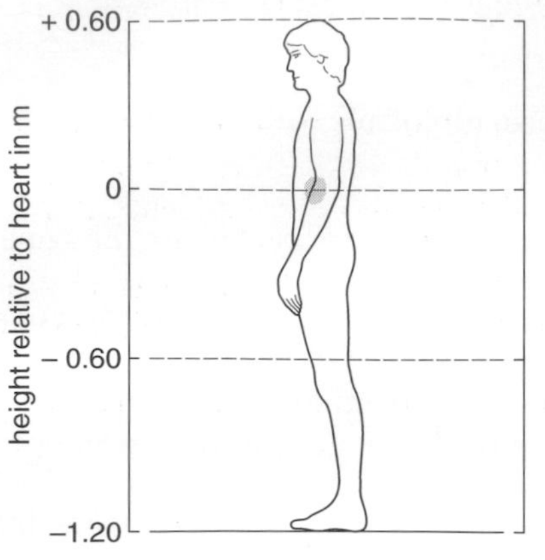

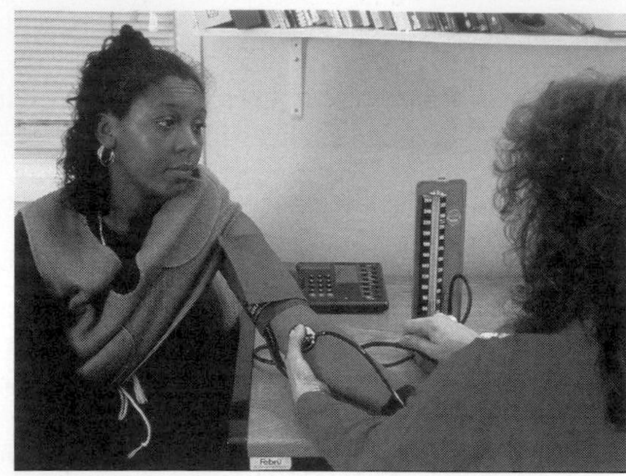

8.20 The blood in the human body flows through capillaries (very narrow paths) from the arteries to the veins. The blood in the veins is therefore at a lower pressure than the blood in the arteries.
In hospitals many fluids (e.g. blood, saline solution, drugs) are given to patients by means of an intravenous drip, i.e. the liquid is passed into the blood stream by means of a needle.
(a) Why would the needle be inserted in a vein, rather than an artery?
(b) If the mean blood pressure in a patient's vein is 20 mmHg (above atmospheric pressure) what is the minimum height at which the bag of liquid must be placed above the patient's arm? Assume that the density of the liquid is the same as that of water. [Use data.]

8.21 A capsule, containing air at a pressure of 150 kPa, is sealed by a valve which can withstand a pressure difference of 650 kPa. How deep in sea water, of density 1020 kg m^{-3}, can the capsule be lowered if no water is to enter? [Pressure at surface of sea = 101 kPa.]

8.22 Oak has a density of about 700 kg m^{-3}. Draw a free-body diagram for a piece of oak floating in water, and explain fully why 0.70 of its total volume is submerged.

8.23 The airship R101 (which burst into flames in 1930 when the hydrogen in it ignited after a crash) had a volume of 1.38×10^5 m^3. Using data given above, calculate
(a) the upthrust on it
(b) the weight of the gas in it if it was filled with **(i)** hydrogen (density 0.0880 kg m^{-3}) **(ii)** helium (density 0.176 kg m^{-3})
(c) the differences between the upthrust and these two weights.

8.24 **(a)** What is the Archimedean upthrust on a body of volume 0.20 m^3 when it is suspended, completely immersed, in a tank of water?
(b) Why is the upthrust exactly the same when it rests on the bottom of the tank?

8.25 A mixture of ice and water is poured into a beaker until it is completely full. Explain whether there will be an overflow when the ice melts.

8.26 A manufacturer wishes to use expanded polystyrene, which has a density of 15 kg m^{-3}, to make buoyancy aids for children learning to swim. What volume of material is needed to produce an upthrust of 20 N?

8.27 In a misguided attempt to measure the density of air a student found the mass of a deflated balloon, blew it up (making sure that the air in it was at atmospheric pressure) and found the new mass. He found that the two masses were identical. Why was this?

8.28 The masses of a table tennis ball and a squash ball were found by placing them on a top pan balance. The readings were 2.39 g and 23.82 g, respectively, and the balls' diameters were 38.6 mm and 40.6 mm.

(a) Using data given above, calculate **(i)** the upthrust of the air on the table tennis ball **(ii)** its true mass.

(b) Repeat **(a)** for the squash ball.

(c) In each case what is the percentage error in the mass?

(d) In what type of situation is it necessary to consider the upthrust of the air on a body?

8.29 The diagram shows the Plimsoll lines which are painted on the sides of ships. They show where the water line should be for different situations.

(a) If the water is at level F for a ship in dock, and then changes to TF, has the ship been loaded or unloaded? Explain.

(b) TF stands for 'tropical fresh water', T stands for 'tropical sea water'. Why is TF higher than T?

(c) S stands for 'summer', W stands for 'winter'. Why is S higher than W?

(d) Why is the weight of a ship often called its displacement?

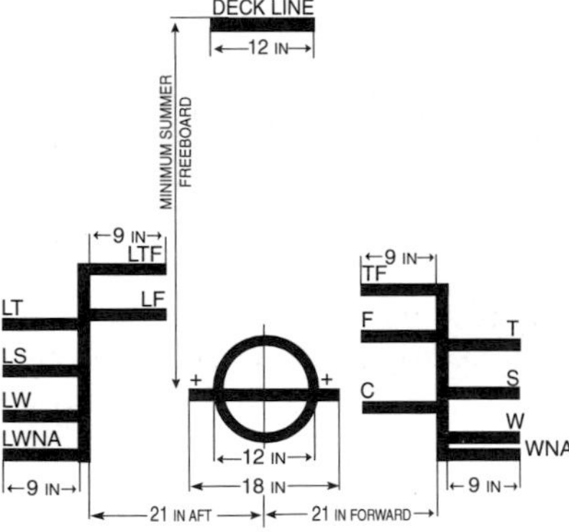

8.30 An aqueduct is a bridge which carries a canal or a river. Is the downward force on the aqueduct greater when a barge is passing over it? Explain.

8.2 Flow

In this section you will need to

- understand what it means to say that a fluid is viscous
- understand the difference between frictional and viscous forces
- understand the difference between streamline and turbulent flow, and the conditions under which each may occur

- be able to use the drag force equation $F = \frac{1}{2}AC_D\rho v^2$ where the symbols have their usual meaning
- understand that in a pipe fluid flows faster where the pipe is narrower, and that the pressure there is less
- understand how the Bernoulli effect is used in situations where air flows over a curved surface (aerofoils, hydrofoils, sails, windsurfing, paragliding)
- be able to use the lift force equation $F = \frac{1}{2}SC_L\rho v^2$ where the symbols have their usual meanings.

8.31 A horizontal pipe A of cross-sectional area 4.0×10^{-4} m^2 narrows into a second horizontal pipe B of cross-sectional area 1.0×10^{-4} m^2. The pipes are full of water, and the speed of the water in pipe A is 0.20 m s^{-1}.

(a) What is the rate of flow of volume in **(i)** pipe A **(ii)** pipe B?

(b) What is the speed of the water in pipe B?

(c) Explain how the water in pipe B has been able to be accelerated.

8.32 Suppose your kitchen tap delivers 500 cm^3 of water in 2.5 s, and at the tap the diameter of the stream of water is 1.2 cm.

(a) At what speed does the water emerge from the tap?

(b) How fast is it moving when it has fallen 0.20 m?

(c) Calculate the diameter of the stream of water when it has fallen that distance.

8.33 Why is a pump needed to keep liquid flowing along a horizontal pipe?

8.34 When you stir a cup of coffee, the circular motion of the coffee ceases after a short time. Why is this?

8.35 Explain the difference between streamline and turbulent flow. Which is more likely to occur when the fluid has a low speed?

8.36 Why would you use viscous, and not frictional, forces to reduce the size of the unwanted oscillations of a body?

8.37 A prototype car is placed in a wind tunnel.

(a) Describe what would be done to enable the designer to tell where the flow of air over the car was streamline and where it was turbulent.

(b) Why is it important that turbulence should be reduced as much as possible?

(c) For a particular car the frontal area is 3.0 m^2, the drag coefficient C_D is 0.30. What is the drag force at speeds of **(i)** 25 m s^{-1} **(ii)** 35 m s^{-1}?

8.38 Water flows along a horizontal tube whose cross-section varies as shown. Sketch a graph to show how the pressure varies with distance along the tube, assuming that the water is viscous.

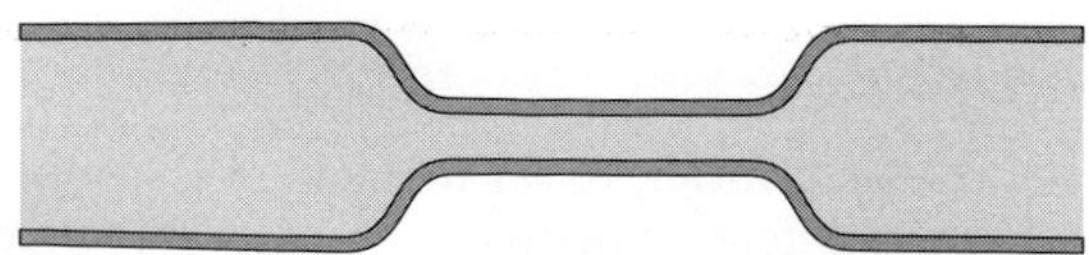

8.39 Photographs of windsurfers often show the sail with a curved shape, but the wind is blowing across the surface of the sail, not into it, as might have been expected: see the photo. Explain why the sail is curved in this way.

8.40 A table tennis ball may be supported on an upward jet of air produced by a hair drier. The molecules of air which hit the ball are pushed downwards, and by Newton's third law, the ball is pushed upwards. But why does the ball return to the centre of the jet if it is pushed sideways slightly?

8.41 Explain why

(a) a long high-sided lorry being overtaken by a similar lorry may be pulled towards each other, especially if there is a large difference in speed

(b) a strong wind blowing along the outside of a garden wall may cause the wall to fall outwards

(c) two table tennis balls, which are hanging side by side and near each other, move closer together when you blow into the space between them.

8.42 **(a)** Draw a diagram which shows a cross-sectional view of an aerofoil, and draw streamlines to show the flow of air past it when it is placed in a wind tunnel.

(b) Hence explain why there is an upward push on the aerofoil.

(c) How does the upward push depend on the area of the surface of the aerofoil, the density of the air, and the speed of the air?

8.43 Use the equation $F = \frac{1}{2}SC_L\rho v^2$ to calculate the upward push of the air flowing past the wings of an air liner which has a total wing area of 500 m^2 and a speed of 280 m s^{-1} in a region where the density of air is 0.025 kg m^{-3}. Take C_L to be 0.50.

9 Mechanical properties of matter

Data: $g = 9.81$ N kg^{-1}

9.1 Materials in tension and compression

In this section you will need to

- understand what is meant by the tension in a stretched wire, spring or rod and that it is the same throughout
- use the equation $F = kx$, which defines the stiffness k of a spring
- calculate the stiffness of two identical springs of stiffness k when arranged in series or in parallel
- use the equation $\sigma = F/A$, which defines tensile stress σ
- use the equation $\varepsilon = \Delta l/l$, which defines tensile strain ε
- understand what is meant by saying that a wire or spring obeys Hooke's law
- understand what is meant by the limit of proportionality, and the elastic limit of a material
- use the equation $E = \sigma/\varepsilon$, which defines the Young modulus E
- understand what is meant by stiffness, and how it differs from strength
- understand what is meant by elastic and plastic behaviour
- explain where tension and compression occur in a beam or cantilever
- draw stress-strain graphs for typical ductile and brittle metals, rubber and other polymers
- describe how to measure the Young modulus for a material in the form of a wire.

9.1 Two tug-of-war teams each pull on a rope with a force of 5000 N. The rope is horizontal. What is the tension in the rope at its mid-point?

9.2 A mass of 6.0 kg is placed at the lower end of a vertical wire; the upper end is fixed to a ceiling. What is the tension in the wire
(a) at its lower end **(b)** at its upper end?
What assumption do you have to make to be able to answer **(b)**?

9.3 The mass of 6.0 kg referred to in the previous question is now supported by a uniform rope which has a mass of 1.0 kg. What is the tension in the rope
(a) at its lower end **(b)** at its upper end **(c)** at its mid-point?

9.4 The graphs in the diagram show how the force F needed to produce an extension x varies for three different springs A, B and C.
(a) Calculate the gradient of each line and hence find the stiffness of each spring.

(b) Use your graph to find the force needed to produce an extension of 0.12 m in **(i)** A **(ii)** B **(iii)** C.

(c) Use your graph to find the extension produced by a force of 16 N in **(i)** A **(ii)** B.

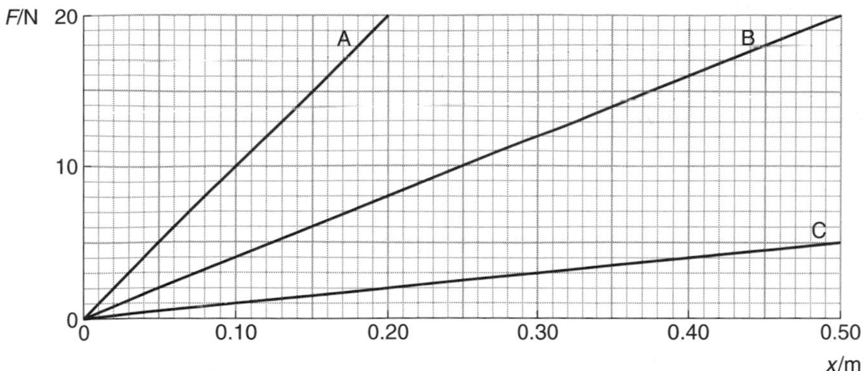

9.5 A spring has a stiffness of 60 N m^{-1}. Assuming that Hooke's law is obeyed, find the force needed to stretch it **(a)** 10 cm **(b)** 50 cm.

9.6 A spring has an unstretched length of 12 cm and a stiffness of 50 N m^{-1}. What force is needed to **(a)** double its length **(b)** treble its length?

9.7 A load of weight 20 N was hung from a vertical spring of stiffness 100 N m^{-1}.

(a) What was the extension?

(b) A second identical spring was hung side by side with the first spring to help support the load, i.e. in parallel. What was then the extension? What was the stiffness of this system of two springs?

(c) This second spring was removed, and instead placed between the first spring and the load, i.e. in series. What was then the extension? What was the stiffness of this system of two springs?

(d) For springs of stiffness k, write down the stiffness, in terms of k, of two springs **(i)** in parallel **(ii)** in series.

9.8 The common 'expendable' springs often found in laboratories have a stiffness of 30 N m^{-1}. What is the stiffness of

(a) three of these springs connected end-to-end

(b) two of these springs connected side-by-side

(c) a spring made by cutting one of these springs into two equal lengths?

9.9 If a wire of a particular material (e.g. copper) is stretched with a particular force, will its extension be greater if

(a) it is made thinner

(b) it is made longer?

9.10 If a particular wire is stretched with a steadily increasing force, will the force at which it breaks depend on the original length of the wire?

9.11 Express these stresses in N m^{-2}, giving the number in standard form:
(a) 101 kPa **(b)** 0.27 MPa **(c)** 35 MPa **(d)** 2.8 GPa **(e)** 235 GPa.

9.12 Copy and complete the following table:

	force	cross-sectional area	stress
(a)	6.0 N	0.10 mm^2	
(b)	12 kN	2.0 mm^2	
(c)		3.4 cm^2	6.0×10^6 N m^{-2}
(d)		$\pi(0.50 \text{ mm})^2$	6.4 MPa
(e)	0.11 kN		0.22 GPa

9.13 A steel wire has a diameter of 0.36 mm.
(a) What is its radius?
(b) What is its cross-sectional area, in m^2?
(c) It is pulled by a force of 3.5 N. What is the tensile stress in the wire?

9.14 The maximum stress in high-tensile steel is about 340 MPa. A particular wire made from this steel has to support a load of 60 kg.
(a) What is the maximum stress, in N m^{-2}?
(b) What force does the load produce? [Use data.]
(c) What is the minimum cross-sectional area of the wire?
(d) What is the minimum diameter of the wire?

9.15 What is the tensile stress in
(a) one of the supporting cables of a suspension bridge which has a diameter of 40 mm and which pulls up on the roadway with a force of 30 kN
(b) a nylon fishing line of diameter 0.35 mm which a fish is pulling with a force of 15 N
(c) a tow rope of diameter 6.0 mm which is giving a car of mass 800 kg an acceleration of 0.40 m s^{-2} (other horizontal forces on the car being negligible)?

9.16 Copy and complete the following table:

	length	extension	strain
(a)	2.0 m	4.0 mm	
(b)	20 cm	50 cm	
(c)	10 m		3.0×10^{-3}
(d)	3.4 m		5.2×10^{-3}
(e)		0.57 mm	1.6×10^{-4}

9.17 What is the tensile strain when
(a) a copper wire of length 2.0 m has an extension of 0.10 mm
(b) a rubber band of length 50 mm is stretched to a length of 150 mm?

9.18 A rectangular strip of polythene is 0.10 mm thick and 10 mm wide (and several centimetres long). When it is stretched it deforms so that the ends still have their original width and thickness, but there is a central section which is still 0.10 mm thick but only 5.0 mm wide. If the force with which each end is being pulled is then 50 N, find the tensions and the tensile stresses in **(a)** the wide **(b)** the narrow part of the strip.

9.19 Copy and complete the following table:

	stress	strain	Young modulus E
(a)	50 MPa	6.0×10^{-4}	
(b)	0.10 GPa	5.0×10^{-2}	
(c)		0.054	0.22 GPa
(d)	0.30 GPa		300 GPa
(e)	1.8 GPa		180 GPa

9.20 The table below gives the corresponding values of load and extension when masses were hung on a wire of length 2.0 m and diameter 0.40 mm.

load/kg	0	0.20	0.40	0.60	0.80	1.00	1.10	1.12
extension/mm	0	1.0	2.1	3.1	4.2	5.4	7.3	9.0

(a) Plot a graph of load (on the y-axis) against extension (on the x-axis).
(b) From the straight part of the graph, where Hooke's law is obeyed, calculate the gradient of the graph (which will have kg/mm as the unit).
(c) From the gradient calculate the ratio of *force* to extension (which will have N/mm as the unit). [Use data.]
(d) What is this ratio in N/m?
(e) What is the cross-sectional area of the wire, in m^2?
(f) In the equation $E = Fl/eA$ you now have the ratio F/e (from the graph), and the values of l and A. Calculate the value of E, the Young modulus, for this material.
(g) What is the advantage of drawing a graph and calculating the gradient, rather than calculating the value of E from one or more pairs of values of load and extension?
(h) Estimate the probable final extension of the wire after the load of 1.12 kg is removed.

[Note: keep your answers for a later question.]

9.21 The diagram shows stress–strain graphs for three metals.
(a) The label on the strain axis is $\varepsilon/10^{-3}$. What is the strain at the point on the strain axis which is labelled '0.5'?
(b) Why does strain not have a unit?
(c) Which of the three metals has the greatest Young modulus?
(d) For each metal calculate the Young modulus.

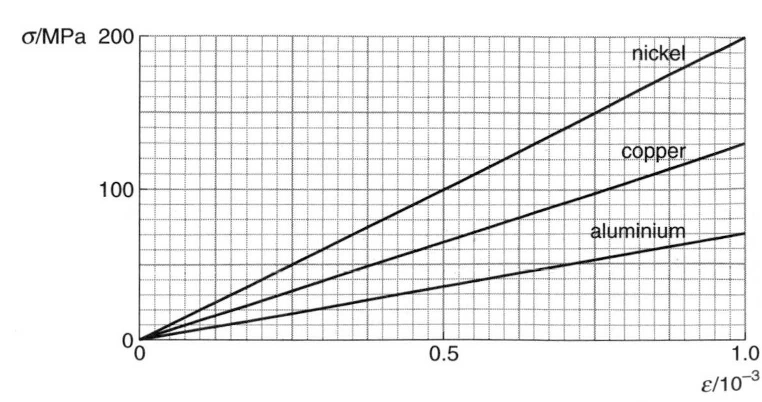

9.22 A copper wire of length 1.2 m and cross-sectional area 0.10 mm^2 is hung vertically; for copper $E = 130$ GPa. A steadily increasing force is applied to its lower end to stretch it. When the force has reached a value of 10 N
(a) what is the stress in the wire
(b) what is the strain in the wire
(c) what is the extension of the wire?

9.23 **(a)** What would be the strain in a wire if a stress equal to its Young modulus were applied to it?
(b) Could this ever occur in practice?

9.24 What is meant by the *ultimate tensile stress* for **(a)** a brittle material such as cast iron **(b)** a ductile material such as copper?

9.25 The diagram shows force–extension graphs for two wires A and B, made from the same material.
(a) Is it possible for A, compared with B, to be
(i) thinner and longer
(ii) thinner and shorter
(iii) thicker and shorter
(iv) thicker and longer?
(b) What quantities should be plotted to get the same graph for both wires?

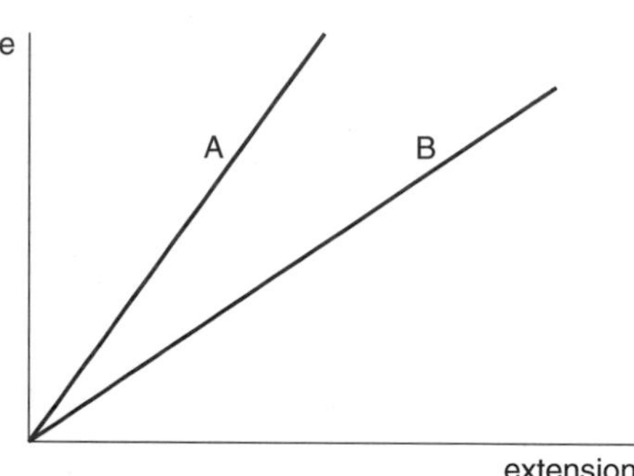

9.26 A garage intends to tow a lorry, using a steel wire. Explain to what extent the garage need consider each of the following factors: the Young modulus of the steel, the ultimate tensile stress of the steel, the length of the wire, the cross-sectional area of the wire.

9.27 Human beings have evolved so that their bones are strong enough to withstand the forces met in normal everyday life, e.g. the cross-sectional area of a leg bone is great enough to be able to support the weight of the body and also the increased forces when walking, running or jumping. Imagine a human being whose dimensions were all twice as great as those of a normal human being.
(a) How many times greater would be **(i)** its volume and its weight **(ii)** the compressive force in a leg bone when it was standing still **(iii)** the cross-sectional area of a leg bone **(iv)** the compressive stress in a leg bone when it was standing still?
(b) What problems would this human being encounter?
(c) Explain why large animals such as elephants and rhinoceroses have very thick legs.

9.28 **(a)** What is the minimum radius of a nylon fishing line which is required to lift a fish of mass 5.0 kg vertically at a steady speed?
(b) If the fish struggles (and therefore accelerates) the force may be increased by a factor of 10. What minimum radius is now required?
[Ultimate tensile stress of nylon filament = 60 MPa.]

9.29 A lift of mass 3000 kg is supported by a steel cable of diameter 20 mm. The maximum acceleration of the lift is 2.5 m s^{-2}. Calculate
(a) the maximum tension in the cable **(b)** the maximum stress in the cable
(c) the maximum strain in the cable. [E for steel = 200 GPa]

9.30 Two wires A and B are made from the same material. A is twice as long as B and has half B's diameter. When the same load is hung from each wire, what is the ratio
(a) (stress in A)/(stress in B) **(b)** (strain of A)/(strain of B)
(c) (extension of A)/(extension of B)?

9.31 Wire A is aluminium; wire B is chromium. B is twice as long as A but has only half the cross-sectional area. The Young modulus of aluminium is $\frac{1}{4}$ that of chromium.
(a) When the same mass is hung from each wire, which will have the larger extension, and how many times larger is it?
(b) Each material has the same ultimate tensile stress. When increasing loads are applied, which wire will break first?

9.32 **(a)** Two wires of the same material but of different lengths and diameters are joined end-to-end and hung vertically to support a load. Explain which of the following quantities must be the same for both wires: tensile force, tensile stress, strain, extension.
(b) Two wires of the same material and of the same length are now hung vertically side-by-side with their ends joined together. Explain which of the following quantities must now be the same for both wires: tensile force, tensile stress, strain, extension.

9.33 A copper wire and a tungsten wire of the same length are hung vertically side by side. The cross-sectional areas of the wires are 0.10 mm^2 and 0.15 mm^2, respectively. Equal steadily increasing forces are applied to each wire. The Young modulus E and ultimate tensile stress σ_u are given in the table:

	E/GPa	σ_u/MPa
copper	130	220
tungsten	410	120

Which of the wires will **(a)** break first **(b)** have the larger extension at that time?

9.34 Two students A and B perform experiments on copper wire. Both wires have the same length, but B's has twice the diameter of A's. Explain which of the following statements are true:
(a) when they apply the same stress they get the same strain
(b) when they apply the same load A's strain is 4 times B's
(c) when they apply the same load A's extension is four times B's.

9.35 There is a risk that thermal expansion might cause railway track to buckle if the temperature rises sufficiently. So the track is stretched before it is laid so that it is in tension. Then, if the temperature rises, all that happens is that the tension in the track is reduced.
Suppose a 100 m length of rail is to be laid. The expansivity of the steel is 1.1×10^{-5} °C^{-1}, i.e. the length increases by this fraction of its original length for each Celsius degree rise in temperature.
(a) What would be the expansion of this length for a 25°C rise in temperature?
(b) The Young modulus for the steel is 200 GPa. What stress would be needed to push the rail back to its original length?

(c) The cross-sectional area of steel rail is 75 cm^2. What force would be needed to push the rail back to its original length, so that a later rise in temperature of 25°C does not put it into compression?

9.36 The diagram shows an experiment in which a copper wire was stretched using increasing loads. The length of the wire was 4.35 m, and its diameter was 0.46 mm. The wire was loaded and unloaded for three cycles, and some of the results are shown in the table.

first cycle		second cycle		third cycle	
load/g	*e*/mm	*load*/g	*e*/mm	*load*/g	*e*/m
0	0	0	0	0	4.3
300	1.0	600	1.9	600	6.0
600	1.9	1200	4.0	1200	7.9
900	3.0	1800	7.9	1800	10.0
1200	4.0	2100	10.9	2100	11.4
900	3.0	1800	10.3	2400	15.9
600	1.9	1200	8.5	2500	18.8
300	1.0	600	6.3		
0	0	0	4.3		

(a) Why was such a long wire used?
(b) What instrument would you have used to measure the diameter of the wire, and what precautions would you have taken?
(c) What safety precautions would you have taken in this experiment?
(d) For which cycle did the wire behave completely elastically? Explain.
(e) Plot a graph of load against extension for the cycle you chose in **(d)**, find its gradient, and hence calculate the Young modulus of copper.
(f) At the end of the second cycle, the wire is 4.3 mm longer than it was. What has happened to the atoms in the copper?
(g) How do the readings in the third cycle show that the copper has been work-hardened?
(h) What was the ultimate tensile stress for the copper in this wire?
(i) In fact the final extension of the wire was not 18.8 mm, because without any further increase in load, the wire stretched to a final extension of 21.8 mm. How was it possible for the wire to continue to stretch without the load being increased?

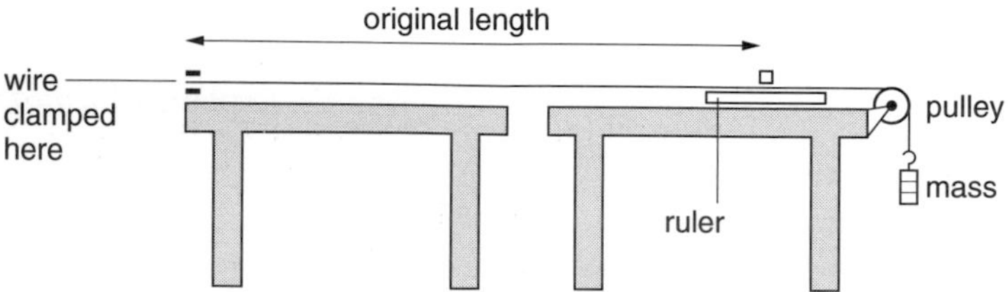

9.37 The diagram shows stress-strain graphs for a glass rod and copper wire.
(a) Which material is stronger?
(b) Why are there two graphs for copper wire?
(c) Does copper obey Hooke's law? If so, to what extent?
(d) Explain which material has the larger Young modulus.
(e) What is meant by the term *nominal stress*?

(f) In **(b)** why does the graph slope downwards, apparently showing the wire continuing to stretch with decreasing loads?

(g) What feature of the graph for the glass rod shows that glass is brittle?

(h) Make a copy of **(c)** and add it to a line which shows what you would expect to happen if the load was gradually removed when a strain of 0.0015 had been reached.

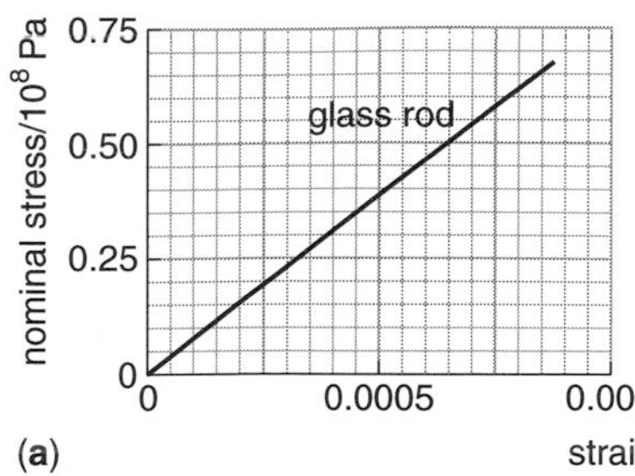

(a)

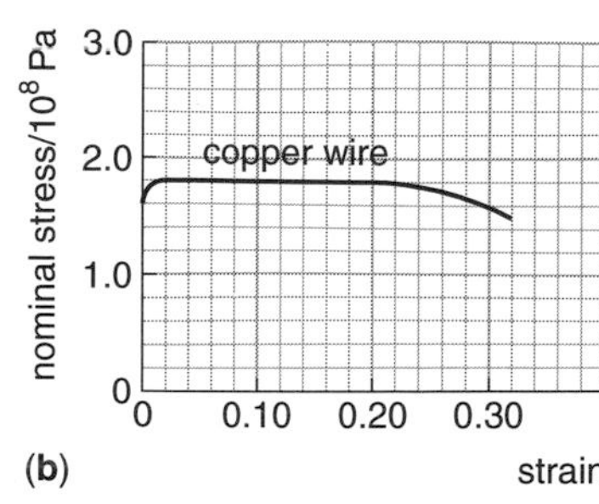

(b)

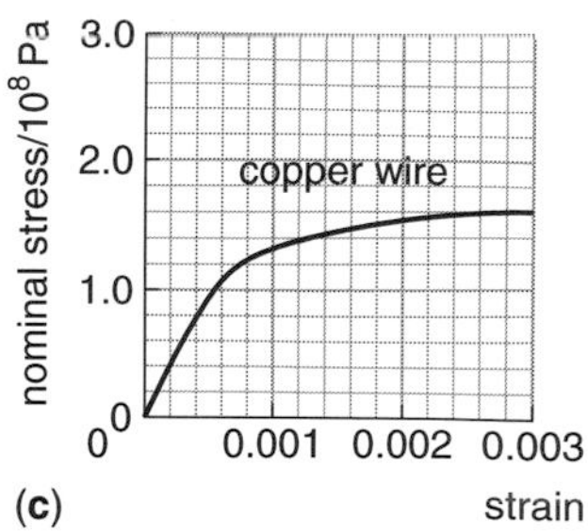

(c)

9.38 Wires of many ductile materials form a neck just before they break, i.e. the wire narrows at one point of the wire. Explain, using terms like force, stress and strain, why the wire must inevitably break where the neck first forms. Sketch stress–strain and force–extension graphs describing the behaviour of the wire before and after the neck forms, and explain why these graphs differ in shape.

9.39 The diagram shows a stress–strain graph for a specimen of polythene (such as you might find holding together a four-can pack of beer or coke).

(a) Describe what the graph shows about the behaviour of the polythene when it is loaded and unloaded.

(b) What is the approximate value of the Young modulus of the polythene?

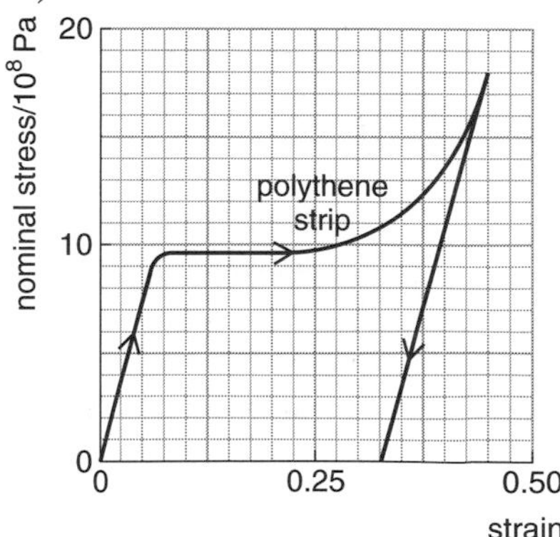

9.40 Describe how you would measure the Young modulus for copper, assuming that you have a choice of copper wires of different diameters and lengths. You should say which instruments you would use to make the measurements, and explain which measurements would have the greatest uncertainty in them.

9.41 A successfully-designed road bridge will be *strong*, *stiff* and *elastic*. Explain why it needs to have each of these properties, making clear the meaning of each of the words in italics.

9.42 In many practical situations the bending of a piece of material (e.g. a concrete beam) produces tension in the material. Draw a diagram of a concrete beam which is supported only at its ends, and mark on it the parts of the beam which have tensile stress in them, and the parts which have compressive stress in them.
The ultimate tensile stress of concrete is about ten times less than the ultimate compressive stress. Mark on your diagram where you would place steel reinforcing bars.

9.43 The Young moduli of copper and aluminium are 130 GPa and 70 GPa, respectively. Bars of the same length and diameter are made from each material. Explain which bar would be the easier to bend.

9.44 The diagram shows the two bones in a human leg and a graph of compressive or tensile stress against strain for the bones of a 30-year old person. The bone breaks at point X.

(a) Use the graph to calculate the Young modulus for the bone (use only the straight part of the graph).

(b) What is the maximum compressive or tensile stress for the bone?

(c) What is the maximum strain for the bone?

(d) In fact bones break more often because they are bent than because of simple compression or tension. Draw a diagram of a bent bone and mark on it the regions where tension and compression would occur.

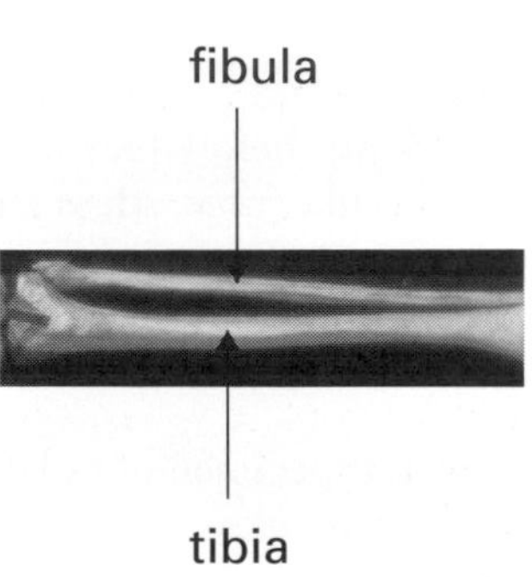

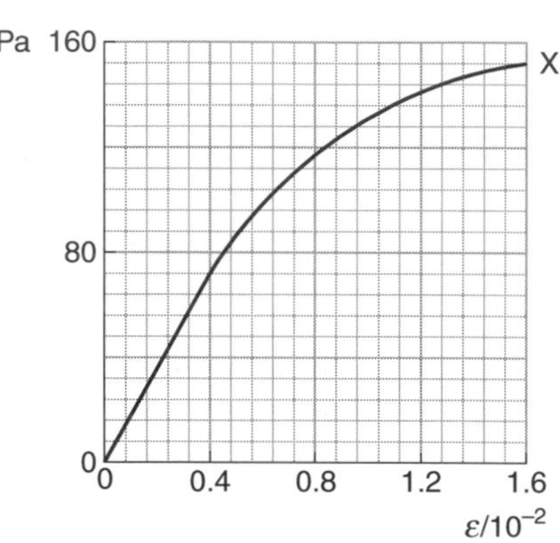

9.2 Energy stored in stretched materials

In this section you will need to

- use the equation $W = \frac{1}{2}kx^2$ to calculate the elastic potential energy W stored in a spring which obeys Hooke's law
- use the equation elastic p.e. per unit volume = $\frac{1}{2}$(stress) × (strain) to calculate the elastic p.e. stored per unit volume in a wire which obeys Hooke's law
- estimate the energy stored in a spring or wire when the force–extension graph is not straight
- understand that the toughness of a material is related to the work that must be done to break it, and that toughness is often a desirable property.

9.45 How much work must be done to stretch a spring of stiffness 30 N m^{-1} by **(a)** 10 cm **(b)** 20 cm **(c)** 30 cm? Assume that the spring obeys Hooke's law. Explain why your answer to **(b)** is not double your answer to **(a)**.

9.46 **(a)** Sketch a graph of force F (on the y-axis) against extension x (on the x-axis) for a spring of stiffness 50 N m^{-1} for values of x from 0 to 1.0 m. Assume that the spring obeys Hooke's law.

(b) Use your graph to find the increase in energy stored in the spring when its extension increases from **(i)** 0 to 0.20 m **(ii)** 0.20 m to 0.40 m **(iii)** 0.80 m to 1.0 m.

9.47 A steel wire of initial length 1.5 m and diameter 0.50 mm is pulled with a force of 45 N.
(a) What is its extension?
(b) How much energy is stored in it?
Assume that Hooke's law is obeyed: E for the steel = 200 GPa.

9.48 Show that the units of both sides of the equation $W = \frac{1}{2}kx^2$ are the same.

9.49 A copper wire is stretched with a steadily increasing force: Hooke's law ceases to be obeyed when the force reaches 80 N. The extension is 5.0 mm when the force is 20 N. How much elastic potential energy is stored in the wire when the force is **(a)** 20 N **(b)** 40 N **(c)** 60 N?

9.50 Refer to question 9.20.
(a) Estimate the work which has been done in stretching the wire up to the point where it carries a load of 1.12 kg.
(b) Estimate how much of this energy has become internal energy, and how much is stored as elastic p.e. (and therefore recoverable).

9.51 The graph shows how the extension varies with the force when a rubber band is stretched. Estimate the work done in stretching it. [Hint: the area of each square represents work done equal to 2.0 N × 100 mm = 0.20 J.]

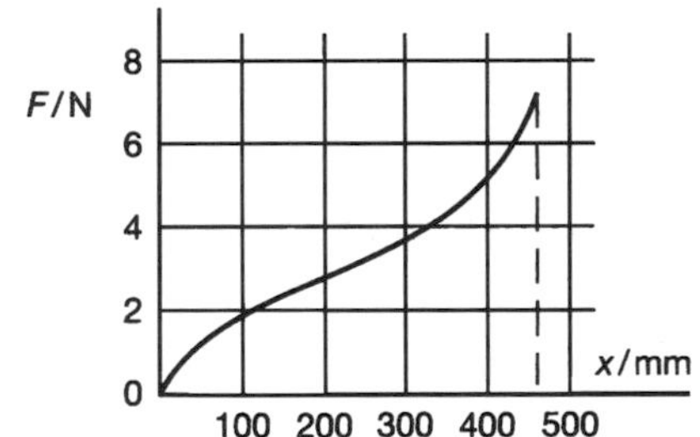

9.52 Consider a wire of length l and cross-sectional area A, whose, material has a Young modulus E, which is being pulled by a force F. Its extension is x, and the energy stored, if Hooke's law is obeyed, is given by $W = \frac{1}{2}Fx$.
(a) Express F in terms of E, l, A and x, and hence show that $W = (EAx^2)/2l$.
(b) Hence show that the energy stored per unit volume in the wire is equal to $\frac{1}{2}$(stress) × (strain) for a wire which obeys Hooke's law.

9.53 The diagram shows idealised forms of the force–extension curve for specimens (of the same shape) of high tensile steel and mild steel up to the point where each specimen breaks.
(a) Explain which is the stronger material.
(b) By considering, in each case, the area between the graph and the extension axis, find the work which must be done to break each specimen.
(c) Explain which is the tougher material.

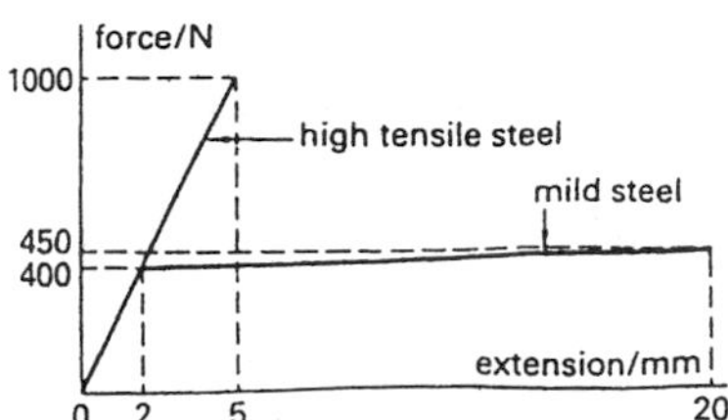

9.54 The diagram shows a force–extension graph for a length of rubber cord which is loaded and then unloaded.

(a) Which curve, the lower or the upper, describes the loading?

(b) What name is given to the fact that the two curves do not coincide?

(c) Explain how this effect causes heating in motor car tyres.

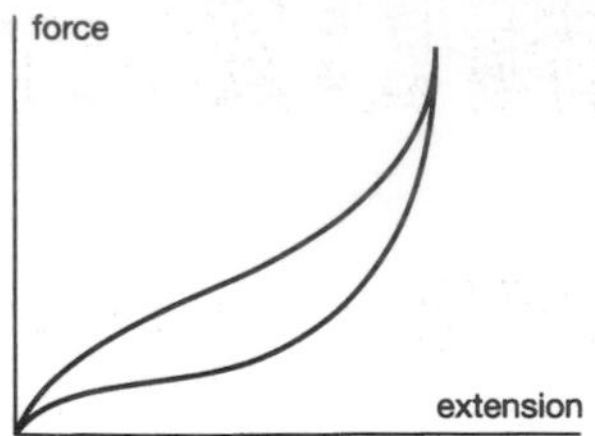

9.55 A locomotive of mass 65 tonnes just fails to stop in a siding and runs into the buffers at a speed of 0.50 m s^{-1}. If the buffers can be thought of as consisting of two compression springs in parallel which obey Hooke's law, what must the stiffness of each spring be if the locomotive is to be stopped in a distance of 10 cm?

9.56 A rubber band was supported so that it hung vertically. The extensions of the band for various loads, when the band was loaded and unloaded, are shown in the table:

load/g	0	200	400	600	800	1000	1200
extension (when loading) /mm	0	70	245	410	530	600	650
extension (when unloading) /mm	0	190	550	610	630	650	650

(a) Plot a graph of load (on the *y*-axis) against extension (on the *x*-axis) for loading and unloading.

(b) Estimate the area beneath each graph.

(c) Hence find **(i)** the work done in stretching the rubber band **(ii)** the internal energy generated in the rubber band.

9.57 A puck of mass 0.10 kg rests in equilibrium at point O on an air table. Frictional forces can be assumed to be negligible. The puck is connected by two springs of stiffness 5.0 N m^{-1} to two posts A and B. Initially the springs are taut but unstretched. The puck is moved a distance of 20 cm towards A and then released.

(a) At what speed does it pass through its original position?

(b) At what point does it first stop?

(c) Sketch a graph to show how the elastic potential energy E_e in the springs varies with distance x from O for values of x from -20 cm to +20 cm.

(d) On the same axes sketch a graph of the kinetic energy E_k of the puck for these values of x.

9.58* Laboratory experiments involving the oscillations of masses supported by springs are common, but similar experiments using masses supported by rubber bands are rarely attempted.

(a) Why is this?

(b) Sketch a graph to show how the amplitude might vary with time for a mass supported by a rubber band.

10 Thermal properties of matter

Data: $g = 9.81 \text{ N kg}^{-1}$

specific heat capacity of water = $4200 \text{ J kg}^{-1} \text{ K}^{-1}$

specific latent heat of fusion of ice = 0.334 MJ kg^{-1}

specific latent heat of vaporisation of water = 2.26 MJ kg^{-1}

10.1 Measuring temperature

In this section you will need to

- understand the principle of measuring a temperature on the centigrade or Celsius scale
- describe several types of thermometer and explain the advantages and disadvantages of each
- understand that temperature scales based on the properties of actual substances will agree only at the fixed points
- understand that the thermodynamic scale of temperature does not depend on the properties of any substance
- understand that 0 K = −273.16°C (though 0 K may be taken as equal to −273°C for most purposes).

10.1 In a mercury thermometer why is
(a) the bore narrow **(b)** the bulb relatively large **(c)** the glass of the bulb thin
(d) mercury preferred to other liquids **(e)** the inside evacuated?

10.2 **(a)** Does a thermometer read its own temperature or the temperature of its surroundings?
(b) Suppose a mercury thermometer was put in a place shaded from direct sunlight, and reached a steady temperature. Another identical thermometer placed in direct sunlight would give a higher steady temperature. What are these two thermometers measuring? Explain, making reference to the rates of emission and absorption of energy by the bulbs, why the readings are different.

10.3 **(a)** The length of a mercury column in a mercury thermometer is 15.3 mm at the ice point and 47.8 mm at the steam point. What is the temperature on the centigrade scale of this thermometer when the length of the column is 21.6 mm?
(b) The resistance of a piece of platinum is 3.254 Ω at the ice point and 4.517 Ω at the steam point. It is used to measure the same temperature as in **(a)** and its resistance is then 3.494 Ω. What is then the temperature on the centigrade scale of this thermometer?
(c) Comment on the discrepancy between your two answers.

10.4 What type of thermometer would you use to measure each of the following? In each case explain the reasons for your choice.

(a) The boiling point of water on a mountain.
(b) The temperature just after ignition in a cylinder of an internal-combustion engine.
(c) The temperature of the filament of an electric lamp.
(d) The normal melting point of zinc.

10.5 **(a)** What is meant by the triple point of water?
(b) Why is this used in defining the thermodynamic scale of temperature rather than, for example, the boiling point of water?
(c) What value is assigned to it, in K?

10.6 What is meant by absolute zero?

10.7 The equation $T/\mathrm{K} = (pV)_T/(pV)_{tr} \times 273.16$ is used to define the thermodynamic scale of temperature, using a constant-volume gas thermometer. Since the volume of the gas in the thermometer is constant the equation can be simplified to

$$T/\mathrm{K} = p_T/p_{tr} \times 273.16.$$

To be precise, the *limiting value*, as the pressure p_{tr} tends to zero, of the ratio p_T/p_{tr} is used.
To measure the temperature T of a substance, the following measurements of the pressure of a gas, kept at the same temperature as the substance, were made:

p_T/mmHg	1621.5	1217.6	812.75	406.88
p_{tr}/mmHg	1000.0	750.00	500.00	250.00

(a) Calculate the ratio p_T/p_{tr} for each of these pairs of values, to 5 significant figures.
(b) What do you notice about the value of the ratio as p_{tr} decreases?
(c) Deduce the value which the ratio would have if p_{tr} were zero.
(d) Hence find the value of T, the temperature of the substance.

10.2 Internal energy

In this section you will need to

- understand what is meant by internal energy
- understand the difference between heating and working
- use the equation $\Delta U = \Delta Q + \Delta W$ where ΔU is the change of internal energy of a body when energy ΔW is supplied to the body by working and energy ΔQ is supplied to the body by heating.

10.8 **(a)** Explain what is meant by internal energy.
(b) Does a block of ice have any internal energy?
(c) Could a lump of iron at 20°C have more internal energy than another lump of iron at 80°C?

10.9 Two copper blocks, placed together in good thermal contact, will soon reach the same temperature. Explain whether the blocks necessarily have the same amount of internal energy.

10.10 What is meant by the internal energy of a body?
If a car is stopped by being braked, in what sense is the increase of energy of the molecules of the brake drums, tyres, road, etc., different from the original kinetic energy of the car?

10.11 Write down a statement of the first law of thermodynamics, using the symbols ΔU, ΔQ and ΔW, and explain what each symbol means.

10.12 Would you describe the following energy exchanges as heating or working? State in each case **(i)** the body losing the energy, and the kind of energy lost **(ii)** the body gaining the energy, and the kind of energy gained.
(a) A can of beer is taken from a refrigerator and put in a warm room.
(b) A man sandpapers a block of wood: its temperature rises.
(c) A night storage heater cools down during the day.
(d) A tennis ball is dropped and after several bounces comes to rest.
(e) The coffee in a mug has just been stirred, and is rotating: later it comes to rest.
(f) A girl pumps air into a bicycle tyre: the pump and air become hotter.

10.13 When a car's brakes are applied frictional forces do 0.20 MJ of work. Because they are hot they lose 0.080 MJ of energy to the surroundings. What are **(a)** ΔW **(b)** ΔQ **(c)** ΔU for this process?

10.14 A battery drives a current through a bulb. During the first few milliseconds, while the filament is still warming up, are ΔU, ΔQ and ΔW positive, negative or zero for the filament?
Are they positive, negative or zero when the filament has reached its steady temperature?

10.3 Heat capacity

In this section you will need to

- use the equation $\Delta Q = cm(\Delta\theta)$, which defines specific heat capacity (s.h.c.) c
- describe how to measure the s.h.c. of a metal in the form of a cylindrical block
- understand that the rate of loss of energy from a body depends on the temperature difference between itself and the surroundings, and on its surface area and the nature of the surface
- understand that a heated body must eventually reach a steady equilibrium temperature when it is losing exactly as much energy as it is gaining
- make simple corrections for the loss of energy from a heated body
- use the equation $\Delta Q = cm(\Delta\theta)$ in the form $\Delta Q = C(\Delta\theta)$ where C is the heat capacity of a body
- use the equation $\Delta Q = cm(\Delta\theta)$ in the form *rate of heating* = *cm* (*rate of change of temperature*)
- understand the significance of water having a particularly high s.h.c.

10.15 In a steel-making furnace 5.0 tonnes of iron have to be raised from a temperature of 20°C to the melting point of iron (1537°C). Find how much energy (in GJ) is needed to do this. [s.h.c. of iron = 420 J kg^{-1} K^{-1}.]

10.16 Make a rough calculation of the cost of using electrical energy to heat water for a bath, if about 0.3 m^3 of water have to be heated from 5°C to 35°C. Assume that the cost of 3.6 MJ of electrical energy is about 10 pence. [Use data.]

10.17 The bit of a soldering iron is made of copper and has a mass of 3.3 g. If the power of its electrical heater is 45 W, how long will it take to raise its temperature from 15°C to 370°C, assuming that there are no energy losses to the surroundings?
[s.h.c. of copper = 385 J kg^{-1} K^{-1}.]

10.18 An instant gas hot water heater is capable of raising the temperature of 2.0 kg of water by 50 K each minute.

(a) What is its power? [Use data.]

(b) What problem might there be in designing an instant *electric* water heater which is to work from the ordinary mains supply and achieve the same rate of heating?

10.19 A power station needs to get rid of energy at a rate of 600 MW and does so by warming up a river which flows past it. The river is 30 m wide, 3.0 m deep, and flows at an average speed of 1.2 m s^{-1}. How much warmer is the river downstream of the power station? [Use data.]

10.20 An electric kettle has a heat capacity of 450 J K^{-1} and an element whose power is 2.25 kW. Ignoring losses of energy to the surroundings, what is the rate of rise of temperature (in K min^{-1}) when the kettle contains 1.0 kg of water? [Use data.]

10.21 A vending machine for serving hot drinks is shown in the diagram. Its water tank has a capacity of 7.5 litres and the manufacturers say that 45 cups of hot water (at 99°C) can be produced at any one time.

(a) What is the volume of a cup?

(b) If 170 cups can be produced in one hour, what is the power of the heater? Assume that the water starts at a temperature of 20°C and that there are no energy losses. [Use data.]

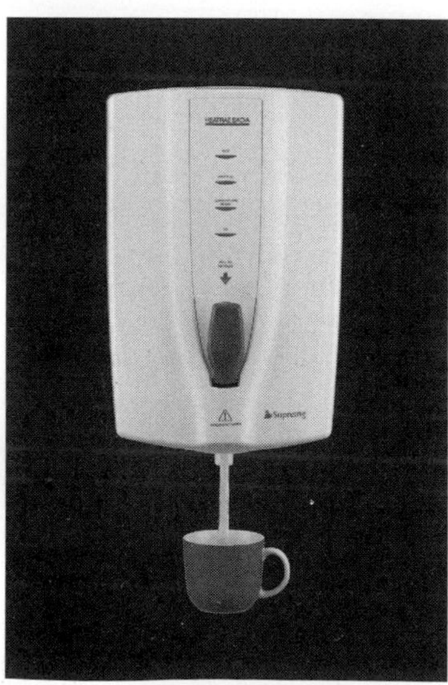

10.22 It is sometimes said that the cost of running an upright freezer is greater than the cost of running a chest freezer because each time the door of an upright freezer is opened cold air falls out and warm air from the room enters and has to be cooled down. Consider an upright freezer of capacity 0.20 m^3 and discuss whether there is any truth in this statement. The temperature inside a freezer may be assumed to be about −18°C, and the s.h.c. of air (under these conditions) is 600 J kg^{-1} K^{-1}. The density of air is 1.3 kg m^{-3}.

10.23 A 100 W immersion heater is placed in 200 g of water in a plastic cup (of negligible heat capacity). Ignore the heating of the surroundings: how fast, in K min^{-1}, does the temperature rise? [Use data.]

10.24 How long would it take a 2.0 kW heater to warm the air in a room which measures 4 m by 3 m by 2.5 m from 5°C to 20°C?
[The s.h.c. of air under these conditions is about 1000 J kg^{-1} K^{-1}.]
Give at least two reasons why in practice it takes much longer (perhaps an hour) for the heater to warm such a room.

10.25 A car of mass 800 kg moving at 20 m s^{-1} is braked to rest 10 times. If 20% of the car's kinetic energy is retained by the steel brake discs, what is their rise in temperature, if each of the four has a mass of 1.5 kg?
[s.h.c. of steel = 420 J kg^{-1} K^{-1}.]

10.26 A squash ball of mass 46 g is struck so that it hits a wall at a speed of 40 m s^{-1}; it rebounds with a speed of 25 m s^{-1}.
(a) What is its rise in temperature? [s.h.c. of rubber = 1600 J kg^{-1} K^{-1}.]
(b) Why is it unnecessary to know its mass?
(c) What will happen to its temperature if the players continue to hit it against the wall?

10.27 James Joule is said to have measured the temperature at the top and bottom of a waterfall as part of his investigation into energy. What temperature rise would you expect there to be at the bottom of a waterfall 50 m high, if the water loses all its kinetic energy on arriving at the bottom of the waterfall? [Use data.]

10.28 Water has a relatively high s.h.c. What effect does this have on
(a) climate
(b) the cost of heating water for baths and showers
(c) the volume of water in a car's cooling system
(d) the running costs of a water-filled central heating system?

10.29 1.0 kg of water at a temperature of 95°C is poured into a copper saucepan of mass 0.70 kg which is at a temperature of 20°C. The water transfers energy to the saucepan and they reach the same temperature before they start to lose energy to the surroundings.
(a) Suppose this temperature is θ. Write down expressions involving θ for
(i) the temperature fall of the water **(ii)** the temperature rise of the saucepan.
[s.h.c. of copper = 385 J kg^{-1} K^{-1}.]
(b) The energy transferred from the water is equal to the energy transferred to the saucepan. Write down an equation involving your answers to **(a)** and hence find θ.
[Use data.]
(c) Explain why the final temperature of the water and saucepan is much closer to the original temperature of the water than it is to the original temperature of the copper.

10.30 A block of copper of mass 400 g is raised to a temperature of 450 K and lowered into 500 g of water contained in a vessel of heat capacity 200 J K^{-1}; these were initially at a temperature of 290 K. The final temperature of the water was 301 K. Equate the energy transferred from the copper to the energy transferred to the water, and hence find the value that these measurements give for the s.h.c. of copper. [Use data.]
Discuss whether this procedure could be used as a method of measuring the s.h.c. of materials in solid form, explaining any precautions you would need to take. Would it matter if the material was a bad conductor of energy?

10.31 A 30 W immersion heater was placed in an aluminium block of mass 1.0 kg, and switched on for 6.0 minutes: the initial temperature of the block and the surroundings was 12.3°C. At successive one-minute intervals, the following temperatures were recorded: 14.1°C, 15.9°C, 17.7°C, 19.5°C, 21.2°C, 22.9°C, 22.8°C, 22.7°C, 22.6°C, 22.5°C.

(a) Look at the initial increases of temperature and estimate the maximum temperature which would have been reached if the block had not been heating the surroundings.

(b) Hence calculate the s.h.c. of aluminium.

10.32 Suppose you were given an immersion heater, a power supply, an ammeter, a voltmeter, a stop clock, a thermometer and a plastic beaker which you could fill with water, and you could ask for any other (simple) apparatus which you might need. Describe how you would measure the s.h.c. of the water, paying particular attention to the precautions you would take to ensure a reliable result.

10.33 Suppose you were asked to design a night storage heater, i.e. a device consisting of a set of blocks (similar to concrete) which is heated at night, when for 7 hours (between midnight and 7 a.m.) electrical energy is relatively cheap, and which releases its energy during the day. Discuss as many aspects of the design as possible, including: drawing a temperature–time graph for a 24-hour cycle, estimating the power of the heater you would need, estimating the mass of the blocks, and drawing attention to any disadvantages of this form of heating.

10.34 A student measures the temperature of a plastic mug of water which is cooling down and draws a temperature–time graph. The mass of water is 153 g and the heat capacity of the mug is negligible. He draws a tangent to the curve at the point where the temperature is 80°C and finds that the line shows a fall in temperature of 25°C in 9.6 minutes.

(a) What is the rate of loss of energy at this temperature? [Use data.]

(b) When the temperature of the water had fallen to 60°C would the rate of loss of energy be larger or smaller? Explain.

(c) Room temperature was 18°C. Assuming that the rate of loss of energy is proportional to how much higher the temperature is than room temperature, estimate the rate of loss of energy when the temperature of the water had fallen to 60°C.

10.35 In an experiment to measure the rate of flow of energy along a copper bar, thin copper tubing carrying water is wrapped round one end of the bar. This keeps that end of the bar cold because energy flows into the water which is passing through the tubing. It is found that the inlet and outlet temperatures of the water are 15.6°C and 34.3°C when the rate of flow of water is such that 430 g are collected in 5.0 minutes. Calculate the rate of flow of energy into the water. [Use data.]

10.36 An immersion heater is fitted to a hot water tank in a house. In the summer months hot water is needed only for washing. Dad says it is better to keep the heater running all the time because if the water is allowed to cool down there will be extra energy needed to warm it up again. Clare says it is best to have the heater switched on only when it is needed. Explain whether you think Dad or Clare is right.

10.4 Latent heat

In this section you will need to

- use the equation $\Delta Q = ml$, which defines specific latent heat (s.l.h.) l
- describe how to measure the s.l.h. of melting of ice
- describe how to measure the s.l.h. of vaporisation of water
- explain in molecular terms why energy is needed to melt a solid or evaporate a liquid
- understand the importance of evaporation in regulating human body temperature.

10.37 Using data given at the start of the chapter, find how much energy must be
(a) given to 2.0 litres of water at 100°C to evaporate it
(b) taken from 0.50 kg of water at 0°C to freeze it.

10.38 Using data given at the start of the chapter, find how long it will take
(a) a 1000 W heater to evaporate 1.0 kg of water which is already at 100°C, its normal boiling point
(b) a refrigerator to freeze 1.0 kg of water which is already 0°C, its normal freezing point, if it can remove energy at a rate of 75 W?

10.39 Using the values of s.h.c. and s.l.h. given at the start of this chapter, draw a graph, with labelled axes, to show how the temperature varies with time when a block of ice, of mass 2.0 kg, is placed in a sealed container and has an immersion heater of power 200 W placed in it. The ice is initially at −10°C; continue the graph until the temperature of the water vapour is 120°C.
[s.h.c. of water vapour under these conditions = 1400 J kg^{-1} K^{-1}.]

10.40 Two lumps of ice, at 0°C, each of mass 20 g, are added to a glass containing a mixture of alcohol and water at a temperature of 15°C. The heat capacity of the glass and its contents is 600 J K^{-1}. When the system has reached equilibrium how much ice is there? [Use data.]

10.41 The mass of liquid nitrogen in an open beaker is found to have decreased by 46.3 g in 10 minutes. If the s.l.h. of vaporisation of nitrogen at its boiling point is 1.99×10^5 J kg^{-1}, at what rate were the surroundings heating the beaker? Why is the heat capacity of the beaker irrelevant?

10.42 A coffee machine in a café passes steam at 100°C into 0.18 kg of cold coffee (s.h.c. the same as that of water) to warm it. If the initial temperature of the coffee is 14°C, what mass of steam must be supplied to raise the temperature of the coffee to 85°C?

10.43 Describe how you would measure **(a)** the s.l.h. of fusion of water **(b)** the s.l.h. of vaporisation of water. In each case

(i) explain how you would calculate the result

(ii) state the factors which lead to uncertainty in the result

(iii) describe the precautions you would take to make your result as accurate as possible.

10.44 Discuss some of the effects on everyday life if we lived on a planet in which the values for water of the following quantities were, separately, one-tenth of their present value:
(a) the s.h.c. **(b)** the s.l.h. of fusion **(c)** the s.l.h. of vaporisation.

10.45 A thermocouple probe connected to a multimeter gave a reading of 20.5°C for room temperature on a particular day. Explain the following observations:

(a) When the probe was placed in some ethanol in a watch glass the recorded temperature fell to 14.8°C.

(b) When the probe was removed from the ethanol and placed in the air again the temperature fell further to 6.8°C, but after a few seconds the temperature began to rise again, eventually reaching 20.5°C.

10.46 An open dish of liquid is very slightly cooler than its surroundings. Why? Your explanation should include an account of why its temperature is steady, and the factors which determine the steady temperature.

10.47 Explain the following:

(a) If you have just finished taking some exercise you should put on a track suit, even if you feel warm. It is particularly important to do this on a windy day.

(b) A bottle of milk is sometimes put under an upturned earthenware pot which is standing in a bowl of water. The pot is porous and water seeps up it. In this way the milk is kept cool.

(c) Snow and ice lie on the ground for some days after the air temperature has risen above 0°C.

(d) A large tub of water placed in a cellar will make it less likely that the temperature in the cellar will fall below 0°C.

10.48 **(a)** A runner of mass 60 kg generates internal energy at a rate of 800 W. Assuming that she loses no energy, and that the average s.h.c. of her body is the same as that of water, at what rate, in K min^{-1}, will her temperature rise?

(b) If she loses energy through conduction, convection and radiation at a rate of 300 W, at what rate, in K min^{-1}, will her temperature rise?

(c) Ideally her temperature should remain constant. Evaporation (from skin and through exhaled air from the lungs) is an additional mechanism by which she can lose energy. At what rate, in g min^{-1}, must she evaporate water in order to keep her temperature constant?

(d) When she stops running she generates internal energy at a rate of 100 W, but continues to lose energy at a rate of 800 W. At what rate, in K min^{-1}, will her body temperature fall?

11 The ideal gas

Data:

Molar gas constant $R = 8.31\ \text{J mol}^{-1}\ \text{K}^{-1}$

Avogadro constant $L\ (= N_A) = 6.02 \times 10^{23}\ \text{mol}^{-1}$

Boltzmann constant $k = 1.38 \times 10^{-23}\ \text{J K}^{-1}$

Density of water $= 1000\ \text{kg m}^{-3}$

$g = 9.81\ \text{N kg}^{-1}$

11.1 The ideal gas law

In this section you will need to

- understand that the temperature used in gas laws is the kelvin temperature, and that $0°\text{C} = 273\ \text{K}$
- understand that amount of substance n is measured in moles
- remember that the number of particles in a mole is called the Avogadro constant L (or N_A)
- use the equations $pV = nRT$, $pV = \frac{m}{M}RT$, and $pV = \frac{N}{N_A}RT$

11.1 Express the following volumes in m^3: **(a)** 1.7 litres **(b)** 6.5 cm^3 **(c)** 3.4 mm^3.

11.2 The diagram shows a graph of p against V, and a graph of p against $1/V$, for a fixed mass of gas kept at constant temperature. Copy the graphs and on each sketch a graph for the following separate changes: **(a)** a lower temperature and **(b)** an increased mass of gas. Label your graphs to make it clear which is which.

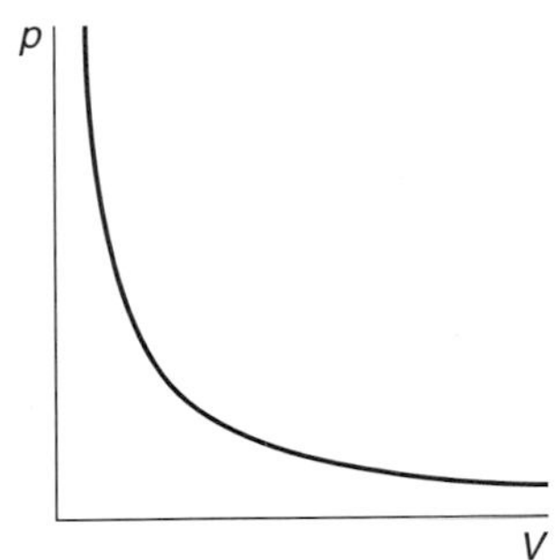

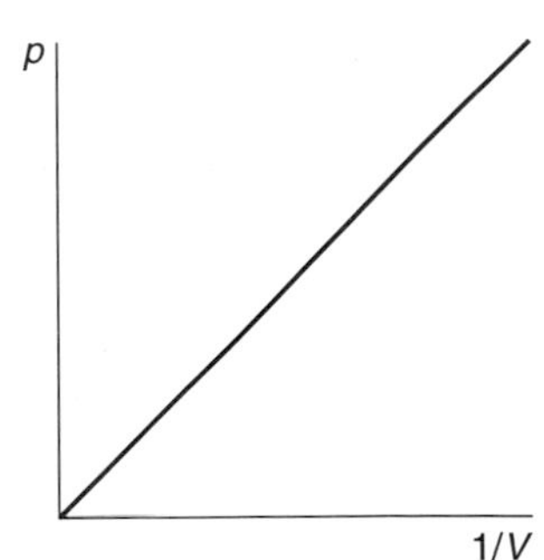

11.3 In an experiment the pressure of a gas and its volume were measured at constant temperature, the following readings were obtained:

p/kPa	102	143	178	200	233
V/cm^3	40.5	28.7	23.4	20.7	17.8

Plot graphs of **(a)** p against V **(b)** p against $1/V$. Does the second graph enable you to say that these measurements show that $pV =$ constant?

11.4 A party balloon has a volume of 1.50 litres and the pressure of the air in it is 128 kPa. It is squashed so that its volume becomes 1.30 litres.
(a) How would you try to measure one of these volumes?
(b) What is the new pressure?

11.5 A cylinder of volume 0.20 m^3 contains gas at a pressure of 200 kPa and a temperature of 290 K.
(a) How many moles of gas are there in the cylinder? [Use data.]
(b) How many molecules of gas are there in the cylinder?
The relative molar masses of hydrogen and nitrogen are 2 and 28 respectively.
(c) What is the mass of gas if it is **(i)** hydrogen **(ii)** nitrogen?

11.6 A balloon is filled with air until its volume is 1.50 litres and the pressure is 110 kPa. The temperature is 290 K. Assume that the volume and temperature remain constant.
(a) How many molecules are there in the balloon?
(b) How many more molecules must be blown into the balloon to increase its pressure to 115 kPa?

11.7 The diagram shows two graphs which show how the pressure of the same fixed mass of gas, kept at constant volume, varies with temperature.
(a) In **(i)** is the pressure proportional to the temperature?
(b) In **(ii)** what would be the label on the x-axis?

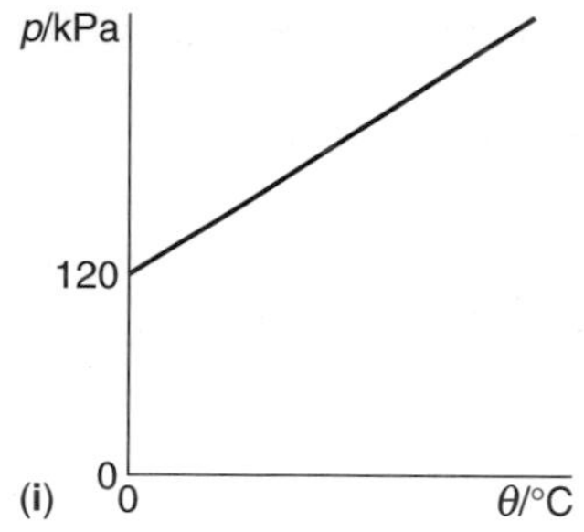

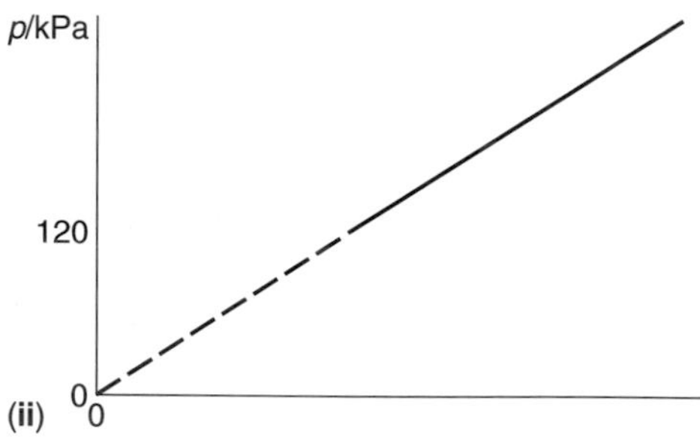

11.8 Graph **(i)** in the previous question contains some numerical information about the gas. Make a copy of graph **(ii)** and transfer this information to it. Hence find the pressure of the gas when its temperature is 100°C.

11.9 At the start of a journey the pressure of the air in a car tyre is 276 kPa and the temperature is 12.0°C. After being driven the pressure is 303 kPa. Assuming that the volume of the air remains constant, what is its temperature now?

11.10 The volume of air in a bicycle tyre is 400 cm^3 and the pressure is 145 kPa.
(a) If the temperature did not change, what volume would the air occupy just before the tyre bursts on a day when the atmospheric pressure is 102 kPa?

(b) In practice the temperature falls: if it falls from 30.0°C to 10.0°C, what volume would the air occupy?

11.11 On a day when atmospheric pressure is 105 kPa, air is pushed into a vehicle tyre until the pressure of the air in it is 360 kPa. If the volume of the inside of the tyre is 0.150 m^3, what volume of air at atmospheric pressure is pushed in, assuming that the volume of the tyre, and the temperature of the air, remain constant?

11.12 The tyre in the previous question later warms up from 15°C to 32°C. What does the pressure in the tyre become, assuming that the volume of the tyre remains constant?

11.13 Some gas occupies a volume of 6.0×10^{-3} m^3 and exerts a pressure of 80 kPa at a temperature of 20°C. What pressure does it exert if, separately
(a) the temperature is raised to 40°C **(b)** the volume is halved
(c) the temperature is raised to 586 K **(d)** the volume becomes 2.5×10^{-3} m^3
(e) the volume becomes 12×10^{-3} m^3 and the temperature becomes 57°C?

11.14 The volume of one cylinder in a diesel engine is 360 cm^3 and the cylinder contains a mixture of fuel vapour and air at a temperature of 320 K and a pressure of 101 kPa. The volume of the mixture is then reduced to 20 cm^3 and at the same time the temperature rises to 1000 K.
(a) Calculate the new pressure in the cylinder.
(b) What assumption have you made?

11.15 An air bubble of volume 3.0×10^{-5} m^3 escapes from a diver's equipment at a depth of 45 m where the water temperature is 5°C. What is its volume as it reaches the surface, where the temperature is 12°C?
[Atmospheric pressure = 101 kPa, density of sea-water = 1020 kg m^{-3}.]

11.16 Describe how you would perform an experiment to investigate how the volume of a gas depends on its temperature while its pressure remains constant. Draw a diagram of the apparatus, describe how you would make the measurements, explain what precautions you would need to take, and sketch a graph to show the sort of results you would expect.

11.17 The diagram shows two graphs of p against $1/V$ for a gas.
(a) What is the gradient of graph A?
(b) If the temperature of the gas was 290 K, what was the amount of gas?
Graph B is for the same amount of gas at a different temperature.
(c) What was the new temperature?

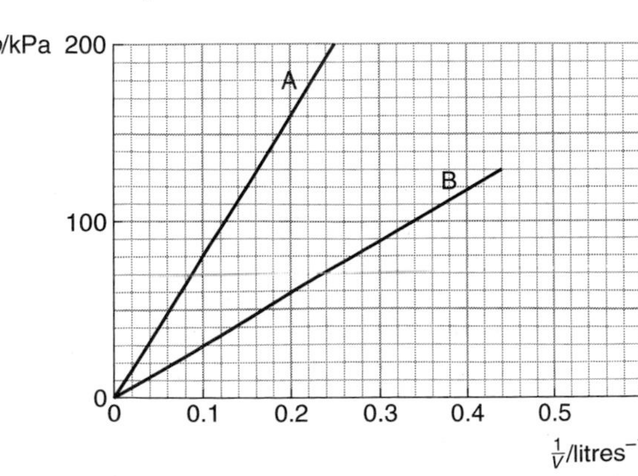

11.18 **(a)** How many moles of gas are there in the following masses: **(i)** 20 g of hydrogen **(ii)** 20 g of helium **(iii)** 200 g of oxygen? (The relative molar masses of hydrogen, helium and oxygen are 2, 4 and 32, respectively.)
(b) If these masses of gas, all at the same temperature, are placed successively in the same container, which will exert the greatest pressure, and which the least?

11.19 A vessel of volume 0.20 m^3 contains a mixture of 2.0 g of hydrogen molecules and 8.0 g of helium molecules. The temperature is 320 K.
(a) Calculate the numbers of moles of each gas. (The relative molar masses of hydrogen and helium are 2 and 4, respectively).
(b) What is the total amount of substance (i.e. the number of moles)?
(c) What is the pressure in the vessel?

11.2 The kinetic theory of gases

In this section you will need to

- understand what is meant by Brownian motion
- interpret $pV \propto nT$ in molecular terms
- remember the assumptions made in the kinetic theory of gases
- understand the steps in the derivation of the equation $p = \frac{1}{3}\rho\langle c^2\rangle$
- understand what is meant by the root mean square (r.m.s.) speed of a collection of molecules
- understand that the average k.e. of the molecules of a gas is proportional to its kelvin temperature
- remember that the average kinetic energy E of a molecule is given by $E = \frac{3}{2}kT$, where k is the Boltzmann constant
- understand that the internal energy of a gas may be changed by either heating or working, and understand the difference between these methods.

11.20 A student is observing the Brownian motion of ash particles in air.
(a) Describe what she sees.
(b) What changes would she notice if, separately, **(i)** larger ash particles were used **(ii)** the temperature of the air were raised?

11.21 What can be deduced about gases from the observation of Brownian motion?

11.22 Consider a cubical box containing air at a temperature of 290 K. Its sides are each 10 cm.
(a) What is its volume?
(b) The pressure in the box is 101 kPa. How many molecules does it contain?
(c) Find the space 'available' to each molecule by dividing the volume of the box by the number of molecules.
(d) The diameter of an atom is about 3×10^{-7} m. Treating it as a spherical body, what is its volume?
(e) How many times larger than this is its available space?

11.23 Consider a rectangular box with sides of 0.30 m, 0.40 m and 0.50 m, respectively, as shown in the diagram. Suppose it contains 1.5×10^{24} molecules, each of mass 5.0×10^{-26} kg. Suppose that each face of the box has one-third of the molecules moving at right angles to it, and that all the molecules have the same speed of 500 m s^{-1}. Consider face X, which measures 0.40 m by 0.30 m. Calculate

(a) the time between successive impacts of a particular molecule on face X

(b) the size of the change of momentum when a molecule strikes face X

(c) using your two previous answers, the average rate of change of momentum when a molecule strikes face X

(d) the average force, caused by all the molecules moving in this direction, on face X

(e) the pressure on face X.

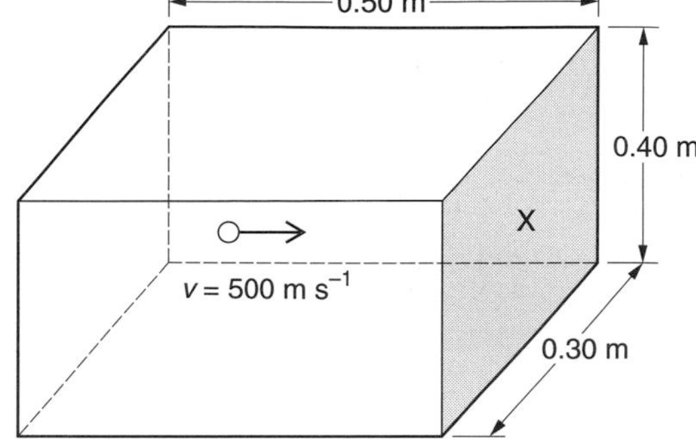

11.24 Referring to question 11.23, explain what the pressure would be

(a) on one of the other faces

(b) if the molecules each had a mass of 8.0×10^{-26} kg

(c) if the molecules each had a of speed of 600 m s^{-1}.

11.25 The kinetic theory of gases predicts how the pressure of a gas depends on properties of the gas. Two forms of this result are $p = \frac{1}{3}\rho\langle c^2\rangle$ and $pV = \frac{1}{3}Nm\langle c^2\rangle$.

(a) Explain what each of the quantities p, ρ, V, N, n, m and $\langle c^2\rangle$ represents.

(b) Give a unit for each of the quantities.

11.26 **(a)** Given that $pV = \frac{1}{3}Nm\langle c^2\rangle$ and $pV = nRT$ derive an expression for the total k.e. $N(\frac{1}{2}m\langle c^2\rangle)$ of the molecules of an ideal gas in terms of n, R and T.

(b) Since $N \propto n$, this could also be written $\frac{3}{2}NkT$, where k is a constant, called the Boltzmann constant. What is the average k.e. of a single molecule in terms of k and T?

(c) R is the *molar* gas constant. Explain why k could be thought of as the *molecular* gas constant.

(d) What is k in terms of R and the Avogadro constant N_A?

11.27 List four assumptions of the kinetic theory of gases, and for each explain why it is necessary to make the assumption.

11.28 How does the kinetic theory of gases explain why the pressure of a fixed mass of gas at constant temperature is inversely proportional to its volume?

11.29 How does the kinetic theory of gases explain why the pressure of a fixed mass of gas at constant volume increases as its temperature rises?

11.30 The measured speeds of 10 vehicles on a motorway are, in m s^{-1}, 42, 32, 28, 40, 33, 32, 35, 34, 32, 25. Calculate **(a)** their mean speed **(b)** their r.m.s. speed.

11.31 The density of argon gas is 1.61 kg m^{-3} at a pressure of 100 kPa.
(a) What is the r.m.s. speed of argon molecules under these conditions?
(b) What would be the r.m.s. speed if the pressure were halved, the temperature remaining the same?

11.32 Estimate the r.m.s. speed of air molecules in the atmosphere at sea level. Assume that the atmospheric pressure is 101 kPa and that the density of air is 1.3 kg m^{-3}.

11.33 **(a)** Find **(i)** the average k.e. **(ii)** the r.m.s. speed of a carbon dioxide molecule at a temperature of 290 K. [Use data.]
(b) Calculate these quantities for a carbon monoxide molecule at the same temperature. The relative molecular masses of carbon dioxide and carbon monoxide are 44 and 28, respectively.

11.34 What is the temperature of a gas if its molecules have an average k.e. of **(a)** 6.21×10^{-21} J **(b)** double that amount?

11.35 What is the average k.e. at 288 K of the molecules of the following gases: **(a)** hydrogen **(b)** nitrogen **(c)** bromine?

11.36 What is the r.m.s. speed at 288 K of the molecules of each of the gases listed in the previous question? Their relative molecular masses are 2, 28 and 160, respectively. [Use data.]

11.37 Air contains a mixture of different gases: nitrogen, oxygen, argon, etc. Explain whether the molecules of the different gases have different **(a)** average kinetic energies **(b)** r.m.s. speeds.

11.38 Two cylinders have the same volume. One contains hydrogen, the other oxygen. The pressures and temperature in the two cylinders are the same. Explain whether the numbers of molecules in each cylinder are the same.

11.39* The figure shows the distribution of molecular speeds v for 1 000 000 oxygen molecules at two temperatures 300 K and 600 K. At any particular whole-number speed the height of the curve represents the number of molecules which have speeds within ±0.5 m s^{-1} of that speed (e.g. for the graph with the higher peak about 2000 molecules have speeds between 319.5 m s^{-1} and 320.5 m s^{-1}).
(a) Explain which graph corresponds to 300 K and which to 600 K.
(b) Estimate the number of oxygen molecules which have speeds within ±0.5 m s^{-1} of 750 m s^{-1} at **(i)** 300 K **(ii)** 600 K.
(c) Is the area beneath the 300 K the same as the area beneath the 600 K curve? What does this area represent?
(d) Estimate the number of oxygen molecules which have speeds of less than 250 m s^{-1} at **(i)** 300 K **(ii)** 600 K.
(e) What is the median speed (i.e., the speed of the greatest number of molecules) at **(i)** 300 K **(ii)** 600 K **(iii)** the ratio of these speeds?
(f) What is the r.m.s. speed of oxygen molecules at **(i)** 300 K **(ii)** 600 K **(iii)** the ratio of these speeds?
(g) Copy the axes, scales and the 300 K graph, and add a graph which shows the distribution of speeds at 1200 K.

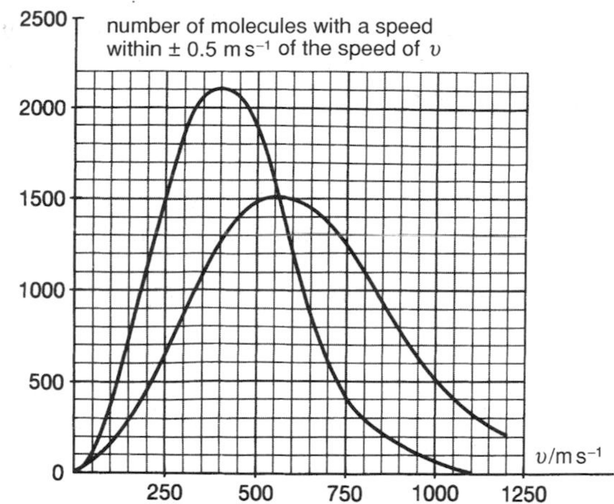

11.40* In questions 11.23 and 11.24 it was assumed that the molecules do not collide with each other: in practice they do. Consider the simple case of a collision between two molecules which are travelling parallel to one edge of the box, and which collide head-on. Explain whether the collision makes any difference to the momentum changes on the opposite faces of the box.

11.3 The internal energy of a gas

In this section you will need to

- use the equations $U = N(\frac{3}{2}kT)$ and $U = \frac{3}{2}nRT$ to calculate the internal energy U of an ideal gas at temperature T
- draw isothermals for different temperatures for an ideal gas
- understand what is meant by isothermal and adiabatic changes
- describe the changes in ΔQ, ΔW and ΔU for constant volume, constant pressure, isothermal and adiabatic changes
- use the equation $\Delta U = \Delta Q + \Delta W$ where ΔU is the change of internal energy of a body when energy ΔW is supplied to the body by working and energy ΔQ is supplied to the body by heating
- remember that the work done ΔW by a gas at constant pressure is given by $\Delta W = p(\Delta V)$ where p is the pressure and ΔV the change of volume
- represent constant-volume, constant-pressure, isothermal and adiabatic processes on a graph of p against V for a gas.

11.41 The diagram shows some gas contained in a cylinder fitted with a piston. Initially the piston is fixed in position, and the gas is heated. On a second occasion, with the gas in the same initial state, the gas undergoes the same amount of heating, but with the piston free to move.

(a) Explain whether the gain of internal energy will be the same on this second occasion.

(b) Will the temperature rise be the same?

11.42 The diagram shows graphs of p against V for a fixed mass of gas, contained in a cylinder fitted with a piston, for two constant temperatures T_1 and T_2.

(a) Which is higher, T_1 or T_2? Explain.

(b) Copy the graphs (but ignore the point '*b*', which is used in a later question), and from the point labelled '*a*' draw lines to show

(i) how p and V vary when the piston is fixed in position and the gas is heated until the temperature is T_2 (call this line X)

(ii) how p and V vary when the piston is free to move and the gas is heated until the temperature is T_2 (call this line Y)

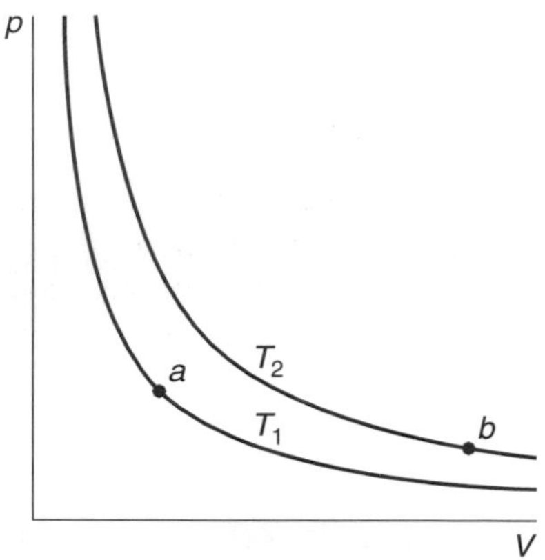

11.43 Refer to the previous question. Explain how it is possible for the gas to be heated without its temperature rising.

11.44 The internal energy U of an ideal gas is given by $U = N(\frac{3}{2}kT)$. What is the internal energy of 1.0×10^{24} molecules of an ideal gas at a temperature of **(a)** 300 K **(b)** 600 K?

11.45 A cylinder is fitted with a frictionless piston and contains gas at a constant pressure of 50 kPa. The area of the piston is 1.0×10^{-2} m^2.

(a) What force does the gas exert on the piston?

(b) What work does the gas do when the piston moves outwards 5.0 mm, the pressure inside the cylinder being kept constant (e.g. by heating)?

11.46 Make a copy of the graph provided for question 11.42. This time the initial state of the gas is given by point '*b*'.

(a) Draw lines to show the variation of p with V when **(i)** the temperature of the gas falls from T_2 to T_1 at constant pressure **(ii)** the temperature of the gas then rises from T_1 to T_2 at constant volume **(iii)** the gas then returns to its original state at constant temperature. Label the lines i, ii and iii.

(b) For each of the three processes, explain whether the internal energy of the gas has risen, fallen or remained the same.

(c) How might these changes have been accomplished, e.g. if the internal energy fell, what happened to the energy?

11.47 The graph shows the variation of pressure with volume for a gas at two temperatures 300 K and 600 K. Use information from the graph to calculate

(a) the number of moles of gas present

(b) the pressure of the gas at *a*

(c) the volume of the gas at *b*.

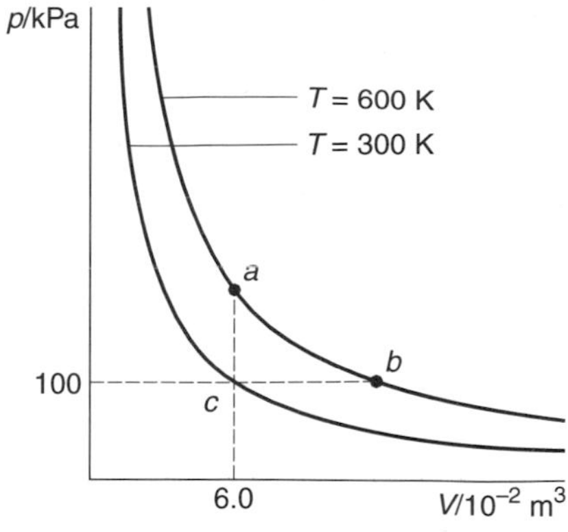

11.48 **(a)** The first law of thermodynamics is sometimes written $\Delta U = \Delta Q + \Delta W$. Why is this said to be a statement of *the principle of conservation of energy*?

(b) Explain the meaning of the symbols ΔU, ΔW and ΔQ.

(c) 'ΔU is about content but ΔW and ΔQ are about processes.' What does this statement mean?

(d) It is not correct to refer to 'the heat in a body'. What is the correct version of this phrase?

11.49 Some soup is warmed up using **(a)** a gas ring **(b)** a microwave cooker. During each of these processes what are the signs of **(i)** ΔW **(ii)** ΔU **(iii)** ΔQ for the soup?

11.50 Refer to question 11.47. The gas is compressed from *b* to *c*. For each answer attach the appropriate sign.

(a) How much work ΔW is done?

(b) If the internal energy U of the gas is given by $U = \frac{3}{2}nRT$, what is the change ΔU in internal energy in this process? (In the previous question you should have found that the amount of gas is 2.4 moles.)

(c) Work is done on the gas, but its internal energy decreases. How is this possible? Calculate the energy ΔQ transferred by heating.

11.51 **(a)** Sketch graphs of p against V for a gas at two different temperatures T_1 and T_2 ($T_2 > T_1$).

(b) Suppose for some gas contained in a cylinder $T_1 = 280$ K and $T_2 = 430$ K and that its initial pressure is 80 kPa and its initial volume is 12×10^{-3} m^3. If the gas is heated at constant volume from T_1 to T_2, use $U = \frac{3}{2}nRT$ to calculate the increase in internal energy. What are ΔQ and ΔW for this process?

(c) If, instead, the gas is heated at constant pressure from T_1 to T_2, what is ΔU? What are ΔQ and ΔW?

11.52 Describe the type of process for which, with the notation of the first law of thermodynamics, **(a)** $\Delta Q = 0$ **(b)** $\Delta U = 0$ **(c)** $\Delta W = 0$, illustrating your answer by drawing lines on a $p-V$ graph for a gas.

11.53 Explain whether the internal energy of a gas always increases when
(a) its temperature rises
(b) it is compressed adiabatically
(c) its volume decreases.

11.54 **(a)** 4.0 mol of an ideal gas is contained in a cylinder at a constant temperature of 300 K. Calculate values of its pressure p corresponding to volumes V given by $V/10^{-2}\ \text{m}^3 = 5.0, 6.0, 7.0, 8.0$ and 9.0, and plot a graph of p against V for this range of values of V. Estimate the area between the graph and the V-axis and hence find an approximate value for the work done ΔW by the gas when it expands at constant temperature from a volume of $5.0 \times 10^{-2}\ \text{m}^3$ to a volume of $9.0 \times 10^{-2}\ \text{m}^3$.
(b) What are ΔU and ΔQ for this process?

11.55* **(a)** Sketch graphs of p against V for a gas at two different temperatures T_1 and T_2 ($T_2 > T_1$) and add a line to your diagram to help you explain what is meant by an adiabatic process.
(b) The table gives corresponding values of p and V for an adiabatic expansion:

$V/10^{-2}\ \text{m}^3$	5.0	6.0	7.0	8.0	9.0
p/kPa	200	155	125	104	88

Plot these values and hence find an approximate value for the work done by the gas when it expands.
(c) Explain whether your answer is greater or smaller than the work which would be done by the gas in an isothermal expansion with the same change of volume.
(d) For the adiabatic expansion what are ΔQ and ΔU?

11.4 Heat engines

In this section you will need to

- understand the principle of a heat engine
- understand that the thermal efficiency of a heat engine is defined by the equation: η = (work done)/(energy supplied by heating)
- understand that the maximum thermal efficiency of a heat engine is governed only by the temperatures between which the heat engine works and use the equation: maximum thermal efficiency $\eta_{max} = (T_1 - T_2)/T_1$
- understand that for a heat engine maximum thermal efficiency is achieved only by a cyclical process of a particular kind (a Carnot cycle) which is not achieved in practice
- understand that the actual (or overall) efficiency of a heat engine is less than the thermal efficiency because of unintentional energy transfers to the surroundings
- understand that a heat pump is a heat engine working in reverse, and that a refrigerator is one example of this.

11.56 The diagram shows a schematic diagram for a heat engine. Energy Q_h is taken from the hot reservoir and energy Q_c is delivered to the cold reservoir (e.g. in a power station the hot reservoir is the furnace and the cold reservoir is the surroundings) and as a result work W is done by the heat engine.

(a) Write an equation for the thermal efficiency of this engine.
(b) Calculate the thermal efficiency if $Q_c = 2.4$ GJ and $Q_h = 3.9$ GJ.

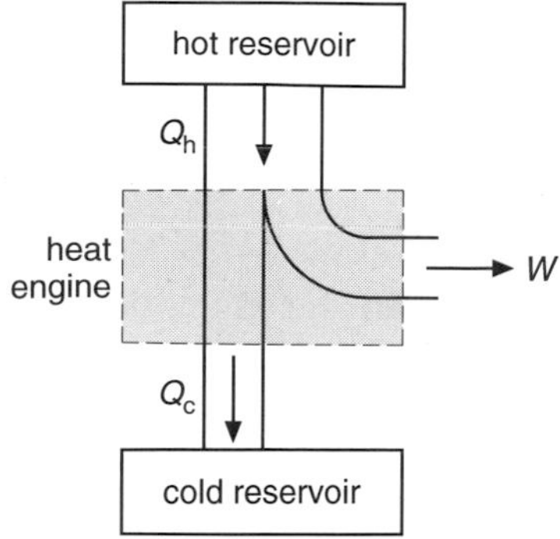

11.57 What is the maximum thermal efficiency of a heat engine working between temperatures of **(a)** 300 K and 550 K **(b)** 300 K and 650 K **(c)** 300 K and 750 K?

11.58 A thermocouple consists of two junctions between two different metals or alloys, as shown in the diagram. A difference of temperature between the junctions can produce an e.m.f. and drive a current round a circuit. This is a heat engine.
(a) What are the hot and cold reservoirs?
(b) Where is the work done?

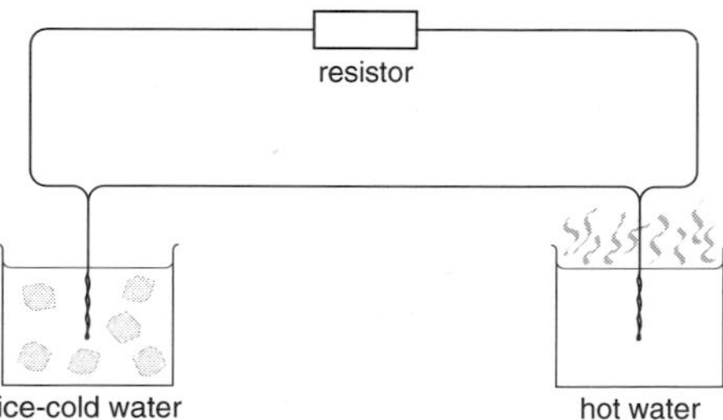

11.59 The figure on the next page shows a cycle *abcda* round which some gas may be taken, starting at *a*. The temperatures of the three isothermal curves are shown. The pressure and volume at *a* are 100 kPa and 1.0 m^3, respectively.
(a) What are the values of pressure and volume at *b*, *c* and *d*?
(b) Work is done by the gas along *bc*, and on the gas along *da*. On a copy of the graph shade areas which represent these two amounts of work done, and explain why the area *abcd* represents the net work done by the gas.
(c) Copy this table and complete the column headed ΔW, remembering to attach signs to the quantities.

path	ΔQ/kJ	ΔW/kJ	ΔU/kJ
ab			
bc			
cd			
da			
abcda			

(d) The energy supplied to the gas by heating along *ab* is 150 kJ. Enter this value in the table, and also the value of ΔU for *ab*.
(e) From *a* to *b* the temperature rose from 300 K to 600 K; from *b* to *c* it rose from 600 K to 1200 K. ΔU for *ab* was +150 kJ, so what was ΔU for *bc*?
(f) Deduce ΔQ for *bc*.
(g) Now calculate ΔU for *cd* and *da*, and hence the rest of the table.
(h) Why, for the complete cycle *abcda*, would you expect ΔU to be zero?
(i) How much energy was supplied by heating?
(j) Use your answers to **(b)** and **(i)** to calculate the thermal efficiency of this cycle.

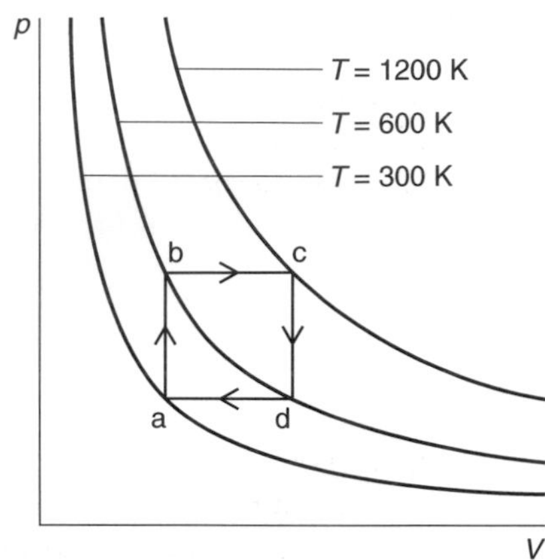

11.60 **(a)** Sketch two isothermals, which need not be to scale, for temperatures of 300 K and 550 K on a p–V graph.
(b) On your graph draw a cycle *abcda* for which *ab* is an isothermal compression at 300 K, *bc* an adiabatic compression to 550 K, *cd* an isothermal expansion at 550 K and *da* an adiabatic expansion to 300 K.
(c) Draw up a table with headings ΔW, ΔU, ΔQ, with one line for each of the paths *ab*, *bc*, *cd* and *da*. Then write '0' in the four spaces for which you know the values must be zero.
(d) Given that $U = \frac{3}{2}nRT$ for the gas, and that on this occasion $nR = 40\ \mathrm{J\ K^{-1}}$, calculate ΔU for the paths *bc* and *da*, and enter these in the table. Also enter the values of ΔW for these paths.
(e) If along path *ab* 7.3 kJ of work is done on the gas, and along path *cd* 13.4 kJ of work is done by the gas, complete the rest of the table.
(f) What is the net work done by the gas, and the energy supplied by heating?
(g) Use your answers to **(f)** to calculate the thermal efficiency of the process.
(h) On your graph shade the area which represents the net work done by the gas.
(i) Use the equation for the *maximum* thermal efficiency of a heat engine to calculate this for a heat engine working between 300 K and 550 K. How does your answer compare with your answer to **(g)** and what does this tell you about the cycle which this question describes? (This cycle is called a Carnot cycle after the French engineer Sadi Carnot who thought of it in 1824.)
(j) How could the thermal efficiency of this kind of cycle be increased still further?

11.61 **(a)** In the year 2000, one electricity company charged 9.90 p during the day for each kW h of energy; in the same area a gas company charged 2.06 p. Assuming that the use of electrical energy for heating in the home is 100% efficient, but the use of gas for heating is only 75% efficient, how many times more expensive is it to use electrical energy rather than gas for heating water?
(b) What is the underlying cause of this difference?

11.62 Question 11.56 showed a schematic diagram for a heat engine. Here is a diagram for a heat pump.
(a) Add labels and arrows similar to those in question 11.56.
(b) Name a heat pump which is found in nearly every home.
(c) For this heat pump what are the 'hot reservoir' and the 'cold reservoir'?

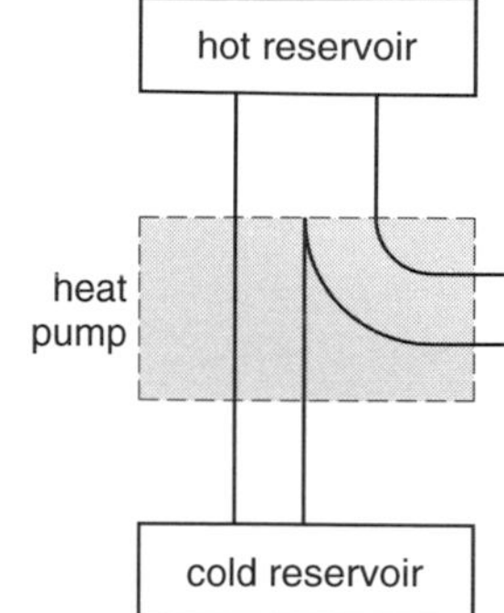

11.63 The coefficient of performance of a heat pump is defined by the equation η_r = (energy transferred from cold reservoir)/(work done) or $\eta_r = Q_c/W$ where W is the work which must be done to transfer energy Q_c from the cold reservoir to the hot reservoir.
(a) Show that it is possible to write $\eta_r = Q_c/(Q_h - Q_c)$ where Q_h is the energy delivered to the hot reservoir.
Heat pumps are in use to warm large buildings such as hospitals and shopping centres. Suppose that a hospital needs an average power of 120 kW for heating, and the cost of electrical energy is 6.0 p per kW h.
(b) What is the cost per day of using conventional electrical heating?
(c) Suppose a heat pump with a coefficient of performance of 2.0 is used to deliver a power of 120 kW to the buildings. What power must be supplied to the heat pump?
(d) Where does the rest of the energy come from?
(e) What is the cost saving per day?

11.64 **(a)** If during the course of a day it requires 2.0 MJ of energy to run a refrigerator, and during that time it transfers 3.0 MJ of energy from the contents of the refrigerator to the room, how much energy is delivered to the room?
(b) What is its coefficient of performance?

11.65 Your friend says 'You're telling me that you can use a power of 1.0 kW and deliver a power of 3.0 kW to a room? I don't believe it – that's something for nothing.' Explain to him or her how this can be done with a heat pump.

11.66 The Peltier effect is the opposite of the effect described in the question 11.58. A modern version of a Peltier device is shown in the diagram over the page. It is 40 mm square. It consists of 127 junctions between p- and n-type semiconductors, arranged in series. An electric current is passed through the device and one junction is cooled and the other heated. A heat sink is fixed to the hot junction so that the energy may be transferred to the surroundings.

(a) Explain why this device is a heat pump.

(b) This device may be used to cool the microprocessor in a computer and therefore enable the microprocessor to run faster. Suppose a particular microprocessor generates 30 W. To keep the temperature of the microprocessor at 5°C when the room temperature is 25°C requires a p.d. of 10.2 V across the device and then the current is 4.9 A. What is the coefficient of performance of this device?

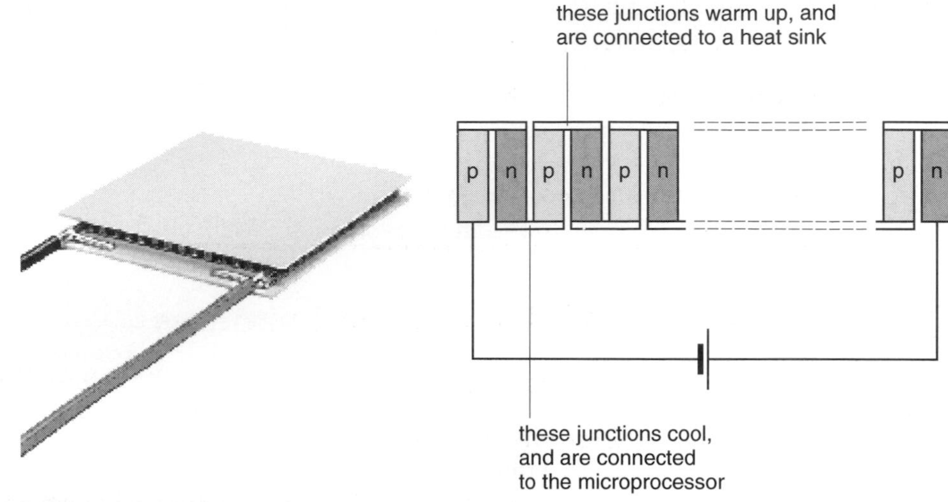

12 Photons and electrons

Data:

speed of light $c = 3.00 \times 10^{8}$ m s^{-1}

the Planck constant $h = 6.63 \times 10^{-34}$ J s

electronic charge $e = 1.60 \times 10^{-19}$ C

mass of electron $m_e = 9.11 \times 10^{-31}$ kg

mass of proton $m_p = 1.672 \times 10^{-27}$ kg

12.1 The energy of electrons and photons

In this section you will need to

- use the equation $W = QV$ for the energy transfer when a charge Q passes through a p.d. V
- understand that the electron-volt (eV) is a unit of energy (though not an SI unit), and remember how it is defined
- understand that the energy E of electromagnetic radiation is quantised
- use the equation $E = hf$ for the energy of a photon, where h is the Planck constant
- use the wave equation $c = f\lambda$
- use the equation $E_k = \frac{1}{2}mv^2$ in solving problems
- use the equation $I = P/4\pi r^2$ to calculate intensity I of radiation.

12.1 The definition of the volt tells us that there is a transfer of energy of 1.0 J when a charge of 1.0 C passes through a p.d. of 1.0 V.
(a) What is the energy transfer when 2.0 C pass through a p.d. of 3.0 V?
(b) What is the energy transfer when an electron passes through a p.d. of 1.0 V? [Use data.]
(c) In TV tubes the electrons may be accelerated through p.d.s of 800 V. How much energy is transferred from electrical potential energy to the kinetic energy of each electron?

12.2 **(a)** What is 1.0 eV in joules? [Use data.]
(b) What is the answer to part **(c)** of the previous question if eV is used as the unit of energy instead of the joule?

12.3 What is the energy transfer, in J, when an electron passes through a p.d. of **(a)** 2.0 V **(b)** 20 V **(c)** 1.0 MV **(d)** 5.0 MV? [Use data.]

12.4 What is the energy transfer, in eV, when an electron passes through a p.d. of **(a)** 2.0 V **(b)** 20 V **(c)** 1.0 MV **(d)** 5.0 MV?

12.5 What is the energy transfer, in eV, when
(a) a proton passes through a p.d. of 1.0 V
(b) a doubly-charged ion passes through a p.d. of 1.0 V
(c) a proton passes through a p.d. of 2.0 MV?

12.6 What is the speed of
(a) an electron with a kinetic energy of **(i)** 6.0×10^{-18} J **(ii)** 50 eV
(b) a proton with a kinetic energy of **(i)** 6.0×10^{-18} J **(ii)** 50 eV?

12.7 In a certain X-ray tube electrons are accelerated through a p.d. of 200 kV. What is their final kinetic energy in **(a)** eV **(b)** J?

12.8 Refer to question 12.5. Suppose the mass of the ion is four times the mass of the proton. What is the ratio (speed of proton)/(speed of ion) when both pass through a p.d. of 1.0 V?

12.9 What is the energy, in J, of a photon of
(a) red light of frequency 4.6×10^{14} Hz
(b) violet light of frequency 6.9×10^{14} Hz? [Use data.]

12.10 Calculate the energies, in J, of a photon of
(a) infrared radiation of wavelength 1500 nm
(b) green light of wavelength 546 nm
(c) ultraviolet radiation of wavelength 365 nm
(d) X-radiation of wavelength 154 pm
(e) γ-radiation of wavelength 2.3×10^{-12} m.

12.11 Suppose a photon has an energy of 1.00 eV.
(a) What is **(i)** its frequency **(ii)** its wavelength?
(b) In what part of the electromagnetic spectrum would you find it?
(c) Repeat part **(a)** for a photon with an energy of 2.00 V.
(d) Repeat part **(b)** for a photon with an energy of 2.00 V.

12.12 **(a)** At what points in a typical house containing the usual electrical equipment would you expect to find electromagnetic radiation of the following wavelengths:
(i) 5×10^{-7} m **(ii)** 3×10^{-10} m **(iii)** 2×10^{-6} m **(iv)** 0.12 m
(v) 1.5×10^{3} m?
(b) What are the energies (in eV) of photons of these wavelengths?

12.13 Copy and complete this table for light from mercury vapour (answers are given for **(a)**, **(b)**, **(c)** only):

colour	yellow	green	violet	ultra-violet
wavelength λ/nm	579	546	435	253
frequency/10^{14} Hz	**(a)**			
photon energy/10^{-19} J	**(b)**			
photon energy/eV	**(c)**			

12.14 Describe the difference between 'red' light (e.g. from a red dress) and the red light emitted by excited hydrogen.

12.15 A laboratory helium-neon laser emits light of wavelength 632.8 nm in a beam of diameter 2.0 mm.
(a) What is the energy of one photon of this light?
(b) If its power is 0.70 mW, how many photons are passing any point in the beam each second?

12.16 A radioactive source is found to emit γ-radiation of wavelength 6.5×10^{-12} m.
(a) What is the frequency of the γ-radiation?
(b) What quantity of energy (in MeV) is carried away from the nucleus by each γ-ray photon?

12.17 A 60 W filament lamp hangs from the ceiling in a room. Assume that 5% of the electrical energy is converted into visible radiation.
(a) Calculate an approximate value for the intensity in W m^{-2} of the visible radiation on a table which is 1.5 m below the lamp. State any further assumptions you make.
(b) The average wavelength of the visible radiation is 550 nm. What is the average energy of each photon? [Use data.]
(c) What is the number of photons striking an A4 sheet of paper which is placed on the table?

12.18* The wavelengths of the two D lines in the spectrum of a sodium lamp are both close to 590 nm.
(a) What is the energy of one photon of sodium light?
(b) A 500 W sodium vapour lamp has an efficiency of 30% (i.e. 30% of the supplied energy is emitted as light in the two D lines). How many photons does it emit per second?
(c) Someone stands 50 m from the lamp. What is the intensity of the D light at that distance?
(d) If his eye pupil in these conditions has a diameter of 3.5 mm, how many photons enter one eye each second?
(e) What is the average distance between photons along the line from the lamp to the eye?

12.19 **(a)** What is the kinetic energy **(i)** in eV **(ii)** in J, of an electron accelerated through a p.d. of 100 kV in a tube designed to produce X-rays?
(b) Photons are emitted when the electron hits the metal target in the tube. If the photon emitted from the tube has the whole of this energy what is **(i)** its frequency **(ii)** its wavelength?

12.20 If a TV tube accelerates electrons through a p.d. of 20 kV, what is
(a) the kinetic energy of the electrons **(i)** in eV **(ii)** in J
(b) the maximum frequency of the X-rays produced
(c) the range of wavelengths of the X-rays?

12.21 Orthochromatic photograph film was in use in the 1940s and 1950s. Its advantage was that it was not sensitive to red light, so could be developed in a darkroom fitted with a red safelight (and the processor could see what he was doing). Why would you have expected it to be less likely that there should be a type of film which was not sensitive to blue light?

12.2 Energy levels

In this section you will need to

- understand that an atom can exist in a few sharply-defined states of energy because there are only certain 'positions' where its electrons may be
- understand that the energy of an atom may be changed when it is bombarded by electrons or photons, or when it emits photons
- calculate the frequency of radiation emitted when an electron moves from one energy level to another.

12.22 The diagram shows some of the energy levels for an atom of hydrogen. Photons are emitted when an electron moves down from one level to another.

(a) When an electron moves from level 2 to level 1, what is
  (i) its loss of energy in eV
  (ii) its loss of energy in J
  (iii) the frequency of the emitted photon
  (iv) the wavelength of the emitted photon
  (v) the part of the electromagnetic spectrum in which this radiation occurs?

(b) Repeat part **(a)** for an electron moving from level 3 to level 2

(c) Repeat part **(a)** for an electron moving from level 4 to level 3.

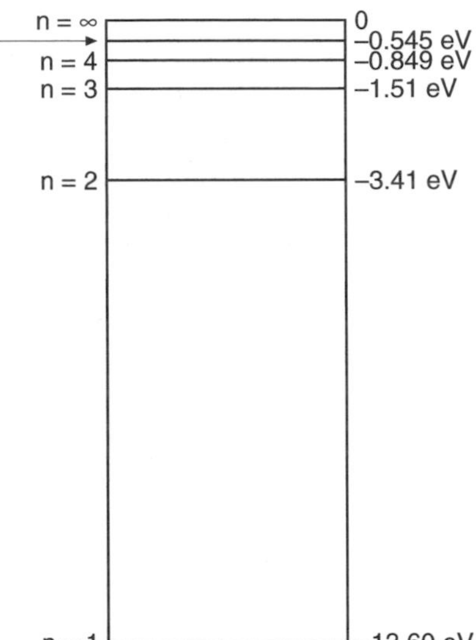

12.23 The two lowest excited states of a hydrogen atom are 10.2 eV and 12.1 eV above the ground state.

(a) Calculate three wavelengths of radiation that could be produced by transitions between these states and the ground state.

(b) In which parts of the spectrum would you expect to find these wavelengths?

12.24 The figure shows an energy level diagram. Sketch a possible line spectrum for the light emitted when electrons make the transitions shown. Label the lines, using the letters shown in the diagram, and indicate on your spectrum diagram which end corresponds to the higher frequency.

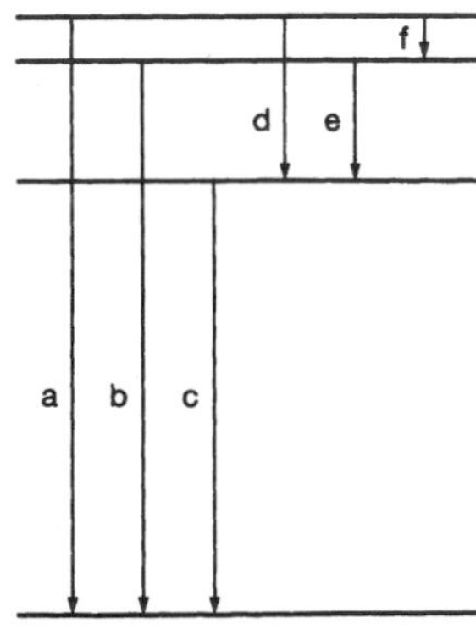

12.25 The four lowest energy levels for an atom consist of the ground state and three levels above that. How many transitions are possible between these four levels?

12.26 The figure shows three energy levels for a particular atom. When an electron moves from level 1 to the ground state the light emitted is blue. In what part of the spectrum would you expect to find the radiation emitted when an electron moves from level 2 to the ground state?

level 2 —— −3 eV

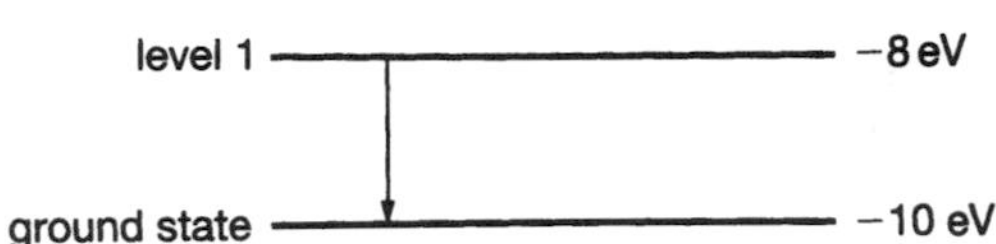

12.27 Suppose an atom has two energy levels E_1 and E_2 above the ground state. Radiation frequencies of f_1 and f_2 correspond to these energies, respectively.
(a) Sketch the energy level diagram of this atom.
(b) What other frequency will be emitted by this atom?

12.28 Refer to the diagram for question 12.22. Plot a graph of the values of energy E on the y-axis against the square of the number of the level (i.e. n^2) on the x-axis. Choose scales so that values of n^2 up to 60 can be plotted on the x-axis, and use a scale of 1 cm per eV on the y-axis.
From your graph deduce the next two highest energy levels above those shown in the previous diagram.

12.29 The ionisation energy of hydrogen is 13.6 eV.
(a) What is the speed of the slowest electron that can ionise a hydrogen atom when it collides with it?
(b) What is the longest wavelength of electromagnetic radiation that could produce ionisation in hydrogen?

12.30 Refer to the diagram for question 12.22.
(a) If the atom is in the ground state, how much energy must be given to it to ionise it?
(b) Suppose an electron of energy 2.2 eV collides with the atom. Explain the possible results if **(i)** the atom is in the ground state **(ii)** its electron is at the −3.41 eV level.
(c) What is the wavelength of the photon which could raise an electron from the −0.849 eV level to the −0.545 eV level?
(d) If an electron returns from the –0.849 eV level to the ground state, what is the wavelength of the photon emitted?

12.31 The table shows the results when electrons of three different energies strike a mercury atom in its ground state. Explain these results.

energy of electron before collision/eV	4.0	4.9	6.0
energy of electron after collision/eV	4.0	zero	1.1

12.32 In an experiment to investigate energy levels electrons were accelerated through a p.d. of 50.0 V and then allowed to strike helium atoms. The energies of the electrons after the collisions were measured.

(a) What is one energy which you might expect some of the electrons to have after the collisions?

(b) Other energies which the electrons had after the collisions included 28.9 eV, 26.8 eV and 26.0 eV. What are the wavelengths of the radiation which might have been observed in this experiment?

12.33* The table gives the frequencies of some of the lines which occur in three groups in the hydrogen spectrum (traditionally the series have the names given at the top of each column):

Lyman $f/10^{14}$ Hz	Balmer $f/10^{14}$ Hz	Paschen $f/10^{14}$ Hz
24.659	4.5665	1.5983
29.226	6.1649	2.3380
30.824	6.9044	2.7399
31.564	7.3084	2.9822

(a) Which series is in the visible part of the spectrum?

(b) Which series is in the infrared part of the spectrum?

(c) Can you see any numerical relationship between the frequencies in adjacent columns? If so, why is that relationship to be expected?

(d) Predict the next two highest frequencies in the Lyman series.

(e) How many different energy levels are responsible for the frequencies shown in the table?

(f) What is the longest wavelength in the Lyman series?

12.34 **(a)** The first energy level for mercury is 4.9 eV above the ground state. When the atom returns from this level to the ground state, what is the wavelength of the radiation emitted, and in what part of the spectrum is this radiation?

(b) How do you account for the following:

(i) Cool mercury vapour strongly absorbs the ultraviolet radiation from a mercury vapour lamp.

(ii) Cool mercury vapour is completely transparent (i.e. non-absorbing) to the visible light from a mercury vapour lamp, provided the ultraviolet light is excluded by a glass filter.

12.35 Some of the energy levels for the sodium atom are –1.51 eV, –1.94 eV, –3.03 eV (two levels very close together) and –5.14 eV, which is the ground state. Draw a labelled diagram for these levels, and describe and explain what might happen if cool sodium vapour (i.e. sodium whose atoms are in the ground state) is bombarded with

(a) electrons whose k.e. is 2.00 eV

(b) electrons whose k.e. is 2.50 eV

(c) light of wavelength 590 nm.

12.36 The wavelengths of visible radiation range from 400 nm to 750 nm approximately. Use the information in the previous question to show that the only light which can be absorbed by cool sodium vapour has a wavelength of 590 nm.

12.37 Light from a filament lamp gives a *continuous spectrum*, i.e. light of all colours in the visible part of the electromagnetic spectrum. This is white light.

(a) Why do excited gases give a *line spectrum*, i.e. light of just a few colours?

(b) An *absorption spectrum* may be obtained when white light passes through a gas. The diagram shows a typical absorption spectrum obtained when white light is passed through sodium vapour: it consists of dark lines in a few particular places superimposed on the continuous spectrum. Explain how these dark lines occur, and whether the number of them would depend on the temperature of the gas.

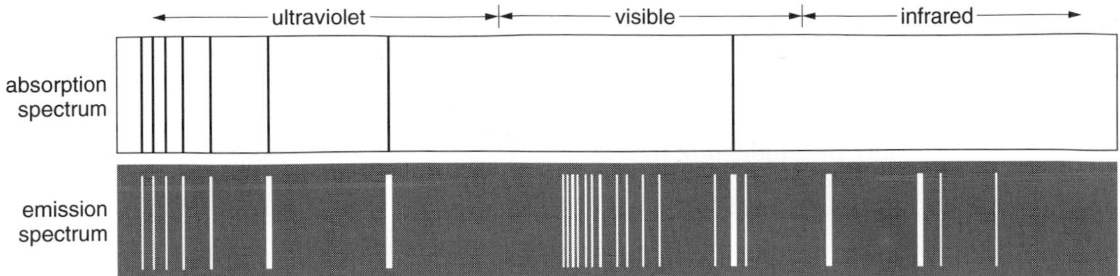

12.3 The photoelectric effect

In this section you will need to

- understand how quantisation explains the photoelectric effect
- use the photoelectric equation $hf = \phi + \frac{1}{2}mv_{max}^2$
- understand that the maximum k.e. $\frac{1}{2}mv_{max}^2$ may be equated to the electrical potential energy eV_s, where V_s is the stopping potential.

12.38 How do the following observations of the photoelectric effect support the idea that light is quantised?

(a) There is no noticeable delay between the arrival of photons and the emission of electrons.

(b) For a particular metal surface, only light of a certain minimum frequency will cause emission of electrons.

12.39 How does the quantum theory explain that

(a) electrons are emitted by a particular metal only when radiation of less than a certain wavelength falls on it

(b) the rate of emission of electrons is proportional to the intensity of the radiation

(c) the maximum speed of the emitted electrons is independent of the intensity of the radiation?

12.40 The minimum frequency of light which will cause photoelectric emission from a lithium surface is 5.5×10^{14} Hz.

(a) Calculate the work function of lithium.

If the surface is lit by light of frequency 6.5×10^{14} Hz, calculate

(b) the maximum energy of the electrons emitted

(c) the maximum speed of these electrons.

12.41 The work function of a freshly cleaned copper surface is 4.16 eV. Calculate

(a) the minimum frequency of the radiation which will cause emission of electrons, and state whether this radiation is visible

(b) the maximum energy of the electrons emitted when the surface is exposed to radiation of frequency 1.20×10^{15} Hz.

12.42 The diagram shows results from an experiment in which two different photocells A and B were exposed to light of different wavelengths. The stopping potential V_s has been plotted against the frequency f of the radiation.

(a) What is the equation which relates V_s and f?

(b) Rearrange the equation so that it reads '$V_s =$ '

(c) Measure the gradient of the two graph lines.

(d) Explain why these two gradients should be the same.

(e) Hence calculate the Planck constant h.

(f) Measure the intercepts on the V_s-axis for the two graph lines.

(g) Hence calculate the work function for the two materials in the photocells.

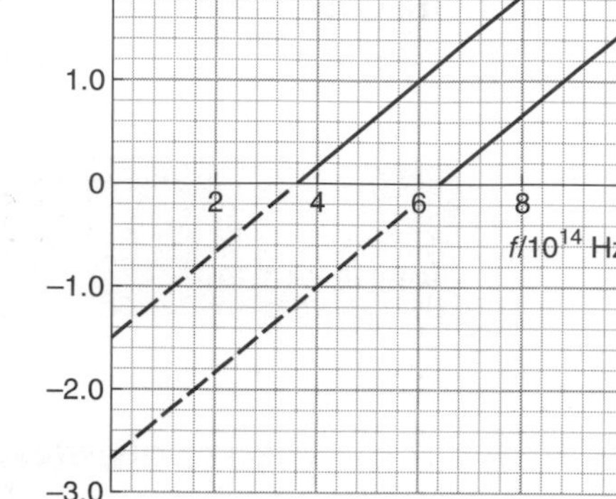

(h) The experiment depends on the photocell being lit by light of several different frequencies. To do the experiment how would you obtain light of particular frequencies?

12.43 Monochromatic radiation of wavelength 546 nm falls on a potassium surface of area 7.5 cm^2 in an evacuated enclosure. The intensity at the surface is 60 mW m^{-2} and it may be assumed that 1% of the photons emit electrons from the surface. What is the photoelectric current?

12.44 The work function of a freshly cleaned zinc surface is 3.6 eV. If it is illuminated with ultraviolet radiation of wavelength 253 nm,

(a) what will be the maximum k.e.(in eV) of the emitted electrons

(b) what p.d. would be needed to stop the electrons being emitted?

If the radiation is shone on the freshly-cleaned zinc cap of an uncharged leaf electroscope, explain

(c) why the emission of electrons will soon stop

(d) whether you would expect there to be any deflection of the leaf.

12.4 Waves and particles

In this section you will need to

- understand that we cannot describe light in terms of other phenomena (i.e. it is not 'like' water waves or bullets)
- understand that just as light has particle properties, so electrons (and other particles) have wave properties, and diffraction experiments are evidence for this
- use the de Broglie equation for momentum $p = h/\lambda$ to calculate the wavelength associated with a particle.

12.45 **(a)** What are the energies (in eV) of the following types of photon: **(i)** a radio wave of wavelength 1500 m **(ii)** infrared radiation of wavelength 5.0×10^{-5} m **(iii)** a gamma ray of wavelength 2.0×10^{-12} m?

(b) Which type would you think of as being most like a wave, and which most like a particle?

(c) Does any of this radiation behave entirely like a wave or entirely like a particle?

12.46 **(a)** Use the equations $E_k = \frac{1}{2}mv^2$ and $p = mv$ for a particle of mass m and speed v to show that its momentum p is related to its kinetic energy E_k by the equation $p = \sqrt{(2mE_k)}$, at speeds which are low compared with the speed of light.

(b) What are the kinetic energy and momentum of, and wavelength associated with, an electron accelerated through a p.d. of 10 V?

(c) What are these quantities when it is accelerated through a p.d. of 100 V?

12.47 Calculate the momentum of, and wavelength associated with, the following particles, when each has an energy of 10 keV:

(a) an electron

(b) a proton

(c) an alpha particle, which consists of two protons bound to two neutrons.

12.48 How could you distinguish, experimentally, between a photon and an electron, if each has a momentum of 5×10^{-23} N s?

12.49 An experimenter wishes to investigate the diffraction of electrons by thin foils. He wants to use wavelengths of 1.5×10^{-10} m.

(a) What momentum should the electrons have?

(b) What kinetic energy do these electrons have?

(c) What p.d. should be used to accelerate the electrons?

12.50 In an electron diffraction tube the electron beam passes through a very thin crystalline foil. The beams diffracted by the crystals form circles on the end face of the tube. The diameter d of one prominent particular circle is measured for a range of values of the accelerating p.d. V, and the following values are obtained:

V/kV	1.5	2.0	3.0	4.5	6.0
d/mm	68	58	48	40	35

The theory of this experiment leads us to expect that d should be proportional to $1/\sqrt{V}$.

(a) Plot a graph of d against $1/\sqrt{V}$, and comment on the result.

(b) What p.d. would be required to give a diffraction circle of diameter 25 mm?

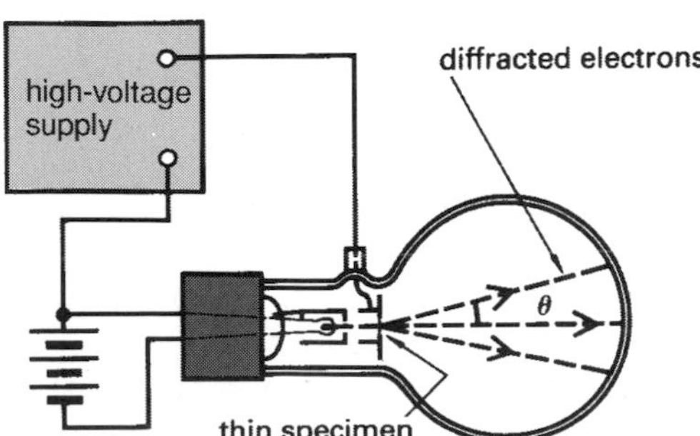

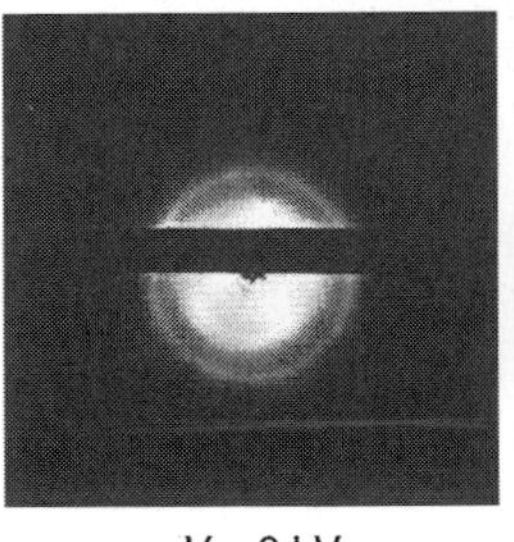

V = 6 kV

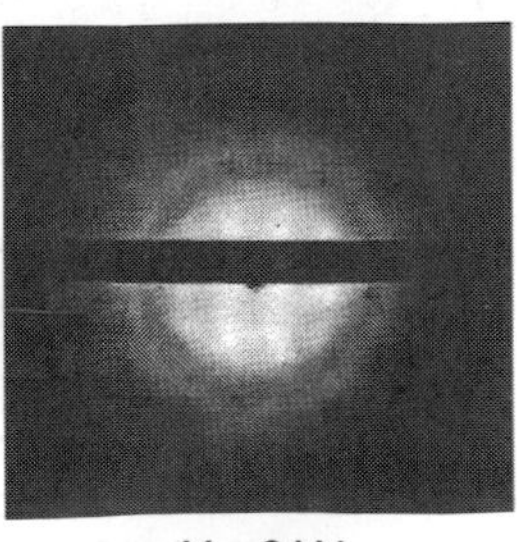

V = 3 kV

13 Radioactivity

Data: Avogadro constant L (N_A) = 6.02×10^{23} mol^{-1}

atomic mass unit u = $1.660\,43 \times 10^{-27}$ kg

relative atomic masses (r.a.m.): sodium 23, iron 56, gold 197

electronic charge $e = 1.60 \times 10^{-19}$ C

13.1 Atoms and nuclei

In this section you will need to

- understand that an atom consists of a nucleus surrounded by one or more electrons
- understand that the nucleus contains protons and neutrons
- use the notation $^{A}_{Z}X$ when referring to nuclides, where Z is the atomic (proton) number and A is the mass (nucleon) number of the nuclide of element X
- use the term neutron number N when describing nuclides
- use the atomic mass unit u and remember that it is equal to one twelfth of the mass of the nuclide $^{12}_{6}C$.

13.1 The element copper comes 29th in a list of the elements. What information does this fact alone give you about atoms of copper?

13.2 The symbol for a nuclide of an element is often written $^{A}_{Z}X$, where A and Z are the mass and atomic numbers of the nuclide, e.g. $^{131}_{53}I$ is one of the nuclides of iodine. How is it possible to abbreviate this to ^{131}I or iodine-131 without there being any confusion?

13.3 The diameters of most atoms are about 3×10^{-10} m. Roughly how many atoms thick is
(a) a piece of gold foil, of thickness 6×10^{-7} m
(b) a strand of the copper wire (diameter 0.10 mm) used to make flexible electrical connecting wire?

13.4 The atomic number Z of carbon is 6; the neutron numbers N of its three common isotopes are 6, 7 and 8. Write down
(a) the mass numbers A of these isotopes
(b) the atomic mass of the most massive isotope.

13.5 The relative atomic mass of gold is 197.
(a) What is the mass of one atom of gold? [Use data.]
The density of gold in the solid state is 1.93×10^{4} kg m^{-3}.
(b) How many atoms of gold are there in a volume of 1.00 m^3?

(c) Treating a gold atom as if it occupied a cubical volume, what is the volume of such a cube?

(d) What is the length of the side of such a cube? This is an approximate measure of the 'diameter' of a gold atom.

Repeat these calculations for an atom of aluminium, for which the relative atomic mass is 26.9 and the density 2.70×10^3 kg m^{-3}. Comment on the two values of diameter.

13.6 Using data given above, find the number of

(a) gold atoms in a gold ring which has a mass of 25 g

(b) iron atoms in a one-kilogram iron mass.

13.7 Tin, symbol Sn and atomic number 50, has more stable isotopes than any other element. The number of neutrons can be 64, 65, 66, 67, 68, 69, 70, 72 or 74. Give a list of the symbols for these nuclides.

13.8 Copper (atomic number 29), has two stable isotopes of mass number 63 (69.2%) and 65 (30.8%), where the numbers in brackets are the relative proportions of the isotopes. Calculate the relative atomic mass of copper.

13.9 The three common isotopes of silicon have mass numbers of 28, 29 and 30. Their percentage abundances are 92.2%, 4.7% and 3.1% respectively. Calculate the relative atomic mass of silicon by finding the weighted average of these mass numbers.

13.10 The mass of an atom of sodium, Na, is 3.82×10^{-26} kg. What is the mass, in grams, of one mole of sodium atoms? Comment on your answer. [Use data.]

13.11 How many atoms are there in a mass of 40 g of each of these nuclides:
(a) $^{40}_{19}K$ **(b)** $^{80}_{35}Br$ **(c)** $^{120}_{50}Sn$? [Use data.]

13.12 This is a list of the commonest nuclides of several different elements with increasing atomic numbers:

$$^{12}_{6}C \quad ^{45}_{21}Sc \quad ^{93}_{41}Nb \quad ^{147}_{62}Sm \quad ^{208}_{82}Pb.$$

For each nuclide calculate the ratio N/Z, where N is the number of neutrons, and comment on your answer.

13.2 Unstable nuclei

In this section you will need to

- remember the nature of the alpha, beta (β^- and β^+) and gamma radiations together with their relative charges and masses
- understand the mechanism of electron capture as another method by which one nuclide changes to another
- write nuclear equations of the form: $^{4}_{2}He + ^{14}_{7}N \rightarrow ^{12}_{8}O + ^{1}_{1}H$ or $^{14}N\ (\alpha,p)^{17}O$ in which both charge and mass are conserved.

13.13 Copy this grid and fill in the gaps to show the properties of these four types of radiation from unstable nuclei:

	α	β^+	β^-	γ
electric charge mass, in u ionising power what it is				

13.14 Samarium-147 (atomic number 62, symbol Sm) decays by α-emission. The following is a list of neighbouring elements with their atomic numbers (in brackets): cerium (58), praseodymium (59), neodymium (60), promethium (61), europium (63), gadolinium (64). Explain what isotope it must decay into.

13.15 Cobalt-56 (symbol $^{56}_{27}$Co) is an isotope that decays by β^+-emission.
(a) What name is given to a β^+-particle?
(b) How many protons, neutrons and electrons are there in each neutral atom of this isotope?
(c) The neighbouring elements in the Periodic Table are $_{26}$Fe and $_{28}$Ni. What is the nuclide into which $^{56}_{27}$Co decays?

13.16 Copy the grid which gives the number of neutrons, *N*, in a nucleus (up) against the number of protons, *Z*, (along), i.e. it is a grid of neutron number against atomic number in the region $N = 81$, $Z = 57$.
(a) Explain why the arrow shows an α-decay and write the nuclear equation for this decay.
(b) Add two labelled arrows to your grid to show
(i) a possible β^--decay
(ii) a possible β^+-decay.
Give the appropriate nuclear equations.
(c) Discuss how the arrows you have drawn for β-decays support the fact that nuclides with an excess of neutrons tend to undergo β^--decay whereas those with too few neutrons tend to undergo β^+-decay.

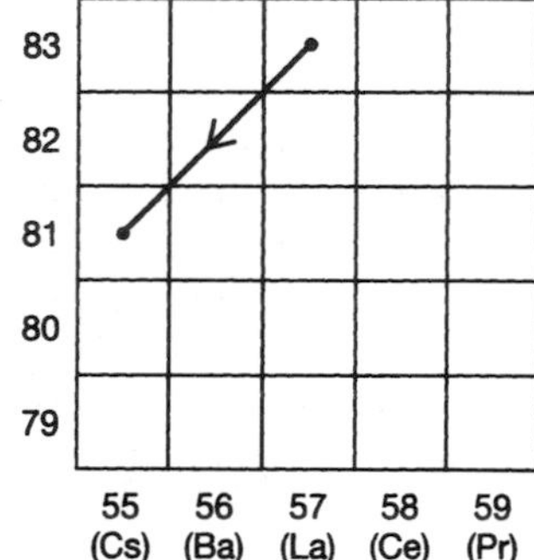

13.17 $^{238}_{92}$U decays by emitting a succession of α-particles and β^--particles to form $^{206}_{82}$Pb. How many of each particle are emitted?

13.18 $^{26}_{13}$Al decays by electron capture.
(a) Write down the equation which describes this decay. The neighbouring elements, with their atomic numbers, are magnesium (12) and silicon (14).
(b) Write down a second equation which shows that the emission of a β^+-particle would produce the same nuclide.
(c) How can electron capture and β^+-particle emission be distinguished in practice?

13.19 When a β^--particle is produced in radioactive decay, is there any change in the number of orbital electrons in the atom **(a)** immediately **(b)** eventually? Explain.

13.20 Answer the previous question for **(a)** α-decay **(b)** β^+-decay **(c)** electron capture.

13.21 The α-particles from a source typically have just one or two particular energies, but the β-particles emitted by a source have a range of energies. For example, β^+-particles from nitrogen-13 may have any energy between zero and a maximum of 1.2 MeV. If a particular β^+-particle from nitrogen-13 has an energy of 0.80 MeV, which two particles share the remaining 0.40 MeV when this β^+-particle is emitted?

13.22 The diagram shows a plot of neutron number N against atomic number Z for stable nuclides. Each small circle shows that there is a stable nuclide with particular values of N and Z. The circles form a 'fuzzy' curve of increasing slope. Unstable nuclides would lie to one side or other of this fuzzy curve.

(a) For $Z = 15$, find a typical value of N, and calculate the ratio N/Z. Repeat this for $Z = 30$, 45, 60, 75 and draw a graph to show how N/Z varies with Z. Write a sentence to describe what the graph shows.

(b) Does the ratio N/Z for a nucleus increase or decrease when **(i)** a β^--particle **(ii)** a β^+-particle is emitted?

(c) Explain why, on the diagram, two regions are labelled β^--emission and β^+-emission.

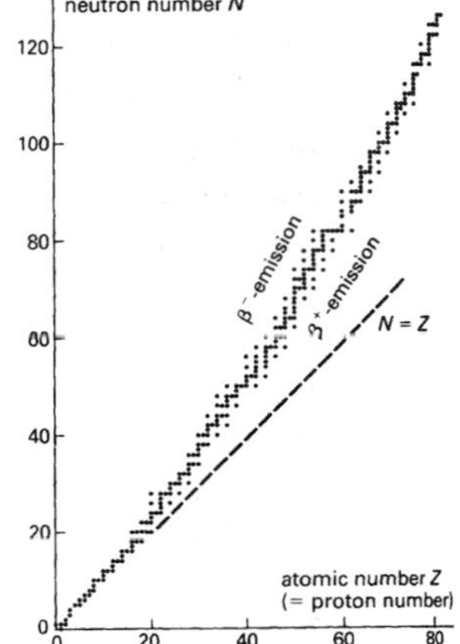

13.23 Just as an atom can exist in different excited states, with its electrons in different energy levels, so a nucleus can exist in different excited states.

^{212}Bi decays to the nuclide ^{208}Tl by emitting an α-particle. If the α-particle has the maximum energy of 6.090 MeV, the ^{208}Tl nuclide will be in its ground state. The diagram shows that the α-particle may have any one of four other different energies, and then the ^{208}Tl nuclide is temporarily in an excited state. To return to the ground state it almost immediately emits one or more γ-rays. The possible transitions are shown in the diagram.

(a) What are, in order of decreasing size, the six different energies of the γ-rays which may be emitted?

(b) What is the shortest wavelength?

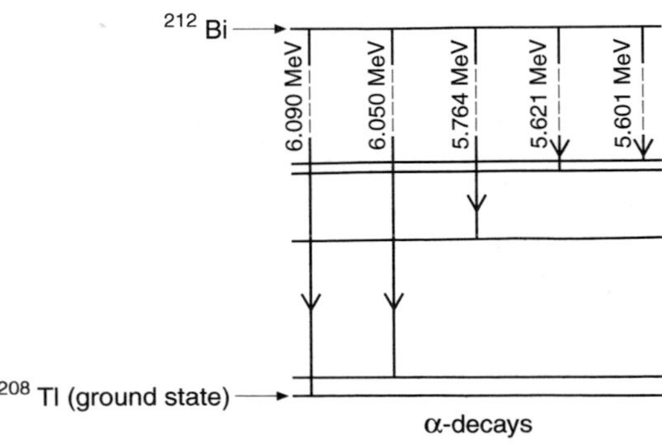

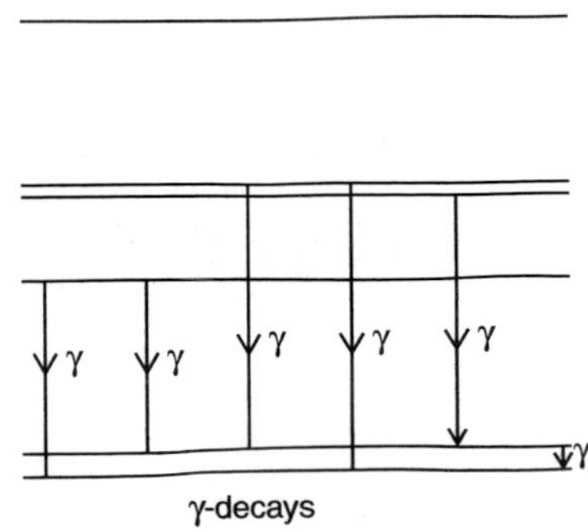

13.3 Properties of the radiations

In this section you will need to

- remember the properties of the three types of radiation
- remember the relative ionising properties of the radiations
- remember the typical penetrating abilities of the radiations
- describe how to use a GM tube and scaler in the study of ionising radiations
- describe how to use a diffusion cloud chamber to study the properties of α-particles
- describe how to use an ionisation chamber to study the properties of α-particles
- understand that magnetic fields may be used to deflect β-particles
- understand what is meant by the half-value thickness of a material absorbing γ-rays.

13.24 The photograph shows tracks left by α-particles in a cloud chamber.

(a) Explain what these white lines consist of.

(b) Each line fades within a few seconds. What has happened to it?

(c) If the source had emitted β-particles or γ-rays there would have been no tracks, or a few very faint tracks. What property of α-particles causes them to leave such obvious tracks?

(d) How can you deduce that these α-particles all have almost the same energy?

(e) What difference would you notice if you covered the source with a very thin sheet of paper?

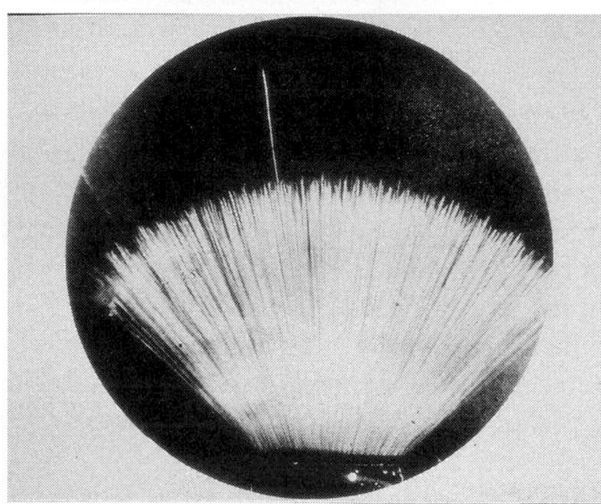

13.25 The graph shows how the range of α-particles in air varies with their initial energy.

(a) Radium-226 emits α-particles which have initial energies of either 4.78 MeV or 4.60 MeV. What are the ranges in air of particles emitted with these energies?

(b) What are the initial speeds of these α-particles? [Use data.]

(c) What is the process by which they lose energy?

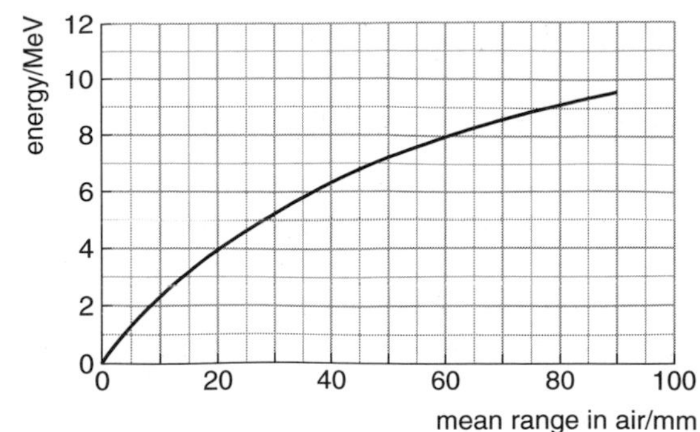

13.26 In 1909 Rutherford and Royds performed an experiment which demonstrated conclusively that α-particles are identical to helium nuclei. Write an account of their experiment.

13.27 The diagram shows one form of spark counter. A p.d. of about 5 kV is connected between the brass plate and the grid of wires. Particles from the source travel into the space between the plate and the wires. If they ionise the air between the plates, a spark passes between the plate and the wires and this event can be registered on a scaler.

(a) Explain what would you expect to happen if two sources, one emitting only β-radiation and the other emitting only γ-radiation, were moved steadily one at a time towards the grid from a distance of 10 cm.

(b) When an α-emitting source is moved steadily towards the grid from a distance of 10 cm, sparks begin to occur when the distance is less than 4.5 cm. Explain

(i) whether, with another source, this distance would be the same

(ii) why the rate of sparking increases as the distance decreases from 4.5 cm

(iii) why the time interval between sparks is not constant, and why successive sparks occur at different places on the grid.

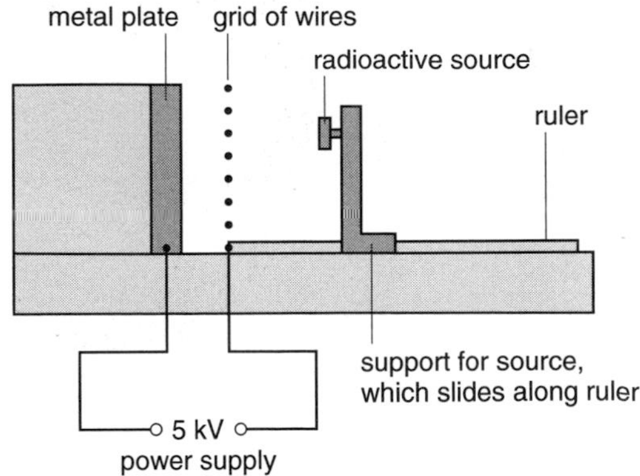

13.28 **(a)** Radon is a radioactive element which is gaseous at room temperature. It emits α-particles with an energy of 6.3 MeV. Each particle loses on average 30 eV of energy for each pair of ions that it creates in a collision. How many ion pairs would you expect an α-particle from radon to create?

(b) What is the charge on each ion? [Use data.]

(c) Some of the gas is pumped into an ionisation chamber, which is shown in the diagram. A power supply is connected so that there is a p.d. between the case and the central electrode. The positive and negative ions therefore move in opposite directions and so there is an electric current which can be measured by a sensitive ammeter. What is the current when the radon is producing 800 α-particles each second?

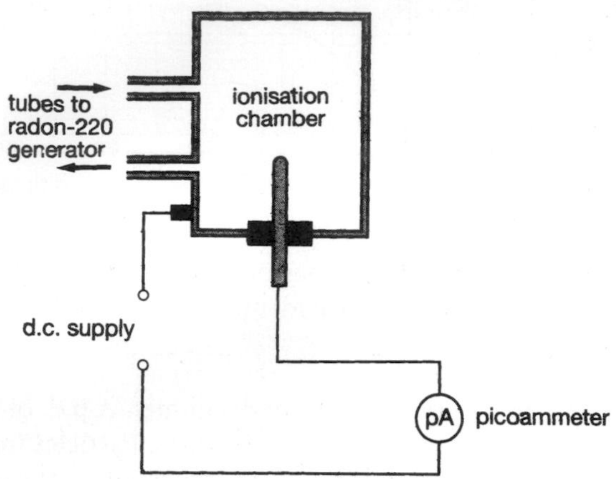

13.29 An α-particle slows down as it travels through air. What eventually happens to it?

13.30 The part **(i)** of the diagram shows the energy levels in a nucleus when an americium-241 nucleus decays to form uranium-237. These are similar to the energy levels in excited atoms.

(a) What are the energies of the emitted α-particles?

(b) When the less energetic α-particle is emitted the uranium-237 nucleus is left in an excited state, but it almost immediately emits a γ-photon and returns to the ground state. What is the energy of this photon, and its wavelength? [Use data.]

(c) 85% of the α-particles are the more energetic. How does part **(ii)** of this diagram convey this information?

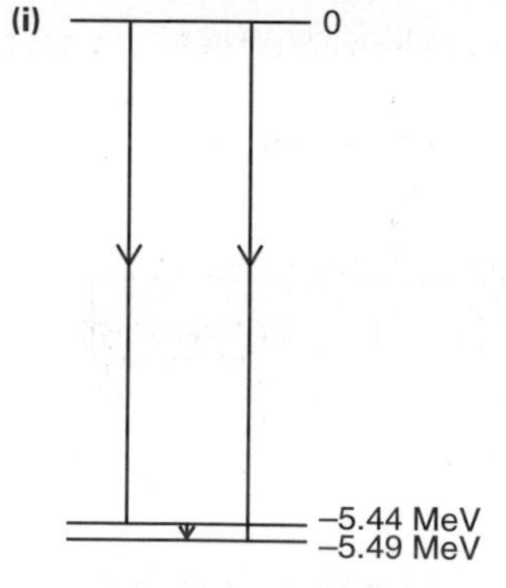

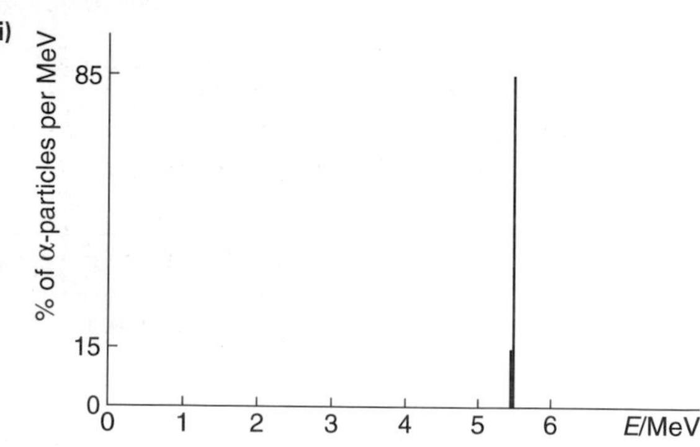

13.31 The α-particles from a source typically have just a few particular energies, but when a radioactive nuclide decays by β-emission, the β-particles have a whole range of energies up to a maximum value, as shown in the diagram for the nuclide $^{35}_{16}S$.

(a) Explain what diagram **(i)** shows. E.g. what does it mean that for 80 keV the percentage is 1.1%?

(b) $^{35}_{16}S$ emits a β^--particle. Write down the nuclear equation for this decay. (Neighbouring elements are $_{15}P$ and $_{17}Cl$.)

(c) Experiments in bubble chambers show that the direction of recoil of the nucleus, and the direction of the β-particle, are not directly opposite. See diagram **(ii)**. Why does this suggest that a third particle is emitted?

(d) This particle does not leave a track in a bubble chamber. What can you deduce from that?

(e) What is the name of the emitted particle which accompanies **(i)** β^--emission **(ii)** β^+-emission?

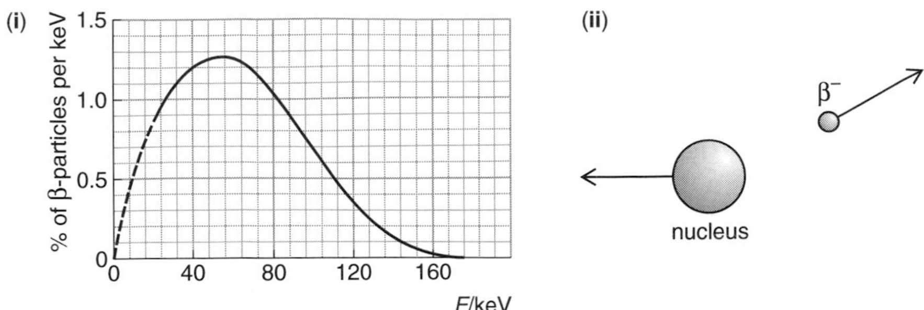

13.32 The 'radium' sources available in school and college laboratories consist of a mixture of nuclides as a result of the decay of radium. They therefore emit α-particles, β^--particles and γ-rays. A student places sheets of different materials of different thicknesses between the radium source and a GM tube placed close to the source, as shown in the diagram, and measures the numbers of events per minute registered by the GM tube. Explain each of the following observations:

(a) When the GM tube is moved a few cm further away from the source the count-rate falls considerably.

(b) When the GM tube is moved back to its original position, and a sheet of paper is placed between the GM tube and the source, the count-rate again falls considerably.

(c) The paper is removed, and the GM tube is moved to a position 10 cm from the source. When a sheet of aluminium of thickness 1.0 mm is placed between the GM tube and the source, the count-rate falls by about 30%.

(d) Each of three more sheets of aluminium of the same thickness cause the count-rate to fall even further.

(e) Two further sheets of aluminium produce little effect on the count-rate, but a sheet of lead of thickness 10 mm has a considerable effect.

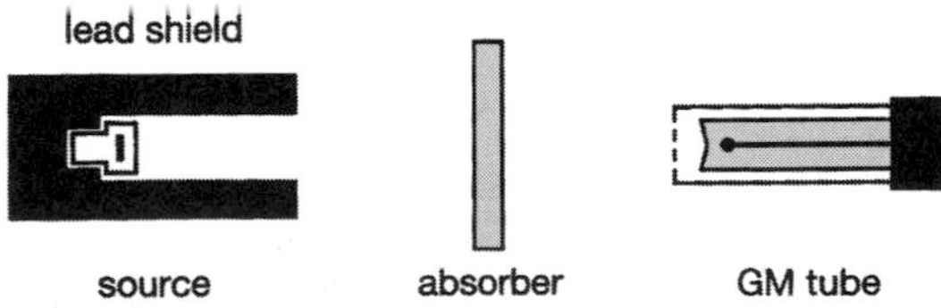

13.33* A student uses the apparatus shown in the diagram for question 13.32 to investigate, using sheets of aluminium, the absorption of β^--particles from a particular source. The sheets of aluminium were stamped with the thickness in mg/cm^2. The results were as follows:

thickness/mg cm^{-2}	0	48	200	304	387	554	637	702
count	19 960	4330	4436	3381	2243	500	443	224
time of count/s	15	30	60	90	120	200	600	600

(a) Calculate the count rate, in counts per minute, for each of the absorber thicknesses.
(b) Why did the student have the counter running for a longer time for the later readings?
(c) Give two reasons why the distance between the source and the counter should be as small as possible.
(d) In this experiment the highest count rate is much greater than the lowest count-rate. Before plotting a graph of the results it helps to compress the count rates by taking their logarithms. Take logarithms, to base 10, of the count rates and plot lg(count min^{-1}) against absorber thickness.
(e) Extend your graph to estimate the thickness of aluminium which would absorb these β-particles completely.

13.34 You know that magnetic fields push on current carrying wires if the wire is at right angles to the field. In fact what they push on is the moving electric charges in those wires, i.e. the electrons. So a magnetic field should push on α-particles and β-particles, since they move and carry electric charge.
The diagram shows how the direction of the magnetic field, the direction of the moving charge, and direction of the force are related.
(a) Will the particle bend to the right or the left if it is **(i)** an α-particle **(ii)** a β^+-particle **(iii)** a β^--particle?
(b) Considering only the different masses of the α-particle and the β-particles, which kind of particle will have the larger deflection?
(c) Considering both the different masses *and* the different amounts of charge of the α-particle and the β-particles, which kind of particle will have the larger deflection?

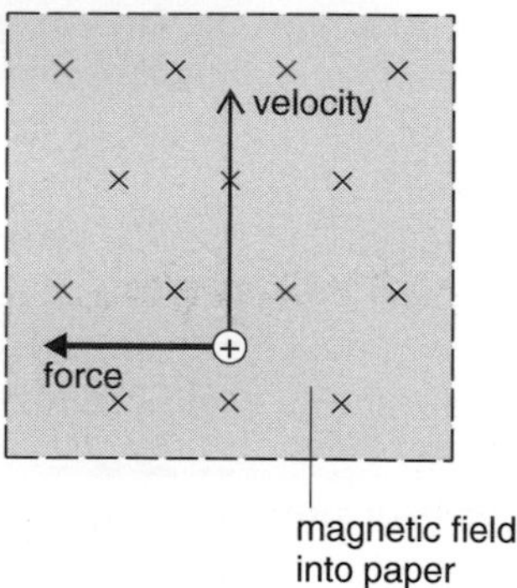

13.35 The half-value thickness for cobalt-60 γ-rays passing through lead is 10 mm.
(a) Draw a graph to show how the intensity of a parallel beam of these γ-rays would vary with distances from zero to 30 mm.
(b) The cobalt-60 γ-ray sources used in school or college laboratories are often contained in lead cylinders whose thickness is about 4 mm. By what fraction is the intensity of the γ-radiation reduced by this thickness of lead?

13.36 A laboratory γ-source is often a tiny quantity of cobalt-60 with a 'strength' of 5 μCi (microcuries). If a one-curie source produces 3.7×10^{10} γ-photons per second, calculate how many γ-photons pass through each of your eyes (taken to be spheres of radius 2.0 cm) every second when you look at a laboratory source from a distance of **(a)** 0.50 m **(b)** 2.50 m.

13.37 A GM tube is used to measure the activity of a source which is emitting γ-rays only. It records an activity of 3200 Bq when it is 30 cm from the source. What would you expect it to record when it is 60 cm from the source? State any assumptions you need to make.

13.4 Laws of radioactive decay

In this section you will need to

- understand that radioactive decay is a random process but that laws govern radioactive decay if the activity is great enough
- remember that the activity A of a radioactive source is equal to the number of disintegrations per second and is measured in becquerel (Bq)
- use the equation $A = -\lambda N$ which relates activity A to the number N of undecayed nuclei, and where λ is the radioactive decay constant
- remember the relationship $\lambda t_{\frac{1}{2}} = \ln 2$
- understand the mathematics in section 21.7
- use the equation describing radioactive decay $N = N_0 e^{-\lambda t}$
- use radioactive decay curves to find the half-life $t_{\frac{1}{2}}$ and the decay constant λ of radioactive nuclides
- 'take logarithms' of both sides of the equation $N = N_0 e^{-\lambda t}$ to give $\ln N = -\lambda t + \ln N_0$
- explain how to measure the half-life of long-lived radioactive nuclides
- apply the laws of radioactive decay to situations where radioactive nuclides are used in practice, for example as tracers or in carbon dating.

13.38 $^{24}_{11}$Na, an isotope of sodium which emits β^- particles, has a decay constant of $1.28 \times 10^{-5}\ s^{-1}$. Suppose a sample initially contains 6.00×10^{10} nuclei.

(a) What is its initial activity?
(b) What is its half-life?
(c) After 30 hours how many undecayed nuclei will there be?
(d) After 30 hours what will be its activity?

13.39 The nuclide phosphorus-32 has a half-life of 14.3 days. It can be added to the fertiliser in the soil in which a plant is growing: the plant's absorption of the phosphate can then be monitored by observing the radioactivity in the stem and leaves.

(a) What is the decay constant, in s^{-1}?
(b) How many atoms of it would be needed to produce an activity of 10 000 Bq?
(c) What mass of the phosphorus isotope would be needed? [Use data.]

13.40 An adult human being has about 250 g of potassium in its body. Ordinary potassium contains 0.012% of a naturally occurring radioactive isotope ^{40}K which has a half-life of 1.3×10^9 years. The potassium in our own bodies contains this proportion and contributes to our exposure to radiation. What is

(a) the decay constant, in s^{-1}
(b) the number of atoms in 250 g of potassium [Use data.]
(c) the number of unstable nuclei in 250 g of potassium
(d) the activity of these nuclei?

13.41 A sample of iodine-131, of half-life 8.0 days, has an activity of 7.4×10^7 Bq. What will be the activity of the sample after 8 weeks?

13.42 Technetium-99 ($^{99}_{43}$Tc) decays by emission of a β^--particle into ruthenium (Ru).
(a) Write down the symbol with superscript and subscript for the ruthenium nuclide.
A sample containing 0.10 μg of technetium-99 is found to emit β^--particles at a rate of 135 Bq.
(b) How many atoms are there in 99 g of ^{99}Tc?
(c) Calculate the decay constant λ of ^{99}Tc.
(d) What is the half-life, in years, of ^{99}Tc?

13.43 Caesium-137 has a half-life of 30 years.
(a) Calculate its decay constant, in s^{-1}.
(b) Calculate the number of nuclei needed to give an activity of 2.0×10^5 Bq.
(c) Calculate the mass needed.

13.44 An exponentially-decaying quantity has a value of 654 at 11 a.m. and a value of 587 at 12 noon.
(a) What is the ratio 587/654?
(b) What will be the value of the quantity at **(i)** 1 p.m. **(ii)** 2 p.m.?
(c) What was its value at 10 a.m.?

13.45 **(a)** Iodine is an element which is selectively absorbed by humans in the thyroid gland, which is in the neck. Its uptake can be monitored by giving a patient some radioactive iodine in the form of a solution of sodium iodide. Two samples of activity 200 kBq are prepared, and one administered to the patient, the other being kept as a reference. The nuclide ^{131}I has a half-life of 8.0 days.
(a) What is its decay constant, in s^{-1}?
(b) How many atoms of ^{131}I are needed to produce an activity of 200 kBq?
(c) What mass of iodine contains this number of atoms?
(d) The reference sample is placed in a 'phantom' neck, a neck which is modelled to give the same dimensions and absorbing properties as the patient's neck. In a normal thyroid gland the uptake of $^{131}_{53}$I might be 30% to 50% after 24 hours. If the activities measured in the sample and the patient are 5.0 kBq and 1.3 kBq, what is the patient's percentage uptake of the iodine?

13.46 Refer to the previous question.
(a) What would the activity of the initial samples have been if half the mass of iodine had been used?
(b) Suppose a nuclide with double the half-life had been used. If the number of atoms was the same, what would the activity have been?

13.47 In the treatment of some cancers needles filled with radioactive material are inserted into the malignant tissue. A typical needle is shown in the diagram. Caesium-137, whose half-life is 28 years, is often used in the needle. What will be its activity if the mass of the caesium is 2.0 mg? [Use data.]

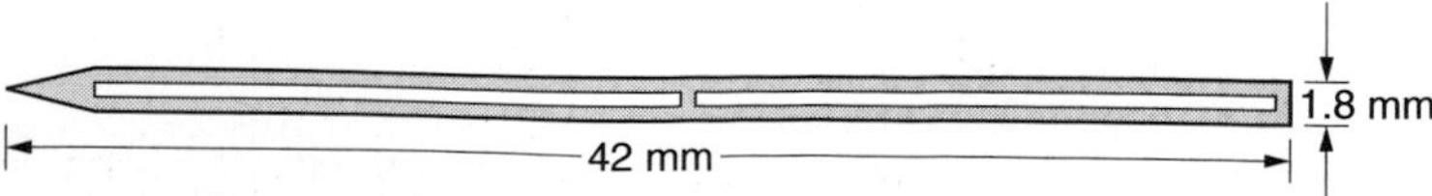

13.48 An initially pure sample of ^{131}I has an initial mass of 4.0 μg. It emits β-particles and its half-life is 8.0 days. What will be the mass of the iodine in the sample after 24 days? Choose one of the following answers and explain your choice: 4.0 μg, 1.3 μg, 1.0 μg, 0.50 μg.

13.49 The half-lives of ^{21}Na and ^{26}Na are 24 s and 60 s, respectively. The graphs in the diagram show the activity of two samples of these nuclides of sodium. Initially each sample contains the same number of sodium atoms.

(a) Which graph refers to the ^{21}Na?

(b) At what time are the activities the same?

(c) At that time what is the ratio (number of undecayed ^{21}Na nuclei)/(number of undecayed ^{26}Na nuclei)?

(d) How would you test whether the graphs, as drawn, show exponential decay?

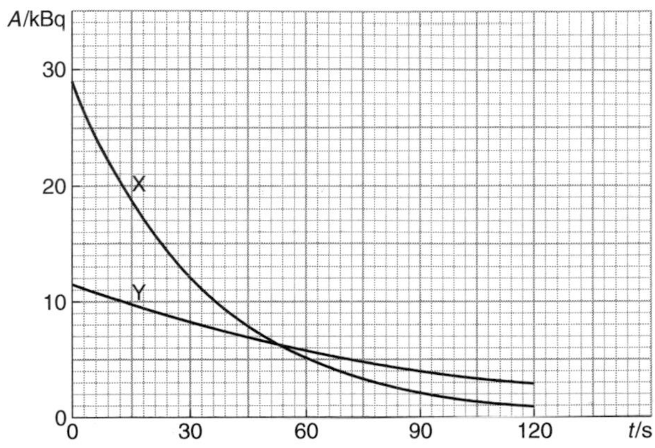

13.50 The graph shows the decay of a nuclide which initially has 200 000 undecayed nuclei.

(a) What is the half-life of the nuclide?

(b) Find the activity A of the source at times t/min = 0, 2, 4, 6, 8, 10, 12 by drawing tangents to the curve at these times and measuring the gradient.

(c) Plot a graph of A against t.

(d) Test this graph to see if it is exponential by measuring three different periods of time in which the activity of the source halved.

(e) Plot a graph of $\ln(A/\text{min}^{-1})$ against t.

(f) For exponential decay this graph is a straight line of slope $-\lambda$. Measure the gradient and hence calculate the half-life of the nuclide.

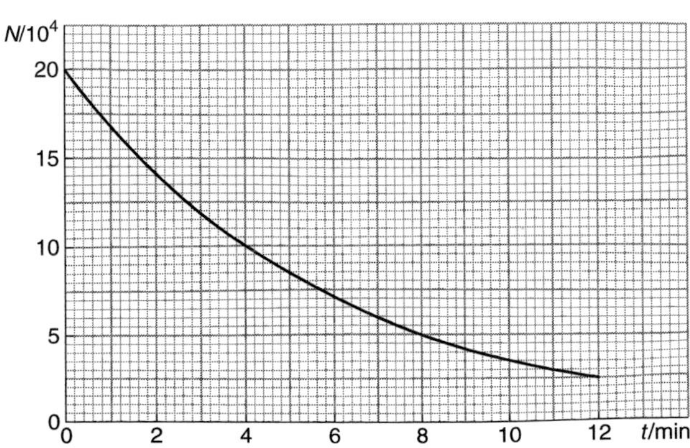

13.51 Suppose a sample of a radioactive nuclide initially has 10 000 undecayed nuclei and that it has a half-life of 1.0 minute.

(a) Plot a graph of the number of undecayed nuclei at 1-minute intervals for the first 5 minutes, using a scale of 1 cm for 1000 nuclei on the y-axis and 2 cm for 1 minute on the x-axis.

(b) Draw tangents to the graph at values of t/min = 0, 1, 2, 3, 4, 5 and calculate the gradients of these tangents, which measure the rates of decay of the nuclide.

(c) *On the same axes* plot a graph of the activity of the sample at 1-minute intervals for the first 5 minutes, using a scale of 1 cm for an activity of 10 Bq on the y-axis and 2 cm for 1 minute on the x-axis.

(d) Describe how the two graphs compare.

13.52* In an experiment to determine the half-life of the nuclide $^{63}_{30}\text{Zn}$, a sample containing some of the radioactive zinc was placed close to a GM tube and the following readings taken:

time/hours	0	0.5	1.0	1.5	2.0	2.5	3.0
count rate/count min^{-1}	259	158	101	76	56	49	37

During the following morning the counter registered 29, 32, 33, 29, 28 and 30 counts over a one-minute period on the hour each hour.

(a) Why was activity still recorded the following morning?

(b) Why did this not appear to be constant?

(c) Allowing for this activity, make a table of the corrected count rate during the experiment on the first day.

(d) Plot a graph of this count rate against time and use it to deduce three different values of the half-life of ^{63}Zn. Find the mean of your values.

(e) Plot a second graph of ln(count rate) against time and use it to deduce a value for the half-life of ^{63}Zn.

(f) Discuss which of **(d)** and **(e)** provides the more reliable value.

13.53 A plastic bottle contains $^{232}_{90}\text{Th}$ in the form of thorium hydroxide and is connected by two tubes to an ionisation chamber (refer to question 13.28). The thorium decays into the radioactive gas radon ($^{220}_{86}\text{Rn}$), so there is radon in the bottle and this can be pumped into the ionisation chamber

The half-life of this radon nuclide is short enough for its rate of decay to be directly measured, and the diagram shows the result of connecting a chart recorder to the ionisation chamber to measure the ionisation current.

(a) Explain how the graph shows both the random and the systematic nature of radioactive decay.

(b) How many α- and β^--particles are emitted when $^{232}_{90}\text{Th}$ decays to $^{220}_{86}\text{Rn}$?

(c) Make measurements on the graph to find the half-life of $^{220}_{86}\text{Rn}$.

(d) The final current would not be zero if the chamber were not cleaned regularly. Explain this.

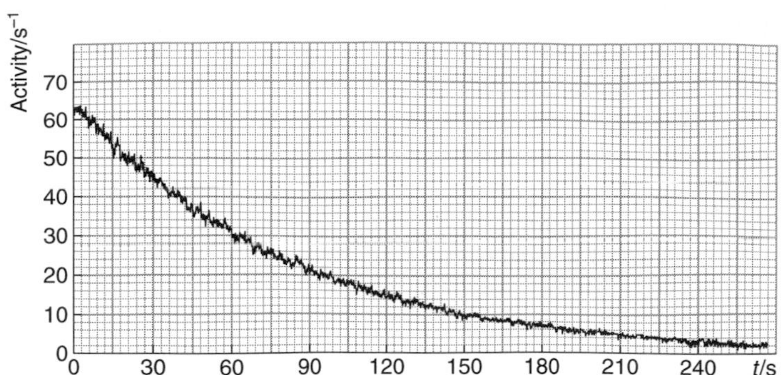

13.54 The equation $N = N_0e^{-\lambda t}$ can be used to calculate the number of undecayed nuclei N there will be after a time t. Explain why there must be a similar equation $A = A_0e^{-\lambda t}$ which can be used to calculate the activity A after a time t.

13.55 Strontium-90 is one of the fission products in nuclear reactors. Its half-life is 28 years. Suppose that initially there are 5.0×10^{11} undecayed nuclei.

(a) What is the decay constant, in y^{-1}?

(b) How many will there be 100 years later?

(c) After what time will the number have fallen to 5.0×10^9?

13.56 In order to find the volume of water in a central heating system a small quantity of a solution containing the radioactive isotope sodium-24 (half-life 15 hours) is mixed with the water in the system. The solution has an activity of 1.6×10^4 s^{-1}.
When 30 hours have elapsed, it is assumed that the sodium-24 has mixed thoroughly with the water throughout the system, and a 100 ml sample of the water is drawn off and tested for radioactivity. It is estimated that the activity of the sample is 2.0 s^{-1}.
What is the total volume of water in the central heating system?

13.57 To measure the volume of red cells in human blood some of the patient's blood is removed and exposed to sodium chromate which contains the radioactive nuclide ^{51}Cr. The ^{51}Cr nuclei attach themselves to the red blood cells. 10 cm^3 of the 'labelled' blood is replaced in the patient. Suppose that at this time the activity of the sample is 0.20 MBq. The labelled blood is allowed to mix in the patient for 10 minutes and 10 cm^3 of blood is then removed. Suppose the activity of this sample is 154 Bq.

(a) The half-life of ^{51}Cr is 28 days. Show that there is no need to allow for the decrease in radioactivity during the time of the experiment.

(b) What is the volume of red blood cells in this patient?

(c) The haematocrit for a typical female patient is 44%, i.e. red blood cells form 44% of the total blood volume. What is the total volume of blood for this patient?

13.58 Caesium-137 was one of the constituents of the fall out from the Chernobyl nuclear reactor disaster in 1986. Its half-life is 28 years. Immediately after the explosion the activity, even outside the 30 kilometre exclusion zone, was 50 kBq per square metre. After what time would this activity be reduced to 1 kBq per square metre?

13.59 In investigations of the blood flow in parts of the brain a patient may be injected with a β^+-particle emitter such as $^{18}_{9}$F, $^{15}_{8}$O, $^{12}_{7}$N or $^{11}_{6}$C. The molecules containing these nuclei then emit positrons and the emission enables doctors to monitor brain activity. This is called positron emission tomography (PET).

(a) Considering the values of N and Z for these nuclides, why might you expect them all to be β^+-particle emitters?

(b) The nuclides are created in a cyclotron on the hospital site because these nuclides all have short half-lives. Name two reasons why short half-lives are desirable?

(c) The emitted positrons are detected because they immediately collide with electrons. Their annihilation then emits two γ-rays, in opposite directions. The detection of these two γ-rays in a scintillation counter enables doctors to pinpoint the source of the positron. Why is it necessary to base PET on the use of positrons, rather than electrons?

13.60* **(a)** Suppose a nuclide X is unstable and decays to a nuclide Y, which is stable. Sketch graphs to show how the numbers of nuclei of X and Y vary with time.

(b) Now suppose that Y is unstable and decays to a nuclide Z which is stable. A model of this situation can be provided by three water tanks, as shown in diagram **(i)**. The tanks have exit pipes of different cross-sectional areas. Initially tank X is filled, but it drains out into Y, which in turn drains into Z. What feature of the model corresponds to **(i)** the number of undecayed nuclei **(ii)** the decay constant of the nuclide?

(c) Diagram **(ii)** shows how the numbers of nuclei of X, Y and Z vary with time. Use the model to help you explain these graphs.

(d) In this example, which has the larger decay constant: X or Y?

(i)

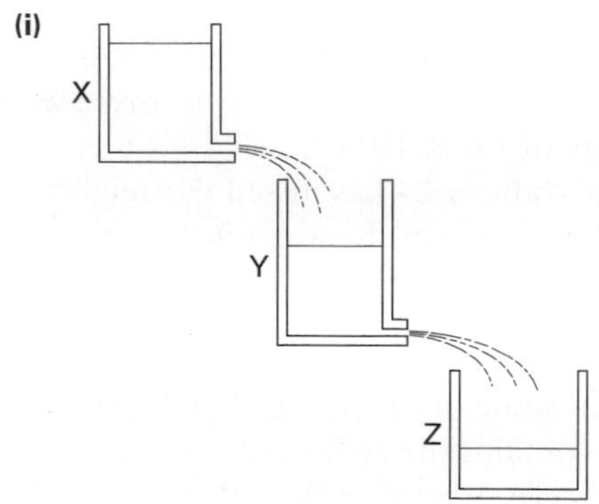

(ii)

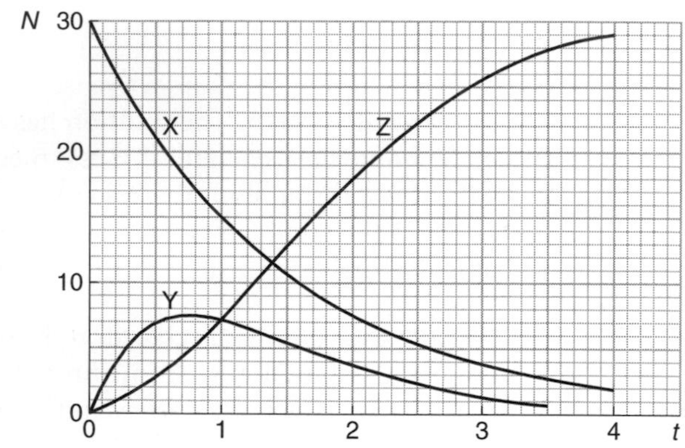

13.61 In a piece of living timber the fraction 1.25×10^{-12} of the total carbon content is in the form of the radioactive isotope carbon-14 (half-life 5730 years). A sample of carbon dioxide containing 2.00 g of carbon from living timber is introduced into a GM tube. This proportion remains constant by exchange with the carbon in the surroundings (as long as the fraction of ^{14}C in the surroundings continues to be 1.25×10^{-12}). When the animal or plant dies this exchange no longer takes place, and therefore the amount of radioactive carbon in the material decreases. Calculate

(a) the number of atoms of carbon-14 in the sample

(b) the decay constant of carbon-14

(c) the average number of disintegrations to be expected in 10 minutes, if the background count obtained with a non-radioactive gas in the GM tube is 10 per minute.

(d) The count rate obtained with an identical sample prepared from an ancient piece of timber is 307 in 10 minutes. Calculate the age of the timber.

13.62 The half-life of carbon-14 is 5730 years and decays by β^--emission.

(a) What nucleus is the product of the decay?

(b) Calculate the decay constant, in s^{-1}, for carbon-14.

(c) Hence find the number of these carbon atoms required to produce 150 β^--particles per second.

(d) If one atom in 8×10^{11} atoms of natural carbon is radioactive, what mass of carbon is required to produce this count rate?

(e) Your answer to **(d)** should suggest a possible difficulty in this method of determining age. What is it?

13.63 The term *half-life* is used to indicate how long the activity of a radioactive source will last. Would it not be simpler to say what its *life* is?

13.64 If radioactivity is a random process, how can there be mathematical laws which can be used to calculate what happens?

14 Nuclear power and nuclear matter

Data: The Avogadro constant L (N_A) = 6.02×10^{23} mol^{-1}

unified atomic mass unit u = $1.660\,43 \times 10^{-27}$ kg

mass of proton = $1.672\,52 \times 10^{-27}$ kg = 1.007 28 u

mass of neutron = $1.674\,82 \times 10^{-27}$ kg = 1.008 67 u

mass of electron = $9.109\,08 \times 10^{-31}$ kg = 0.000 549 u

1 u ≡ 931.5 MeV

speed of electromagnetic radiation = 3.00×10^{8} m s^{-1}

electronic charge $e = 1.60 \times 10^{-19}$ C

14.1 Exposure to radiation

In this section you will need to

- explain that the cumulative effect of background radiation varies from place to place
- understand the precautions that need to be taken to minimise our exposure to radiation.

14.1 It is suggested that, in order to reduce exposure to background radiation, it is preferable to live in a house built of granite which is well sealed against draughts than in a large tent designed to allow cool breezes to flow through it. Explain what you think of this suggestion.

14.2 Suppose a worker in the nuclear industry is working near a radioactive source which has an activity A.

(a) Explain why, if t is the time for which she is exposed to the radiation and d is her distance from the source, the dose D of radiation received is proportional to At/d^2.

(b) The damage done also depends on the nature of the radiation. The *dose equivalent* H is given by $H = QD$ where Q is the quality factor for the radiation. Explain why the quality factor for α-particles (20) is much higher than the quality factor for γ-rays (1).

14.3 The nuclide $^{99}_{43}\text{Tc}^{\text{m}}$ is a nuclide whose nucleus is in an excited state: the nucleus returns to the ground state by emitting a γ-ray, and is then $^{99}_{43}\text{Tc}$. There is no other change in the nuclide. It is often used in medical diagnosis within the body because it is a pure γ-ray emitting nuclide.

(a) Explain why it is preferable to use γ-emitting nuclides rather than β-emitting nuclides.

(b) Why would an α-emitting nuclide be useless and dangerous?

14.4 People living near Chernobyl when the explosion took place absorbed ^{131}I from the fall-out into their thyroid glands. ^{131}I has a radioactive half-life of 8 days, but as well as decaying it is also removed from the body by excretion. In 15 days half is excreted.
(a) What is the decay constant, in d^{-1}, for **(i)** radiation **(ii)** excretion?
(b) What is the total decay constant?
(c) What is the effective half-life for ^{131}I in the thyroid gland?

14.5 The diagram shows how the average annual radiation dose equivalent of 2 mSv (millisievert) is made up for people living in the United Kingdom.
(a) What percentage of the average dose comes from artificial sources?
(b) The dose received varies considerably over the British Isles. Suggest why some regions are regions of high dose.
(c) In the year following the Chernobyl accident the average dose over the whole of Britain increased by about 0.1 mSv. Express this as a percentage of the average dose from **(i)** natural sources **(ii)** artificial sources.

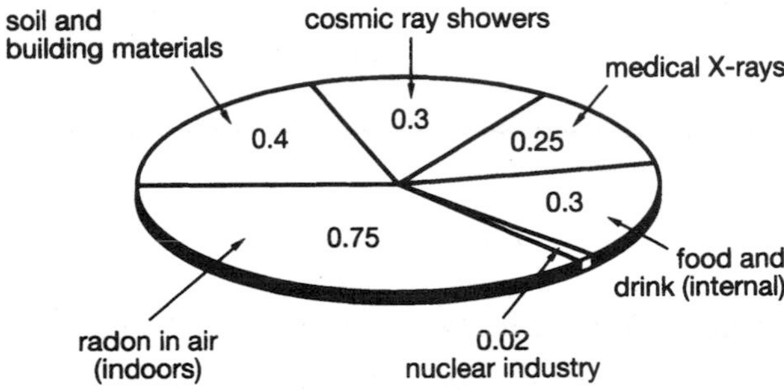

14.6 A radioactive source is to be mixed with some mud in preparation for a study of the movement of sediment in an estuary. The team involved wear protective clothing, handle the source with very long pincers and complete the task as quickly as possible after removing the γ-source from its lead-lined box.
Explain how they are reducing the dose they receive to as small a value as practicable.

14.7 The diagram shows a film badge which would be used by someone who is likely to be more exposed to nuclear radiation than the average person.
(a) Name two areas of work where such a badge might be used.
(b) What is the simplest way of reducing our exposure to γ-radiation?

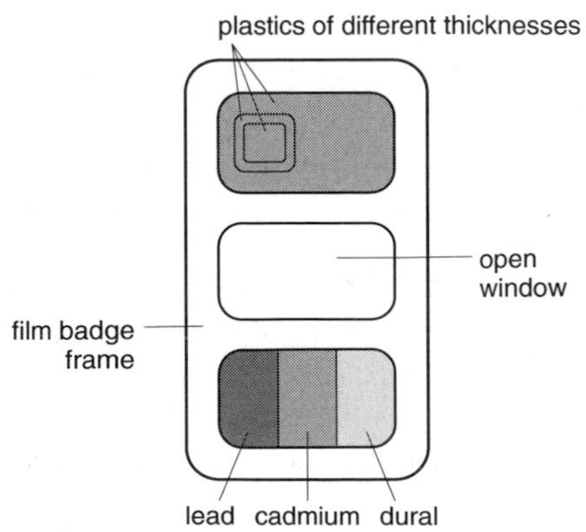

14.8 Suppose you are the Radiation Protection Officer for the Physics Department in your school or college. You have to deal with several sealed sources of radiation. What precautions would you take?

14.2 Mass and energy

In this section you will need to

- use the mass–energy equivalence equation $\Delta E = c^2\Delta m$
- understand that all changes in energy result in changes of mass, and vice versa
- remember how to convert mass–energies between the following units: u, kg, J, eV
- explain the terms binding energy and mass defect of a nucleus
- explain β^-- and β^+-decay in terms of a surplus or deficit of neutrons
- understand what is meant by antimatter, that anti-particles may annihilate each other, and that pairs of anti-particles can be formed.

14.9 Express the following masses in atomic mass units, u: **(a)** $1.674\,82 \times 10^{-27}$ kg **(b)** $9.109\,08 \times 10^{-31}$ kg **(c)** $1.992\,51 \times 10^{-26}$ kg. [Use data.]

14.10 The mass of a proton is 1.007 276 u.
(a) Express this mass in kilograms.
(b) What is the energy equivalent of this mass, in joules?
(c) Express this energy in MeV.

14.11 Copy and complete this table, using data from the start of the chapter. Some have been done for you as examples.

	u	kg	J	MeV
1 u =		$1.660\,43 \times 10^{-27}$	**(a)**	**(b)**
1 MeV =	**(c)**	**(d)**	1.60×10^{-13}	

14.12 An α-particle emitted from a particular nucleus has a kinetic energy of 5.48 MeV.
(a) Express this energy in J.
(b) What additional mass, does the moving α-particle have as a result of this energy? Give your answer **(i)** in kg **(ii)** in u. [Use data.]
(c) The rest mass of the α-particle is 4.001 506 u. What is its additional mass as a fraction of its rest mass?
(d) Now consider a β-particle with the same kinetic energy. The answers to **(a)** and **(b)** will be the same. What is the additional mass of this β-particle as a fraction of its rest mass? [Use data.]

14.13 It requires 1.75 GJ to raise the temperature of a steel bar of mass 5.0 tonnes from 300 K to 1100 K.
(a) What is the mass equivalent of this energy?
(b) What is its additional mass as a fraction of its rest mass?

(c) Compare this answer with the answer to **(c)** of the previous question. What would you say to someone who said 'This mass–energy equivalence is nonsense – whoever felt the mass of something increase just because it had more energy? We all know the mass of something is constant.'

14.14 The diagram shows $^{15}_{6}C$, a nuclide whose atomic mass is 15.010 60 u.

(a) Add up the masses of its protons, neutrons and electrons. [Use data.]

(b) Your answer to **(a)** is larger than the atomic mass of the atom. How much larger?

(c) There is more mass because the particles have more energy when they are separated (it takes energy to pull them apart). What is the *binding energy* for this nuclide i.e., how much more energy do the particles have when they are separate? Give your answer in MeV.

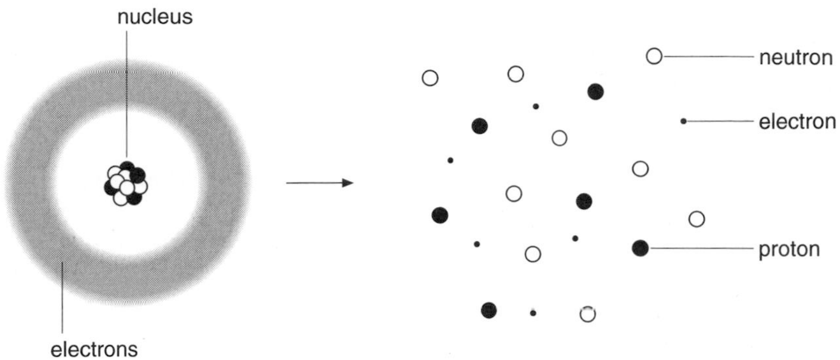

14.15 $^{15}_{7}N$ and $^{15}_{6}O$ are atoms with the same mass number as $^{15}_{6}C$; the three nuclides are called *isobars*. These nitrogen and oxygen atoms have masses of 15.000 10 u and 15.003 07 u, respectively.

(a) Repeat parts **(a)**, **(b)** and **(c)** of the previous question for these two atoms.

(d) Which of the three atoms would require most energy to separate it into separate particles?

(e) So which of the three nuclides is most stable?

14.16 **(a)** Copy the table and complete it to calculate the binding energies of the five nuclides, and then the binding energies *per nucleon*. [Use data.]

nuclide	$^{7}_{3}Li$	$^{27}_{13}Al$	$^{56}_{26}Fe$	$^{138}_{56}Ba$	$^{238}_{92}U$
mass of protons	3 × 1.007 276 u				
mass of neutrons	4 × 1.008 665 u				
mass of electrons	3 × 0.000 549 u				
total mass of particles	7.058 135 u				
mass of neutral atom	7.016 005 u	26.981 535 u	55.934 93 u	137.905 0 u	238.050 76 u
binding energy	39.24 MeV				
binding energy per nucleon	5.606 MeV				

(b) Do you see any pattern in the way in which the binding energy per nucleon varies with atomic number?

(c) On the basis of these results, in which part of the atomic table would you expect to find the most stable nuclides?

14.17 When a $^{224}_{88}$Ra nucleus emits an α-particle and decays to a $^{220}_{86}$Rn nucleus, there is a decrease of 0.006 22 u in the rest masses of the particles.

(a) How much kinetic energy, in MeV, do the particles have after the decay?

(b) The kinetic energy of the α-particle has been measured to be 5.68 MeV. What other particle has kinetic energy?

14.18 Under certain circumstances γ-radiation can cause the production of pairs of electrons and positrons, the energy of each γ-ray photon being converted into the mass of an electron–positron pair. Take the rest mass of an electron as equivalent to an energy of 0.511 MeV.

(a) What is the smallest energy (in MeV) that a γ-ray photon must have in order to give rise to an electron–positron pair?

(b) What is the wavelength of such a γ-ray?

(c) If a γ-ray of half this wavelength produces an electron–positron pair and its energy is equally shared between the particles, what is the energy (in eV) of each of the particles produced?

14.19 The positron is the anti-particle of the electron. When they meet they annihilate each other and two γ-rays are produced.

(a) Why must *two* γ-rays be produced? What are their directions relative to each other?

(b) How much energy is available for the creation of the two γ-rays? [Use data.]

(c) What is the wavelength of each γ-ray?

(d) Why will the emission of a β^+-particle always result in the creation of two γ-rays, but it is extremely unlikely that this will happen when a β^--particle is emitted?

14.20 The rest mass of an electron is 9.11×10^{-31} kg.

(a) Calculate the minimum energy of a γ-photon which could produce an electron–positron pair.

(b) What will be the speed of recoil of the particles if a γ-photon with 1% more than this minimum energy causes pair production? State any assumptions which you make.

14.21 The table gives the masses of some silicon atoms and some phosphorus atoms. For each silicon nuclide an equation for β^--decay could be written down to show it decaying into the phosphorus nuclide, e.g. $^{28}_{14}\text{Si} \rightarrow {}^{28}_{15}\text{P} + {}^{0}_{-1}\beta$. Considering the difference in masses before and after the suggested decay, say which of the decays **(a)**, **(b)**, **(c)**, **(d)** and **(e)** are possible.

(a)	$^{28}_{14}$Si	27.976 93 u	? →	$^{28}_{15}$P	27.992 u
(b)	$^{29}_{14}$Si	28.976 49 u	? →	$^{29}_{15}$P	28.981 82 u
(c)	$^{30}_{14}$Si	29.973 76 u	? →	$^{30}_{15}$P	29.978 32 u
(d)	$^{31}_{14}$Si	30.975 35 u	? →	$^{31}_{15}$P	30.973 762 u
(e)	$^{32}_{14}$Si	31.974 0 u	? →	$^{32}_{15}$P	31.973 908 u

14.22 These two equations could be written down for the β^+ and β^- decay of the scandium nucleus: ${}^{47}_{21}\text{Sc} \rightarrow {}^{47}_{20}\text{Ca} + {}^{0}_{1}\beta$ and ${}^{47}_{21}\text{Sc} \rightarrow {}^{47}_{22}\text{Ti} + {}^{0}_{-1}\beta$.
The masses of the scandium, calcium and titanium atoms are 46.952 40 u, 46.954 50 u and 46.951 76 u respectively.
Will the scandium decay by β^+ or by β^- emission?

14.23 The diagram shows an atom of thorium (${}^{232}_{90}\text{Th}$) decaying by α-emission to ${}^{228}_{88}\text{Ra}$. The numbers of protons, neutrons and electrons are shown at the moment that the decay takes place.

(a) At this moment are the two decay products neutral, or are they charged?

(b) How do their masses compare with the masses of the neutral atoms?

(c) To find the kinetic energy available in this decay you would want to add up the masses of the particles after the decay. You would be given the masses of the *atoms*. Explain why you would not have to allow for the fact that the radium is an atom with two extra electrons, and the helium nucleus is an atom which is missing two electrons.

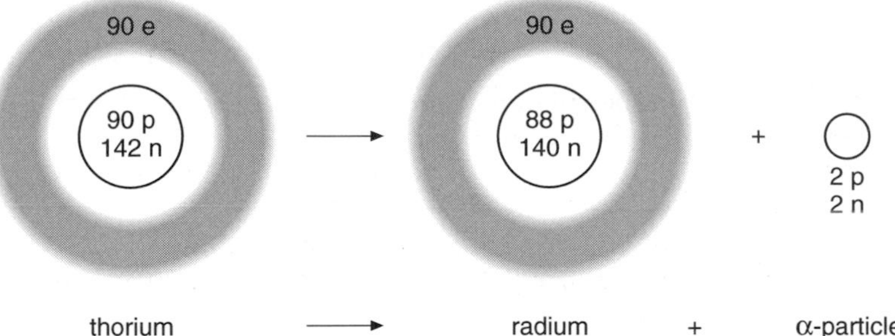

14.24 The diagram shows an atom of bismuth (${}^{212}_{83}\text{Bi}$) decaying by β^--emission to ${}^{212}_{84}\text{Po}$. The numbers of protons, neutrons and electrons are shown at the moment that the decay takes place.

(a) At this moment is the polonium neutral, or is it charged?

(b) How does its mass compare with the mass of the neutral atom?

(c) To find the kinetic energy available in this decay you would want to add up the masses of the particles after the decay. You would be given the mass of the polonium *atom*. Explain what you would do to find the total mass of the decay products.

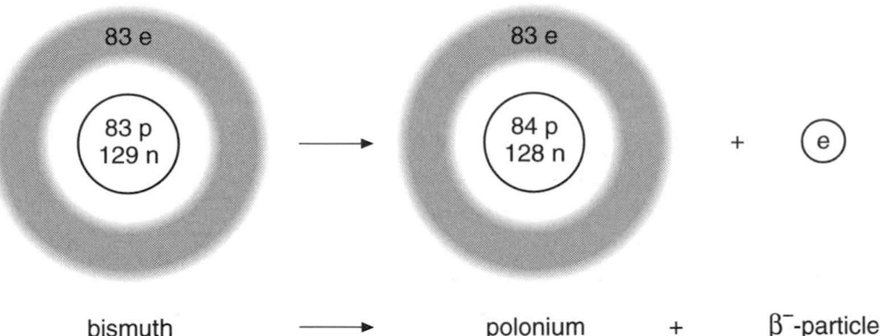

14.25 The last two questions showed you that in α- and β^-- decays you can use just the masses of the atoms when adding up the masses of the decay products. It is not quite as simple for β^+-emission.

(a) Draw a diagram similar to the last two diagrams for the following decay:

$$^{23}_{12}\text{Mg} \rightarrow\ ^{23}_{11}\text{Na} +\ ^{0}_{1}\beta.$$

(b) When calculating the total mass of the decay products, how many electron masses would you have to add to the mass of the sodium atom?

14.26 Draw a diagram similar to the last two diagrams for this electron capture decay:

$$^{57}_{27}\text{Co} +\ ^{0}_{-1}\text{e} \rightarrow\ ^{57}_{26}\text{Fe}.$$

Explain why in calculating the change in mass you can simply subtract the mass of the iron atom from the mass of the cobalt atom.

14.27 Uranium-233 decays by α-particle emission to give thorium-229 according to this equation:

$$^{232}_{92}\text{U} \rightarrow\ ^{229}_{90}\text{Th} +\ ^{4}_{2}\text{He}$$

The masses of the atoms involved are given in the table

atom	^{233}U	^{229}Th	^{4}He
mass/u	233.039 50	229.031 63	4.002 604

(a) What is the change in mass of the particles involved in this reaction?
(b) How much energy is available as kinetic energy?
(c) The energy of the α-particles is measured to be 4.82 MeV. Why is this less than your answer to **(b)**?

14.28 **(a)** $^{14}_{6}$C is a radioactive isotope of carbon that decays by the emission of a β^--particle to give a nuclide of nitrogen. Write down the equation for this nuclear reaction.
(b) The mass of the nitrogen atom is 14.003 074 u. What is the total mass of the products of this decay?
(c) The mass of the carbon atom is 14.003 242 u. How much less mass is there after this decay?
(d) What is the maximum energy for the emitted β^--particle in MeV? [Use data.]

14.29 An atom of phosphorus-32 decays by β^--emission into an atom of sulphur-32, and in the process 1.71 MeV of nuclear energy is transferred to the kinetic energy of the particles.
The atomic mass of sulphur-32 is 31.972 07 u. Calculate the atomic mass of phosphorus-32. [Use data.]

14.30 $^{15}_{8}$O decays by emitting a β^+-particle to form a nuclide of nitrogen.
(a) Write down the equation for this reaction.
(b) The masses of the atoms of oxygen and nitrogen are 15.003 072 u and 15.000 108 u, respectively. What is the mass of the products of this decay?
(c) What four particles are involved in this decay?
(d) How much energy is available for the particles?

14.31 $^{37}_{18}$Ar decays by electron capture to give an isotope of chlorine (Cl).
(a) Write down the equation for this reaction.
(b) The masses of the atoms of argon and chlorine are 36.966 77 u and 36.965 90 u, respectively. Calculate the energy released in this reaction.
(c) A neutrino takes away nearly all of this energy. What happens to the rest?

14.3 Nuclear power

In this section you will need to

- understand that when both nuclear fusion and nuclear fission take place there is an increase in the binding energy per nucleon of the nuclei involved in the reaction
- understand that the absorption of slow neutrons by uranium-235 results in the fission of the resulting nuclide
- understand what is meant by a chain reaction and that it is possible if a reaction induced by one neutron results in the emission of several neutrons
- remember that only 0.7% of naturally-occurring uranium is uranium-235 and that the rest is uranium-238
- understand the use of moderators and control rods in a nuclear pile and which materials might be used for each
- calculate the power available from a nuclear reactor, given the masses of the nuclides involved
- understand that the bombardment of nuclei by fast-moving particles can result in the creation of nuclides which do not occur naturally.

14.32 Question 14.16 asked you to calculate the nuclear binding energy per nucleon for five different nuclides. A graph of how this varies with mass number *A* is shown in the diagram. (Not all nuclides lie exactly on the line.)

(a) Consider the nuclides with a binding energy per nucleon of less than 8 MeV per nucleon. Are these more stable, or less stable, than the nuclides with a binding energy per nucleon of more than 8 MeV per nucleon?

(b) Here are three ranges of mass number: 40–50, 50–60, 60–70. In which range are the most stable nuclides?

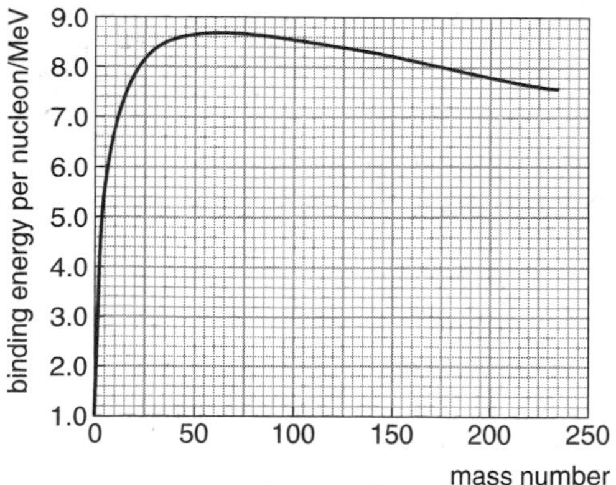

14.33 **(a)** In a fission process a uranium-235 or a plutonium-239 nuclide splits into two less massive fragments. How does the binding energy per nucleon for these fragments compare with the binding energy per nucleon of the original nuclides?

(b) Why does your answer to **(a)** suggest that fission is probable, at least from the point of view of the energy involved?

(c) The diagram overleaf is an enlarged section of the previous graph. Suppose a ^{235}U nucleus splits into fragments which have mass number 91 and 141. Use the graph to find the binding energy per nucleon for the ^{235}U nucleus and the fragments.

(d) Use these figures to calculate **(i)** the binding energy of the ^{235}U nucleus **(ii)** the sum of the binding energy of the two fragments.

(e) How much energy would this fission make available?

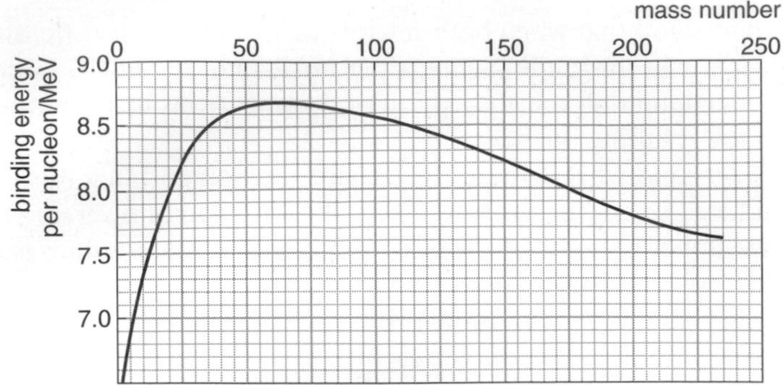

14.34 When a uranium-235 nucleus absorbs a neutron the resulting nuclide splits into two massive fragments and also emits at least one neutron. See the diagram. One possible neutron-induced fission reaction for uranium-235 is $^{235}_{92}\text{U} + ^{1}_{0}\text{n} \rightarrow ^{141}_{56}\text{Ba} + ^{92}_{36}\text{Kr} + 3^{1}_{0}\text{n}$ The masses of the atoms involved are given in the table

atom	$^{235}_{92}\text{U}$	$^{141}_{56}\text{Ba}$	$^{92}_{36}\text{Kr}$
mass/u	235.043 9	140.914 3	91.926 3

(a) Calculate the change of mass when one of these reactions occurs.

(b) Hence calculate the energy released in MeV.

(c) In 1.00 kg of uranium-235 how many moles are there?

(d) In 1.00 kg of uranium-235 how many atoms are there?

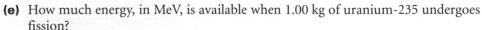

(e) How much energy, in MeV, is available when 1.00 kg of uranium-235 undergoes fission?

(f) Express this energy in **(i)** J **(ii)** kW h.

(g) Suppose a power station is transferring energy at a rate of 1000 MW. How long could it do this using 1 kg of ^{235}U?

14.35 Two other possible pairs of fission fragments are **(a)** $^{140}_{54}\text{Ru}$ and $^{94}_{38}\text{Sr}$ **(b)** $^{137}_{55}\text{Cs}$ and $^{95}_{37}\text{Rb}$. If both pairs result from the capture of a neutron by $^{235}_{92}\text{U}$, explain how many neutrons are emitted in each reaction.

14.36 **(a)** If fission results in the production of neutrons, are the fission fragments likely to be neutron-rich or proton-rich? Explain.

(b) Calculate the ratio N/Z for the six fission fragments mentioned in the previous two questions.

(c) Refer to the diagram for question 13.22 and explain whether these are stable nuclides and, if not, whether they decay by emitting β^--particles or β^+-particles.

(d) If one of these fission fragments decays by β-emission, is a single β-emission likely to be enough to make the nucleus stable? Describe what is likely to happen.

(e) Consider the first two fission fragments mentioned, i.e. $^{141}_{56}\text{Ba}$ and $^{92}_{36}\text{Kr}$. They eventually decay to the stable nuclides $^{141}_{59}\text{Pr}$ and $^{92}_{40}\text{Zr}$, respectively. How many β^--particles are emitted in each of these decay processes?

14.37 Some fission fragments have quite long half-lives, e.g. $^{137}_{55}Cs$ has a half-life of 30 years. What difficulties is this likely to cause?

14.38 **(a)** Referring to the nuclear equation in question 14.34, explain what is meant by a chain reaction.

(b) The diagram shows a fission event occurring in a sample of uranium-235. Six neutrons are emitted. Draw a similar diagram and use it to illustrate your answer to **(a)**.

(c) The possibility of a chain reaction depends on whether the neutrons emitted in a fission event cause more than one fission in another uranium nucleus. For a given mass of uranium what shape is most likely to allow a chain reaction to occur?

(d) Explain what is meant by *critical mass.*

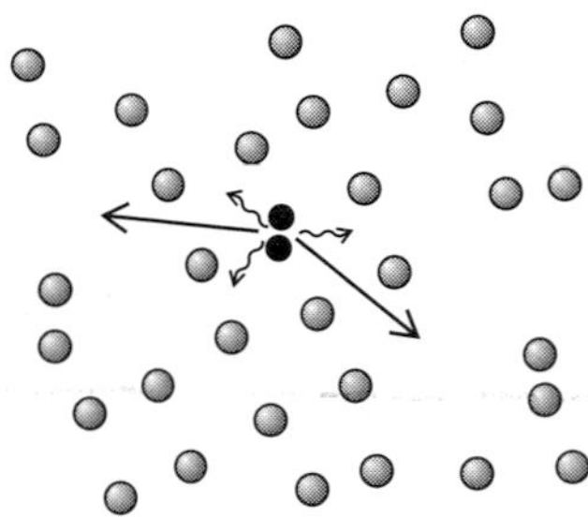

14.39 Naturally-occurring uranium contains only 0.7% of the nuclide ^{235}U. The rest is ^{238}U, which does not undergo fission, but does capture fast neutrons. ^{235}U is more likely to capture a neutron if the neutron's energy is relatively low, e.g. less than 0.05 eV.

(a) Neutrons emitted in fission may have energies of at least 1 MeV. Name two reasons why the emitted neutrons must be slowed down if a chain reaction is to proceed.

(b) It is important to increase the proportion of ^{235}U. Why is it difficult to separate ^{235}U from ^{238}U?

14.40 **(a)** Neutrons could be slowed down in a nuclear fission reactor if they made a series of collisions with other particles. A particle loses most energy if it collides head-on with a particle of the same mass. What kind of atom would therefore be suitable for slowing down, or *moderating*, the fast neutrons?

(b) Neutrons can be captured by hydrogen nuclei, and as a result a γ-ray is emitted. Write down the equation which describes this nuclear reaction. What name is given to the product?

(c) Hydrogen is not a suitable element for moderation, since it captures neutrons. What element is usually used instead of hydrogen to slow down neutrons, and in what form is it used?

14.41 **(a)** What is the purpose of the control rods in a nuclear reactor?

(b) Name two elements which could be used to make them.

(c) For one of these elements find out the values of A and Z and write down the equation which describes the nuclear reaction that takes place.

14.42 The diagram on the following page shows a section of a nuclear pile; the fuel, the moderator and the control rods are shown. Draw a diagram, similar to that for question 14.38, which shows the function of the moderator and the control rods.

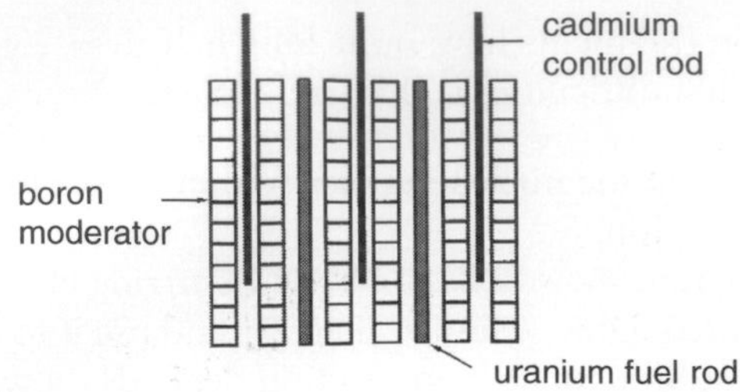

14.43 Two possible hazards of operating nuclear power stations are the emission of radiations from the pile, and the disposal of the fuel rods when their uranium ceases to be useful. Describe what is done to overcome these hazards.

14.44 It is important to be able to detect whether neutrons are escaping from the pile into the surroundings.

(a) Why is it more difficult to detect neutrons than it is to detect α- and β-particles and γ-rays?

(b) Neutrons are absorbed by boron-10 nuclei which then decay. An α-particle is emitted. Write down the equation for this nuclear reaction. (You may find it helpful to have the following list of atomic numbers: lithium 3, beryllium 4, boron 5, carbon 6.)

(c) How could this reaction be used to help detect neutrons escaping from the pile?

14.45 In an earlier question it was said that $^{238}_{92}U$ captures energetic neutrons.

(a) Write down the equation for this event.

(b) The new isotope of uranium decays by β^--emission into neptunium (Np), which itself decays by β^--emission into plutonium (Pu). Write down the equations for these two decays.

^{239}Pu undergoes fission and is therefore a suitable fuel for a nuclear reactor.

(c) Why is it easier to separate ^{239}Pu from ^{238}U than it is to separate ^{235}U from ^{238}U?

(d) A reactor in which the production of plutonium is encouraged is called a *fast breeder* reactor. Why is this?

14.46 In fossil-fuel power stations it is easy to see that the energy comes from the burning of coal, oil or gas. But what process heats the water in a nuclear power station?

14.47 The fission of one atom of uranium-235 releases about 200 MeV of energy. Suppose a nuclear power station that uses uranium-235 has an output of 1000 MW and is 40% efficient. Calculate

(a) the rate at which energy is converted in the reactor

(b) the number of atoms of uranium-235 that it uses per hour

(c) the quantity of uranium-235 atoms (in mol) that it uses per hour

(d) the mass of uranium-235 that it uses per hour.

14.48 Plutonium-238 decays by emitting α-particles of energy 5.5 MeV with a half-life of 86 years.

(a) Explain the principle of using a radioactive source to generate power.

(b) Why is plutonium-238 a suitable isotope to power a heart pacemaker?

(c) If the initial power is to be 50 mW calculate the minimum activity of the source.

(d) How many plutonium-238 atoms would be needed to produce this activity?

(e) What would be the mass of the source?

14.49 Refer to the graph for question 14.32.

(a) Splitting up massive nuclides releases energy. What would need to be done to nuclides of low atomic mass (e.g. < 20) to enable them to release energy?

(b) Use the graph to estimate the binding energy per nucleon that would be expected for a helium nucleus.

(c) In fact the binding energy per nucleon for a helium nucleus is 7.10 MeV per nucleon. Does this mean that a helium nucleus is more or less stable than would have been expected?

(d) What would you say to someone who said 'I'm surprised that an α-particle emerges from a nucleus – I'd have thought that in such an violent event it would disintegrate'?

14.50 The equation describes the fusion of two deuterium nuclei:

$$^{2}_{1}\text{H} + ^{2}_{1}\text{H} \rightarrow ^{3}_{2}\text{He} + ^{1}_{0}\text{n}$$

(a) Using data from the previous question, calculate **(i)** the mass defect in this reaction **(ii)** the energy available from this reaction.

(b) How many such fusions per second would be needed to produce a power output of 200 MW?

(c) What mass of deuterium would be needed each day?

14.51 The solar constant (the mean intensity of solar radiation arriving at the Earth) is 1.4 kW m^{-2}. If the mean distance of the Earth from the Sun is 1.5×10^{11} m what is

(a) the power of the Sun **(b)** the rate of loss of mass of the Sun?

14.4 Nuclear matter

In this section you will need to

- understand that α-particle scattering provided evidence that the atom had a nucleus
- understand that high-energy electron scattering provided evidence that protons and neutrons have a structure
- understand that accelerators are needed to give energy to particles and that greater amounts of energy provide more information
- understand that the four fundamental forces are gravitational, electromagnetic, weak nuclear and strong nuclear and that these have different ranges and have different effects on different groups of particles
- remember that electrons, muons and neutrinos (and their anti-particles) form a group called leptons, and that these are fundamental particles
- understand that exchange particles are a mechanism by which the four fundamental forces operate, and remember which particles mediate the different kinds of force
- draw Feynman diagrams to illustrate the operation of the four fundamental forces
- remember that protons and neutrons (and their anti-particles) are baryons, that pions and kaons (and their anti-particles) are mesons, and that these are all hadrons, and are not fundamental particles
- understand that the up and down quarks have charge and baryon number and that the strange quark has charge, baryon number and strangeness
- remember that quarks are fundamental particles and have an electric charge which is a multiple of $\pm\frac{1}{3}e$
- remember that charge Q, baryon number B and strangeness S are conserved with the exception that strangeness may vary by 1 in weak interactions
- use information about Q, B and S to test whether a reaction is possible.

14.52 In a series of experiments Geiger and Marsden fired α-particles through thin metal (e.g. gold) foils. Most of the particles were almost undeflected, but a few were deflected through large angles.

(a) What did Rutherford deduce about the nature of atoms?

(b) The diagram shows a few of the paths which α-particles would take when they approach a nucleus. They have been labelled 'a', 'b', 'd' and 'k'. Copy the diagram and add the paths which the α-particles would take if they passed through the other unlabelled points.

(c) A diagram like this gives a false impression of what happens. Why? (Consider the scale of the diagram.)

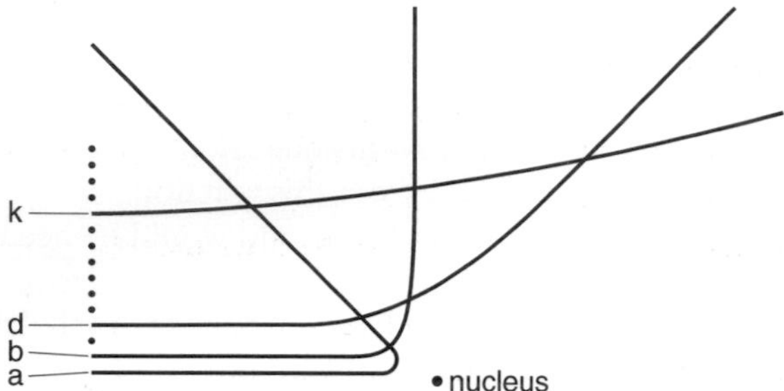

14.53 The diagram shows an α-particle being deflected by the nucleus of an atom. Copy the diagram and mark on it

(a) the point in the α-particle's path where its speed is least (label it X)

(b) the direction of the push of the nucleus on the α-particle at points A, B and C.

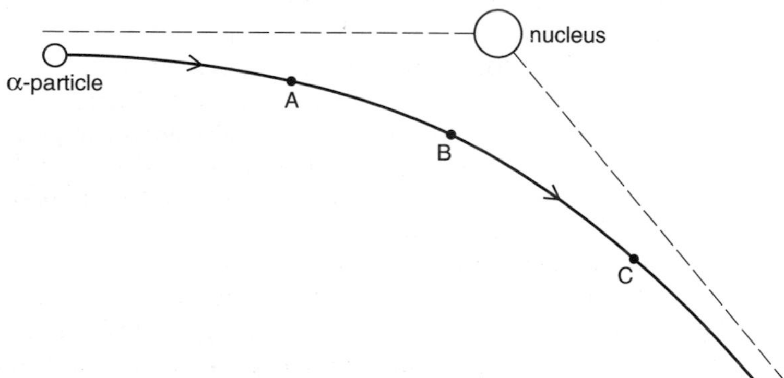

14.54 An atom of uranium-238 has 92 protons, 146 neutrons and 92 electrons.

(a) What fraction of the mass is not in the nucleus? [Use data.]

(b) If the nucleus of this atom may be thought of as a sphere of radius 7.4×10^{-15} m, what is the average density of the material of the nucleus?

14.55 **(a)** Consider how much volume several spheres will take up when they are closely packed together. For example, compared with a single sphere, what will be the width of a group of **(i)** 8 spheres **(ii)** 27 spheres?

(b) What relationship could you write down between the width *w* of the group and the number *n*?

14.56 Electron scattering at 'low' energies (i.e. less than 500 MeV) provides information about the radii of nuclei. The table shows how the radius r varies with mass number A assuming that the charge is distributed uniformly through the nucleus.

nuclide	H	O	Ca	In	Sr	Au	Bi
mass number A	1	16	40	88	122	197	209
radius $r/10^{-15}$ m	0.83	2.84	3.78	4.83	4.98	5.73	5.94

(a) Plot a graph of r on the x-axis against $\sqrt[3]{A}$ on the y-axis.
(b) Your plot should show that $r \propto \sqrt[3]{A}$, or $r = r_0 \sqrt[3]{A}$. What is the value of r_0?
(c) Calculate the radii of these nuclei **(i)** ${}^{23}_{11}\text{Na}$ **(ii)** ${}^{107}_{47}\text{Ag}$.

14.57 **(a)** Estimate the size of the full stop at the end of this sentence.
(b) If this dot represents the nucleus of a gold atom, how large, on this scale, would an atom be? (Take the diameter of a typical atom to be 3×10^{-10} m: use the radius of the gold nucleus given in the previous table.)
(c) What is in the space around the nucleus?

14.58 The average density of the matter in a nucleus is given by its mass divided by its volume.
(a) In the nuclide ${}^{A}_{Z}\text{X}$ what is the number of nucleons?
(b) If each nucleon has a mass m, what is the total mass of the nucleus?
(c) Suppose the radius of a nucleon is r_0. What is sum of the volumes of the separate nuclei?
(d) The volume of the nucleus will be similar to the sum of the volumes of the separate nuclei. Use your answers to **(b)** and **(c)** to write down an approximate expression for the density of the nuclear matter.
(e) Why does your answer to **(d)** show that the density of nuclear matter is the same for all nuclei, whatever their mass and charge?
(f) Use the answer to **(d)** to calculate the value of the density of nuclear matter. (Use data, and take the value of r_0 from question 14.56.)

14.59 A neutron star is a dead star consisting entirely of neutrons. Its mass may be about the same as that of the Sun (2×10^{30} kg). A typical diameter is 18 km.
(a) Calculate the density of the material.
(b) Neutron stars are formed from stars which have masses which are less than three times the mass of the Sun. Dying stars with more than this mass collapse to form black holes. Assuming that these have the same density as neutron stars, what is the diameter of a black hole which has 20 times the mass of the Sun?

14.60 The Stanford Linear Accelerator (SLAC) can accelerate electrons to an energy of up to 20 GeV. When fired at protons the electrons *rebound with different energies and at different angles.*
(a) This is called deep inelastic scattering. Which part of the phrase in italics shows that the scattering is inelastic?
(b) The experiment shows that the proton has a structure. How is this experiment similar to Rutherford's α-particle scattering experiment?
(c) What did physicists mean when they said that the experiment showed that the proton was not a *fundamental particle*?

(d) The particles which make up the proton are called the up (u) and down (d) quarks. They have electric charges of $+\frac{2}{3}e$ and $-\frac{1}{3}e$, respectively, where e is the electronic charge. What is the only possible way in which three of these quarks can form a proton?
(e) Three of these quarks also form a neutron. What is the only possible combination of quarks?

14.61 **(a)** What are the anti-particles of the following particles: **(i)** electron **(ii)** proton **(iii)** neutrino?
(b) Name one property which is **(i)** the same **(ii)** different, for a particle and its anti-particle.
(c) What happens when a particle meets its anti-particle?
(d) Does a neutron have an anti-particle?

14.62 Outside the nucleus a neutron decays to form a proton.
(a) Which other two particles are emitted as a result of this decay?
(b) What happens to the quarks when a proton is formed from a neutron?
(c) Is the proton a stable particle?

14.63 The table shows six leptons. (There are another six leptons which are the anti-particles of these.)

e^-	μ^-	τ^-
ν_e	ν_μ	n_τ

(a) Copy the table and add the names of these particles.
(b) Are leptons fundamental particles?
(c) Which of the four fundamental forces affect leptons?
(d) What is the symbol and name of the anti-particle of e^-?
(e) What are the lepton numbers of each of these particles?
(f) What are the lepton numbers of each of their anti-particles?

14.64 Is electric charge conserved in these reactions?
(a) $\nu_e + n \rightarrow p + e^-$ **(b)** $n \rightarrow p + e^- + \bar{\nu}_e$
(c) $\nu_\mu + n \rightarrow p + \mu^-$ **(d)** $\mu_e \rightarrow e^- + \bar{\nu}_e + \nu_\mu$.

14.65 Is lepton number conserved in the reactions listed in the previous question?

14.66 The diagram shows examples of Feynman diagrams for each of the four fundamental forces.
(a) For each diagram describe the event which is taking place, name the fundamental force, and for **(i)** and **(ii)** name the exchange particle.
(b) Are exchange particles fundamental?

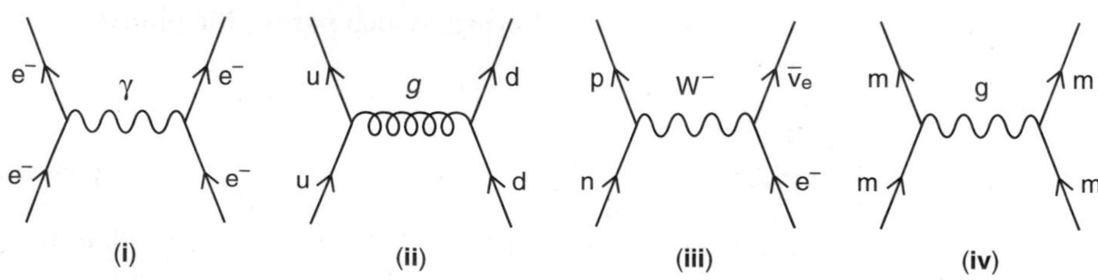

14.67 **(a)** What family of particles do baryons and mesons both belong to?
(b) How many quarks or antiquarks do **(i)** all baryons **(ii)** all mesons have?

14.68 The table shows the symbol and charge for three quarks.

name	symbol	charge
up	u	$+\frac{2}{3}e$
down	d	$-\frac{1}{3}e$
strange	s	$-\frac{1}{3}e$

(a) Make a similar table for the antiquarks.
(b) What would be the charge of particles which had the following combinations of quarks: **(i)** $u\,\bar{d}$ **(ii)** $\bar{u}\,d$ **(iii)** $u\,\bar{u}$ **(iv)** $d\,\bar{d}$?
(c) What would be the charge of particles which had the following combinations of quarks: **(i)** $u\,\bar{s}$ **(ii)** $\bar{u}\,s$ **(iii)** $d\,\bar{s}$ **(iv)** $\bar{d}\,s$?
(d) What family do all these particles belong to?
(e) Why would it not be possible for these combinations of quarks to exist: **(i)** uu **(ii)** us?

14.69 The first table lists the charge *Q*, baryon number *B* and strangeness *S* for some quarks and antiquarks. The second table lists these quantities for some particles.

type	*Q*	*B*	*S*
u	$+\frac{2}{3}e$	$+\frac{1}{3}$	0
d	$-\frac{1}{3}e$	$+\frac{1}{3}$	0
s	$-\frac{1}{3}e$	$+\frac{1}{3}$	−1
$\bar{u}$	$-\frac{2}{3}e$	$-\frac{1}{3}$	0
$\bar{d}$	$+\frac{1}{3}e$	$-\frac{1}{3}$	0
$\bar{s}$	$+\frac{1}{3}e$	$-\frac{1}{3}$	+1

symbol	*Q*	*B*	*S*
p	+1	+1	0
n	0	+1	0
π^+	+1	0	0
π^-	−1	0	0
π^0	0	0	0
K^+	+1	0	+1
K^-	−1	0	−1
K^0	0	0	+1
$\bar{K}^0$	0	0	−1

Use the first table to work out which quarks and antiquarks make up the last seven particles in the second table.

14.70 **(a)** What name could be given to all the particles in the second table in the previous question?
(b) What name could be given to **(i)** the first two **(ii)** the last seven?

14.71 **(a)** By considering the conservation of charge and baryon number explain whether the following reactions are possible:
(i) $p + p \rightarrow p + p + \pi^0$ **(ii)** $p + p \rightarrow p + p + \pi^0 + \pi^0$ **(iii)** $p + n \rightarrow \pi^+ + \pi^0$
(iv) $\pi^- + p \rightarrow n + \pi^0$ **(v)** $p + p \rightarrow p + \pi^+ + \pi^- + \pi^0$.
(b) Compare **(i)** and **(ii)**. In **(ii)** an additional π^0 was produced. What might have caused this to happen?

14.72 The diagram shows an event recorded in a bubble chamber, in which streams of K^- particles are entering from the left. A K^- particle collides with a proton at A to produce a pion and a Σ^+ (sigma-plus) particle. Shortly afterwards the Σ^+ particle decays at B to produce a second pion and another particle, which leaves no track.

(a) The K^- particle is curving anti-clockwise in the magnetic field in the bubble chamber. Which way is the pion (produced at A) curving? What is therefore the sign of its charge?

(b) Write down an equation for the event which occurs at A.

(c) Use the information in the tables in question 14.69, and the idea of conservation of Q, B and S to find Q, B and S for the Σ^+ particle.

(d) How can you tell that at B a second particle must result from the decay of the Σ^+ particle? What can you deduce from the fact that the second particle leaves no track?

(e) The decay of the Σ^+ particle is a weak interaction in which strangeness is not conserved. Use the conservation of Q and B to deduce which of the particles in the second table this particle is.

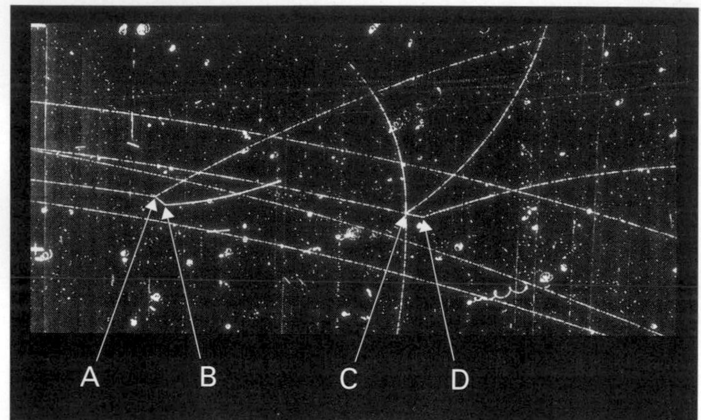

14.73 The diagram for question 14.72 also shows a second event. Another K^- particle collides with a proton at C to produce three pions and a Σ^- particle. Shortly afterwards this particle decays at D to produce another pion and another particle, which leaves no track.

(a) Write down an equation for the event which occurs at C.

(b) Use the information in the tables in question 14.68, and the conservation of Q, B and S to find Q, B and S for the Σ^- particle.

14.74 From your answers to the last two questions find what quarks make up the Σ^+ and Σ^- particles.

14.75 The table lists the four fundamental forces: the headings show some of their properties.

type of force	range/m	relative strength	exchange particles
gravitational			
weak nuclear			
electromagnetic			
strong nuclear			

(a) Copy the table and complete the column headed 'range', using the following values (one of them more then once): 10^{-15} m, 10^{-18} m, infinity.
(b) Complete the column headed 'relative strength', using the following values: 10^{2}, 10^{-3}, 1, 10^{-36}.
(c) Complete the column headed 'exchange particles'.

14.76 Here are three groups of particles: hadrons, leptons, quarks.
(a) Which two groups of particles are affected by the weak nuclear force?
(b) Which group of particles is not affected by the strong nuclear force?
(c) What kind of particles are affected by the electromagnetic force?

14.77 **(a)** Is strangeness conserved in strong reactions?
(b) Is strangeness conserved in weak reactions?
(c) By considering the conservation of charge *Q*, baryon number *B* and strangeness *S* in these strong reactions, explain whether they are possible:
(i) $p + p \rightarrow p + n + K^{+} + \bar{K}^{0}$
(ii) $p + K^{-} \rightarrow \pi^{+} + \pi^{+} + \pi^{-} + \Sigma^{-}$ (Σ^{-} contains dds quarks).
(d) In these weak reactions *Q* and *B* are conserved. Is *S*?
(i) $K^{+} \rightarrow \pi^{+} + \pi^{+} + \pi^{-}$
(ii) $\Sigma^{+} \rightarrow \pi^{+} + n$ (Σ^{+} contains uus quarks)
(iii) $\Sigma^{-} \rightarrow \pi^{-} + n$ (Σ^{-} contains dds quarks)

15 Waves

Data: speed of electromagnetic radiation $c = 3.00 \times 10^8\ \text{m s}^{-1}$

speed of sound in air = $340\ \text{m s}^{-1}$

speed of sound in water and sea water = $1500\ \text{m s}^{-1}$

15.1 Wave properties

In this section you will need to

- understand that waves transmit energy from place to place
- use rays or wavefronts to describe the transmission of wave energy
- use the wave equation $c = f\lambda$
- describe waves as transverse or longitudinal
- understand that transverse waves, e.g. all electromagnetic (e-m) waves, can be plane polarised but that longitudinal waves, e.g. sound, cannot
- remember the order of magnitude of the wavelength of different parts of the electromagnetic spectrum
- remember that all e-m waves travel at $3.00 \times 10^8\ \text{m s}^{-1}$ in a vacuum.

15.1 A fishing vessel uses ultrasonic sound pulses to search for shoals of fish.

(a) Draw a diagram to illustrate the principle involved.

(b) Calculate the times for a pulse to return from 35 m and 40 m below the vessel. [Use data.]

(c) Suggest a maximum time for the length of each pulse.

15.2 Light takes 8 minutes 20 seconds to travel from the Sun to the Earth. Calculate the Sun–Earth distance in metres.

15.3 Ocean waves, which travel in the open sea at $5\ \text{m s}^{-1}$, arrive at a beach once every 4 s.

(a) Calculate the distance between wave crests in open sea.

(b) Calculate how long these waves would take to reach the beach if they were formed in a storm 1200 km away. Give your answers in hours.

15.4 A recording station observed that there was an interval of 68 s between the reception of P (push or primary) waves and S (shake or secondary waves) from an underground nuclear test explosion.
The speed of P and S waves in the Earth's crust are $7800\ \text{m s}^{-1}$ and $4200\ \text{m s}^{-1}$, respectively. Calculate the distance of the test site from the explosion.

15.5 The diagram shows a transverse wave on a rope. It is moving to the right. Several particles on the rope have been labelled.

(a) Copy the diagram and draw arrows to show the direction in which particles P, R and T are moving.

(b) What can you say about the motion of Q and S?
Mark on your diagram two particles which are **(i)** in phase, i.e. moving together – call them P and P′ **(ii)** in antiphase, i.e. moving oppositely – call them A and A′.

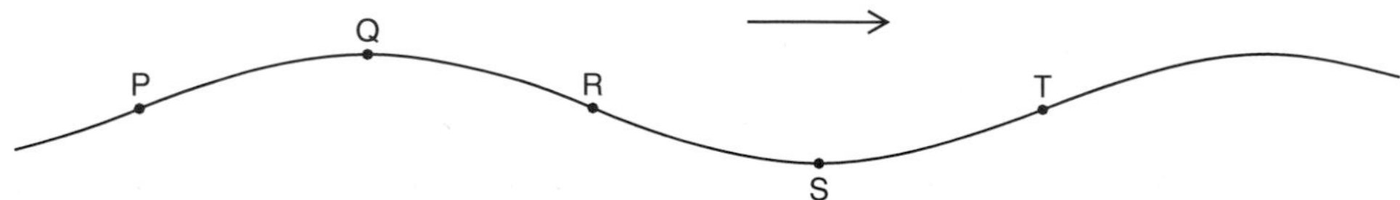

15.6 Dolphins communicate by emitting ultrasonic waves of frequencies in the range 100 kHz to 250 kHz. What range of wavelengths in water does this represent? [Use data.]

15.7 **(a)** A 50 Hz power line emits electromagnetic radiation. What is the wavelength?
(b) A microwave oven operates at 2.4 GHz. What is the distance between wave crests in the oven?

15.8 The following data represent some frequencies and wavelengths used in radio and television broadcasting:

200 kHz/1500 m (AM radio)
92.6 MHz/3.24 m (FM radio)
0.516 GHz/0.581 m (UHF TV)

Show that the waves associated with these broadcasts all travel with the same speed.

15.9 Careful measurements in a ripple tank produced the following data:

f/Hz	22	38	62
λ/mm	10	7.0	5.0

Discuss whether or not these waves all travel at the same speed.

15.10 The figure shows a sinusoidal wave travelling to the right along a rope. The solid line represents the wave at $t = 0$ and the dashed line the wave at $t = 0.25$ s.

(a) What is
(i) the amplitude y_0
(ii) the wavelength λ of this wave?

(b) Calculate
(i) the speed c
(ii) the frequency f
(iii) the period T of the wave.

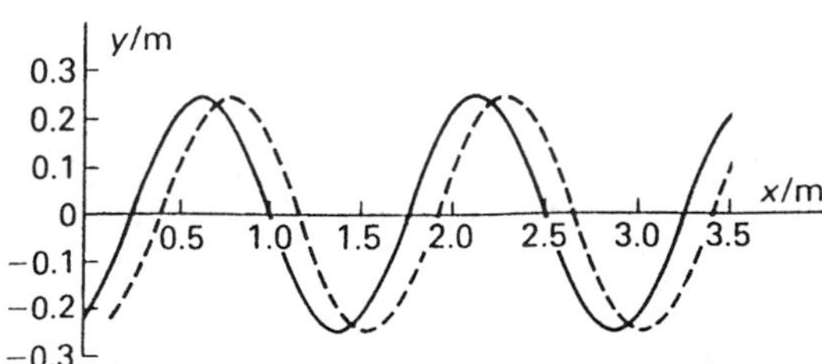

15.11 The diagram shows an idealised wave pulse travelling to the right along a heavy rope at 5.0 m s^{-1}.

(a) Sketch a graph to show how the displacement, y_P, of P varies with time t during the next 0.50 s.

(b) Add a second graph using the same time axis to show how the displacement, y_Q, of Q varies with time t.

(c) Explain how your graphs would differ if the wave pulse was travelling at 1.0 m s^{-1}.

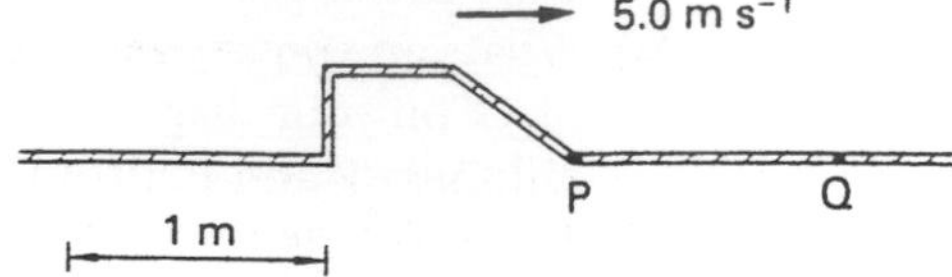

15.12 At large sports meetings the crowd sometimes produces what is called a 'Mexican wave'.

(a) What would be a better name for this manoeuvre?

(b) Describe what the crowd would need to do to produce a wave and suggest values for its frequency and amplitude.

15.13 The speed of a transverse wave along a heavy stretched spring is given by $c = \sqrt{(T/\mu)}$ where T is the tension in the spring and μ its mass per unit length.

(a) Show that the units of the right hand side are equivalent to m s^{-1}.

(b) The speed of a wave pulse on a spring of mass 2.6 kg and stretched length 8.0 m is found to be 5.5 m s^{-1}. Calculate the tension in the spring.

15.14 From which parts of the electromagnetic spectrum do waves of the following wavelengths come?

(a) 2×10^{-10} m **(b)** 5×10^{-7} m **(c)** 0.10 m.

15.15* The diagram shows an idealised wave pulse, a 'step', travelling to the right along a long rope.

(a) Draw a graph to show how the velocity v_P of the point P varies with time t. Mark scales on your graph.

(b) Draw a second graph to show the transverse velocity of points on the rope against distance x from O at the instant shown in the diagram. Mark scales on your graph.

(c) How would your graph in **(b)** differ if the wave pulse was travelling at 2.0 m s^{-1}?

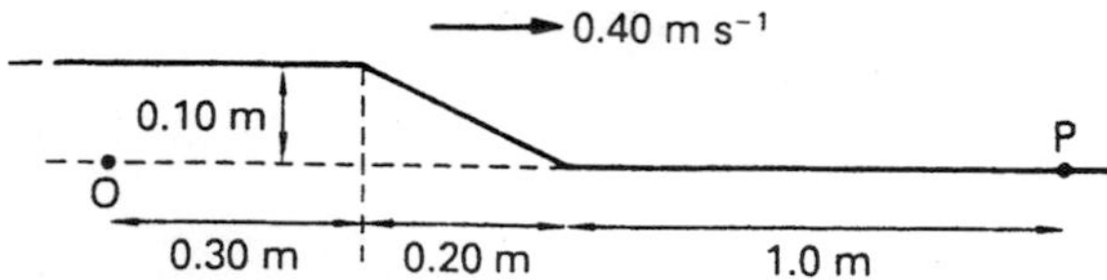

15.16 A student rotates a sheet of Polaroid through which light is passing. He does this with light from three different sources A, B and C and finds that the transmitted light

(a) from A shows no change of intensity

(b) from B varies slightly in intensity twice per revolution

(c) from C is cut off completely twice per revolution.

Describe the state of polarisation of the incident light in each case.

15.17 A laboratory source T of 3 cm microwaves and a microwave receiver R are set facing each other a few metres apart. A strong signal is detected by R.

(a) Describe what R detects as T is rotated 360° about a horizontal axis while still facing R.

(b) Describe what R detects as it is rotated about a horizontal axis while still facing T.

15.18 The diagram shows a TV aerial. The signal is picked up by the quarter wave dipole. The other rods help to aim the aerial at the transmitting station and to give the received signal a greater strength. By looking at local TV aerials and using the data

(a) estimate the wavelength of the broadcast signals and deduce their frequency

(b) explain how the signal from the transmitter is polarised.

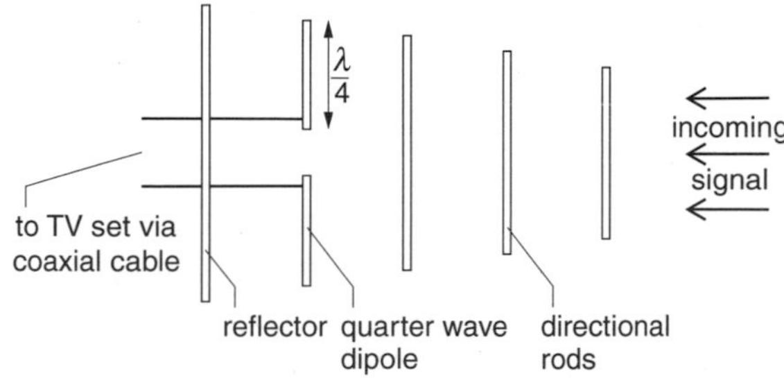

15.19 Light reflected from the surface of water is partially plane polarised. Explain what this means and how it enables polaroid spectacles to reduce glare.

15.2 Point sources and moving sources

In this section you will need to

- understand the inverse square law for the intensity of a wave (its power per unit area) and use the relationship $I = P/4\pi r^2$
- understand that there is a change in the received frequency (and wavelength) of waves from a moving source and use the Doppler shift relationship $\Delta\lambda/\lambda \approx v/c$
- explain the change in pitch or frequency heard when an observer and a source of sound are moving relative to one another.

15.20 A light with an output in the visible range of 50 W is switched on at night at the top of a high tower. Calculate the intensity of the light from the tower at distances of

(a) 100m **(b)** 200 m **(c)** 300 m.

15.21 In listening to a person talking to you who is standing 4.0 m away the intensity of the sound at your ear is 1.2 μW m^{-2}. What is the power output of the speaker's voice?

15.22 A radio-operated garage door opener responds to signals with an intensity greater than 20 μW m^{-2}. For a 250 mW transmitter unit which broadcasts equally in all directions, what is the maximum distance from the garage at which the transmitter will open the door?

15.23 The sound intensity 8.0 m from a rock band is 0.25 W m^{-2}. Calculate the intensity at distances of **(a)** 24 m **(b)** 17 m.

15.24 Suppose the page of your book is 1.7 m from a 75 W light bulb, and the light is just adequate for you to read comfortably. How far from a 40 W bulb would the page have to be to get the same light intensity? [Treat both bulbs as point sources of energy.]

15.25 The intensity of the Sun's radiation at Earth orbit is 1370 W m^{-2}. If the radius of the Earth's orbit is 149×10^6 km, calculate the Sun's total power output.

15.26 A human face emits infra-red radiation (with a peak wavelength of about 10 mm). The total power radiated is 12 W. Treating the head as a point source, calculate the intensity of the radiation at a distance of **(a)** 20 cm **(b)** 3.0 m from the centre of the head.

15.27 The minimum intensity that can be detected by a small portable radio receiver is known to be 2.2×10^{-5} W m^{-2}. Calculate the maximum distance that the receiver can be from a 15 kW transmitter so that it is just able to detect the signal.

15.28 An object is vibrating vertically in a water surface producing waves with a wavelength of 12 mm which travel away from the object at a speed of 120 mm s^{-1}. The object is then made to move in a straight line over the water surface at a speed of 20 mm s^{-1}.

(a) Draw a diagram showing the positions of the waves emitted during half a second.

(b) Calculate the approximate wavelengths of the waves **(i)** directly ahead of the moving object **(ii)** directly behind the moving object.

15.29 A police car travels with uniform speed along a long straight road (represented by the horizontal axis on the diagram) sounding its siren. A microphone connected to a frequency meter is placed close to the side of the road and records how the received frequency f varies with the distance x of the car. Explain the graph.

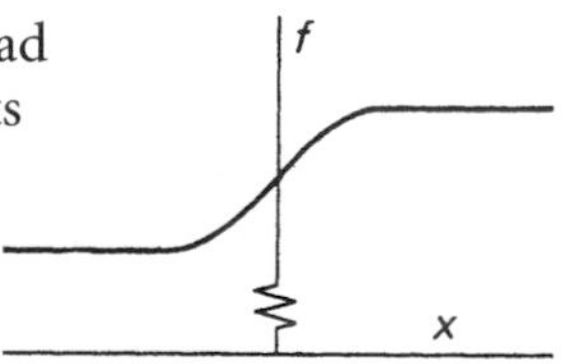

15.30 The spectrum of atomic hydrogen contains a bright line at a wavelength of 486 nm. When the spectrum of light from a distant galaxy was analysed this same hydrogen was measured to have a wavelength of 497 nm. Calculate the speed of recession of the galaxy from the Earth.

15.31 The Sun has a period of rotation about its own axis of 2.3×10^6 s and a mean diameter of 1.4×10^9 m.

(a) Calculate the Doppler change in wavelength for light of wavelength 5.0×10^{-7} m which comes from the edge of the Sun that is moving towards us.

(b) Hence explain why a sharp line in the Sun's spectrum is found to be split into two lines when it is examined very closely.

15.32 **(a)** Explain how measurements of the Doppler shift in the light from galaxies beyond our own have led to the Big Bang theory of the origin of our universe.

(b) Hubble's law states that the speed of recession v of a distant galaxy is directly proportional to its distance d from us, the constant of proportionality being 2.2×10^{-18} s^{-1}.

For a galaxy which shows a red shift of 4.0 %, calculate

(i) its speed of recession **(ii)** its distance from us.

15.3 Refraction

In this section you will need to

- draw diagrams showing the behaviour of rays and wavefronts as they reflect and refract
- use the law of refraction: that for a ray refracted from medium 1 to medium 2

$$n_1 \sin\theta_1 = n_2 \sin\theta_2$$

- understand that the ratio of the speeds of light in two media is equal to the inverse ratio of refractive indexes of the media:

$$c_1/c_2 = n_2/n_1$$

- remember that the refractive index of a given material varies with the wavelength or colour of the light and that this variation is called dispersion
- explain that total internal reflection can occur only when light meets the boundary of a medium with a lower refractive index than the medium in which it is travelling
- use the critical angle condition for total internal reflection:

$$\sin\theta_c = n_1/n_2$$

- understand how light (and infrared radiation) travels along an optical fibre.

15.33 A narrow beam of light in air strikes a glass block of refractive index 1.52 at angles of incidence $\theta_a/° = 0, 5, 10, 15, 20, 25$.
(a) Calculate **(i)** the angle of refraction θ_g in each case **(ii)** the angle of deviation $\varphi = \theta_a - \theta_g$ in each case.
(b) Plot a graph of **(i)** θ_a against θ_g **(ii)** θ_a against φ.
(c) Predict the rough shape of your two graphs as θ_a rises from 25° to 90°.

15.34 The diagram shows how a narrow beam of light strikes a layer of oil on the surface of a tank of water at an angle of 58.0°. If the refractive index of oil is 1.28 and that of water 1.34 calculate
(a) the angle of refraction in the oil
(b) the angle of refraction in the water
(c) the angle of refraction in the water if the layer of oil was removed.

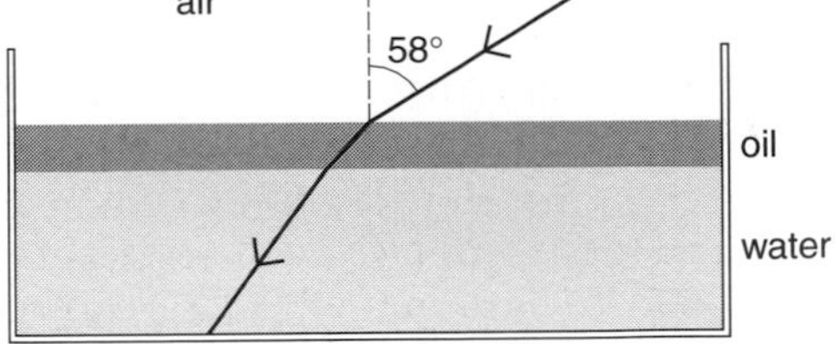

15.35 The refractive indexes of the glass of a prism for red and blue light are 1.538 and 1.516, respectively. Find the angle between the emerging blue and red light when the white light is incident normally on one face of a prism of angle 40° as shown in the diagram.

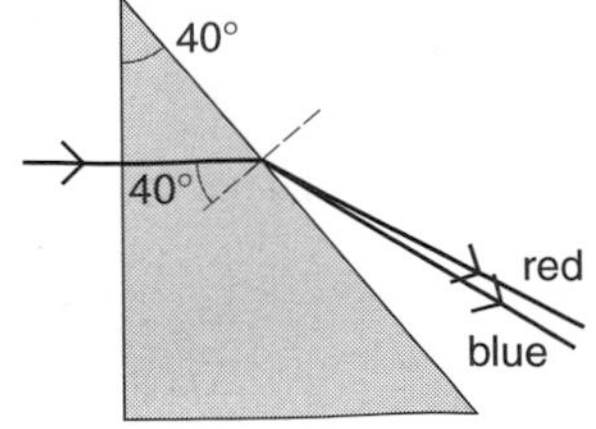

15.36 Repeat the previous question for the case where the angle of the prism is **(a)** 30° **(b)** 20° **(c)** 10°.

15.37 The diagram shows rays representing the transmission of energy by S waves (secondary or shake or shear waves) from an earthquake with its epicentre at E.

(a) What does the existence of the S-wave shadow zone tell you about the nature of the S waves?

(b) Suggest why the ray paths are not straight lines.

(c) Copy the diagram and add wavefronts. [Hint: they are not circular.]

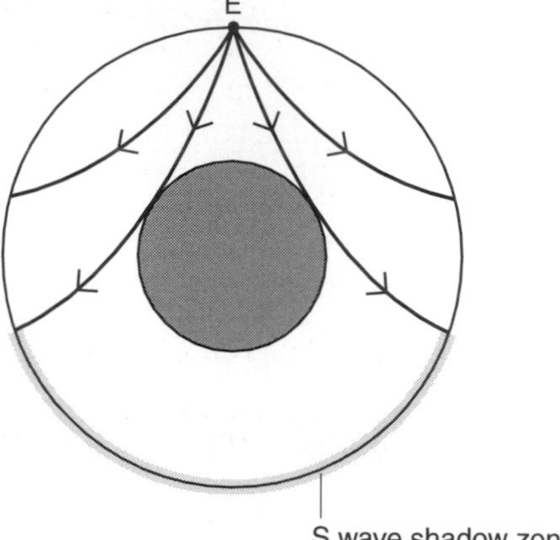

15.38 Draw a scale diagram showing wavefronts in air and in water to illustrate the refraction of sound of frequency 500 Hz. Take the angle of incidence of the sound energy in air to be 45°. [Use data.]

15.39 In an experiment to measure the speed of light, a light pulse is reflected up and down a straight tube which is 1200 m long. How much longer will the light take to make 20 journeys there and back if the tube is full of air rather than if it is evacuated? Take the refractive index of air to be 1.000 29.

15.40 **(a)** What are the critical angles for **(i)** ice of refractive index 1.309 **(ii)** ethyl alcohol of refractive index 1.361 **(iii)** common salt of refractive index 1.544?

(b) The critical angle for diamond is 24.42°. What is the refractive index of diamond?

15.41 **(a)** What is the difference between the critical angles of red light and blue light for a glass for which $n_{blue} = 1.639$ and $n_{red} = 1.621$?

(b) How could you demonstrate that the critical angles were different?

15.42 The output of an LED contains energy in the infrared part of the e-m spectrum. The refractive indexes of a glass fibre for the shortest and longest of these wavelengths are 1.479 and 1.482, respectively. Calculate the times taken by infrared waves of these two wavelengths to travel 1000 m along the fibre. What is the time difference? [Take $c = 2.997 \times 10^8$ m s^{-1}.]

15.43 A step-index optical fibre has a core of refractive index 1.48 and cladding of refractive index 1.46. A narrow beam of light in air enters the (flat) end of the core at an angle of 12° to the axis of fibre.

(a) Will the beam be propagated along the fibre or not? Support your answer with a diagram and appropriate calculations.

(b) What is the maximum angle to the axis for propagation to occur?

15.44 An optical fibre core is made of glass of refractive index 1.472 and is surrounded by a cladding of refractive index 1.455.

(a) Calculate the critical angle for the fibre.

(b) How long does it take energy to travel 3000 m along the axis of the fibre (this is called monomode propagation). [Take $c = 2.997 \times 10^8$ m s^{-1}.]

(c) The diagram shows a ray which bounces from side to side at the critical angle (this is called step-index propagation). Calculate how long energy takes to travel along a 3000 m fibre in this mode of propagation.

15.4 Communications

In this section you will need to

- list the types of noise affecting communications links
- understand the different ways in which radio waves travel from transmitter to receiver: ground waves, sky waves and space waves
- understand the term bandwidth
- understand the difference between analogue and digital information signals
- understand how pulse code modulation is achieved
- remember that bit rate is calculated as:

$$(\text{number of bits per sample}) \times (\text{sampling frequency})$$

- understand the advantage of transmitting digital rather than analogue signals
- explain the nature of time division multiplexing
- understand the advantages of using optical fibres for information transfer.

15.45 When you listen to the radio you sometimes hear a crackling sound.

(a) What is this sound called?

(b) Why would turning up the amplitude on the radio not help?

15.46 Radio waves from a transmitter at T reach a receiver which is over the horizon at R.

(a) Explain with the aid of a diagram how the signal gets from T to R.

(b) Discuss whether there is a limit to the distance on the Earth's surface between T and R.

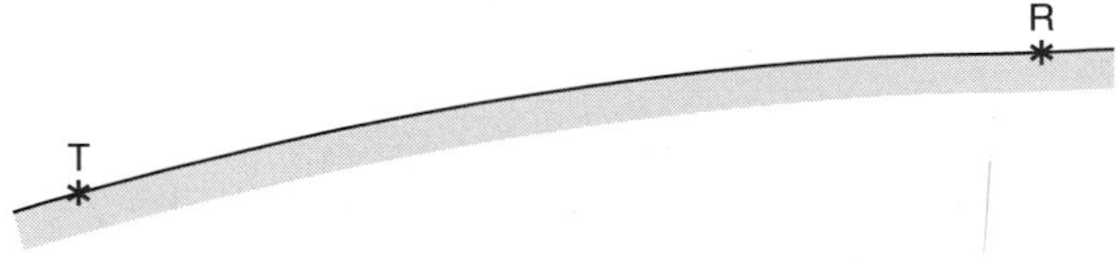

15.47 Explain the role of the ionosphere in the transmission of sky waves in the frequency range of 3–30 MHz which are received beyond the visible horizon.

15.48 A channel of 500 MHz wide is used by a satellite communication system. For TV signals an agreed bandwidth of 8 MHz is used. Different bands are used for the up signal and the down signal.

(a) How many TV signals can be handled by this one satellite?

(b) Why are different bands used for the up and the down signals?

15.49 In the MF band the allowed range for radio broadcasts is 526 kHz to 1606 kHz. A carrier wave of 1053 kHz is amplitude modulated with a signal in the range 300 Hz to 3400 Hz, the range used for voice communications.

(a) Explain what is meant by amplitude modulation.

(b) What are the signal bandwidth and the channel bandwidth in this case?

(c) If a bandwidth of 1.4 times the signal bandwidth must be allowed for transmission, how many amplitude modulated signals could be transmitted in this MF band?

15.50 The diagram shows an analogue signal which has been sampled at regular time intervals to produce a pulse amplitude modulated (PAM) signal.

(a) If the frequency of the analogue signal is 1800 Hz, suggest a value for the minimum useful sampling frequency?

(b) For your suggested sampling frequency, what is the time interval between the sampled pulses?

(c) Explain how the sampled pulses can be encoded as a digital signal.

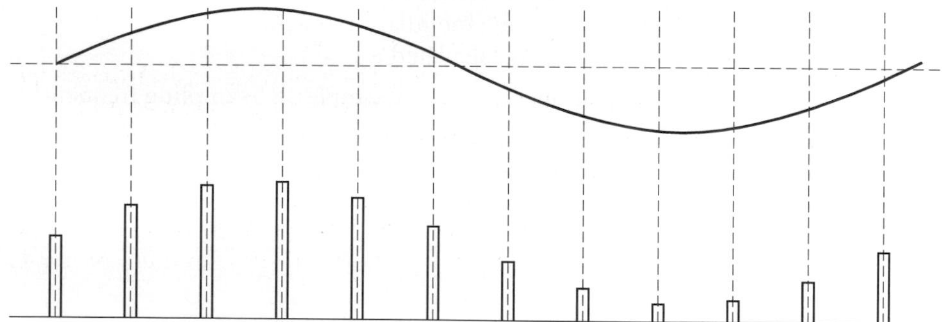

15.51 **(a)** Draw a sketch to show the result of amplifying an analogue signal to which noise has been added.

(b) Draw a second sketch to show the result of amplifying and reshaping a digital pulse to which noise has been added.

(c) Use your sketches to explain why the transmission of digital signals is preferable to the transmission of analogue signals.

15.52 An audio signal is sampled at intervals of 125 μs. The samples are digitally encoded as a series of 3-bit numbers and the following sequence is obtained:

100 010 001 010 100 110 111 110 100

(a) Reconstruct the analogue signal as it would appear at the receiving end of a transmission line.

(b) Estimate the frequency of the audio signal.

15.53 The analogue signal shown in the diagram is to be transmitted using PCM with eight sampling levels and a sampling frequency of 10 kHz.

(a) Write down the quantised levels at each of the eight sampling points.

(b) What are the 3-bit binary numbers for these quantised levels?

(c) Draw the transmitted pulse train for the 0.7 ms period shown.

(d) What is the bit rate for this transmission?

(e) Reconstruct, on graph paper, the received signal. Is it a faithful representation of the original analogue signal?

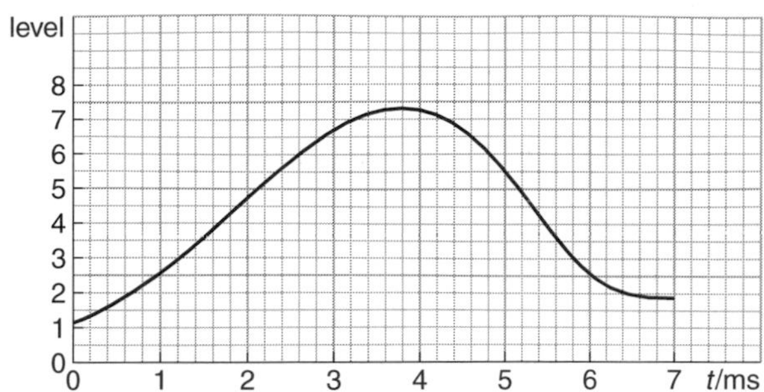

15.54 An optical fibre system operates at a wavelength of 1.3 μm and at a bit rate of 480 kbit s^{-1}.

(a) Where in the electromagnetic spectrum do such waves occur? Why are such wavelengths chosen for 'optical' fibre links?

(b) Calculate the sampling frequency if there are 16 bits (2 bytes) required to encode each sample such as might be required for high-quality music.

(c) What is the maximum frequency in the transmitted music signal which this sampling frequency can successfully handle?

15.55 Some of the secondary advantages of optical fibre links for communication are: no externally radiated signals, the use of common natural materials, small and light cables.

(a) Explain these secondary advantages.

(b) Give a list of primary advantages.

15.56 An analogue signal is sampled every 125 μs and is turned into a digital signal using PCM. The resulting digital signals, consisting of 8-bit pulse trains, are then fed into a telephone transmission line at the rate of 1.0 Mbit s^{-1}.

(a) How long does it take to feed each 8-bit pulse train into the line?

(b) How long elapses after each pulse train is fed in before the next pulse train is fed in?

(c) Use your answers to explain the concept of TDM – time division multiplexing.

(d) If the sampling frequency is increased, what must be changed so that the same number of channels can still be transmitted in such a TDM system?

16 Interference patterns

Data: speed of electromagnetic radiation $c = 3.00 \times 10^8$ m s^{-1}

speed of sound in air = 340 m s^{-1}

16.1 The principle of superposition

In this section you will need to

- understand that the displacement at any point in the path of a wave is the sum of the displacements affecting the point at that instant
- draw the result of superposition as a sequence of displacement – time graphs
- understand the phrases *in phase* and *in antiphase* for two sinusoidal waves arriving at a point.

16.1 The photographs show the position at equal time intervals of a stretched rope as two wave pulses moving in opposite directions travel 'through' each other.

(a) Sketch the rope at each stage and add an explanation of what is happening in terms of wave superposition.

(b) Comment on where the energy of the pulses is at each stage. [Hint: the more blurred a piece of the rope is in the photograph, the faster that piece of the rope is moving sideways.]

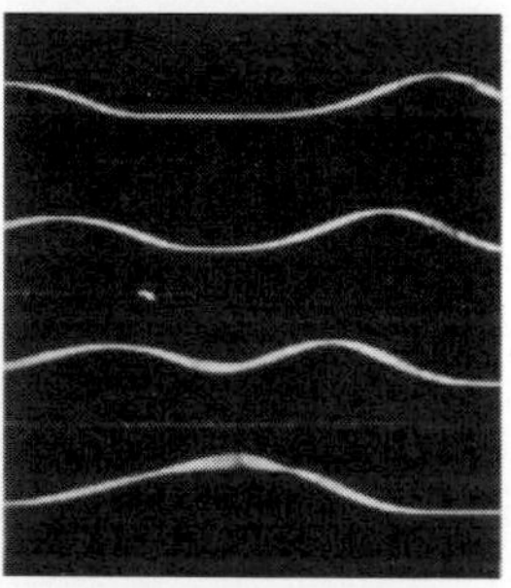

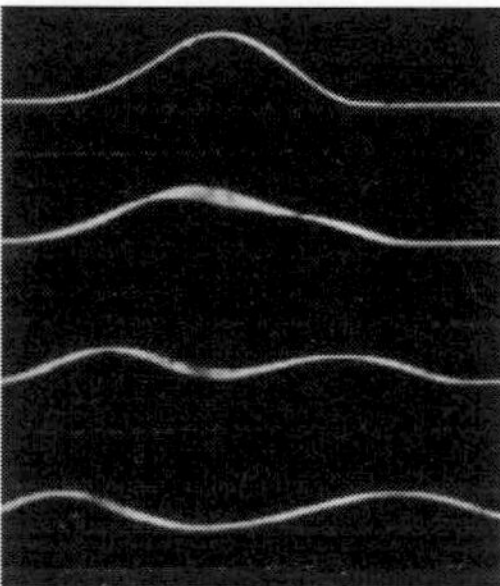

16.2 The diagram shows two wave pulses at time $t = 0$. The markings on the x-axis are 1 m apart.

Draw a series of y–x graphs at one second intervals to illustrate their superposition until after they have crossed. [Hint: sketch the two pulses in lightly and then add the displacements to find the resultant pulse.]

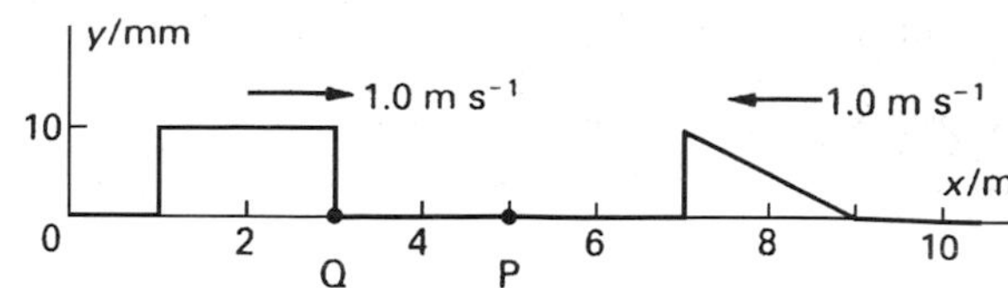

16.3 Draw a diagram to describe the result of superposing two sinusoidal waves of equal wavelengths. They are moving along the same line and are in phase, but one has twice the amplitude of the other. Make your diagram cover at least two wavelengths.

16.4 Use a diagram to help you explain how two sinusoidal waves travelling in opposite directions along a rope can, for an instant, 'cancel each other out'.

16.5 Consider the two points P and Q marked on the diagram of question 16.2.

(a) Draw a graph showing the displacement of the point Q for the next 6 s as the wave pulses pass and separate.

(b) Draw a second graph to show how the displacement of P varies during the next 6 s.

16.2 Stationary wave patterns

In this section you will need to

- explain how two waves travelling in opposite directions produce a stationary wave
- describe how to demonstrate stationary waves on stretched strings and with microwaves
- remember that the distance between adjacent nodes in a stationary wave pattern in $\lambda/2$
- understand that stationary waves can only be set up at certain frequencies (and with certain wavelengths) on strings and in air columns and that these frequencies are the resonant frequencies of the string or air column.

16.6 **(a)** Sketch three stationary patterns which could be formed on the cord in the diagram. Mark the nodes and antinodes in each sketch.

(b) If in one of your sketches the distance between nodes is 0.55 m when the signal generator frequency is 40 Hz, what is the speed of the waves on the string?

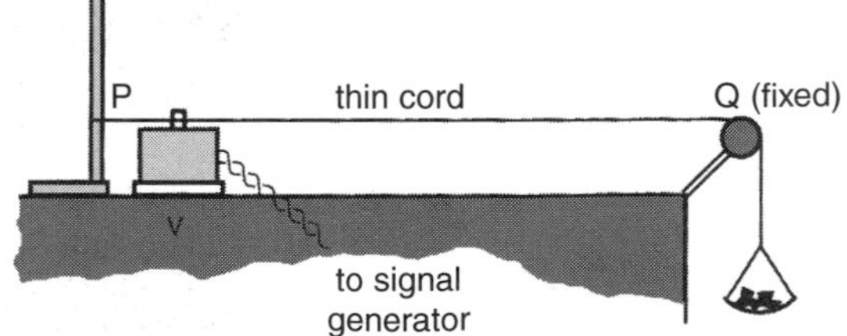

16.7 The natural frequencies f of vibration of the string in the previous example can be found by putting $n = 1, 2, 3$, etc. in the formula

$$f = \frac{n}{2l}\sqrt{\frac{T}{\mu}}$$

where l is the length of the string, T the tension in the string and μ its mass per unit length.

(a) Show that the units of the right hand side of the formula are s^{-1} or Hz.

(b) Using the arrangement in the diagram explain how you would show experimentally that T was inversely proportional to l for a fixed value of n, e.g. $n = 2$.

16.8 Two loudspeakers face each other at a separation of about 30 m. They are connected to the same audio oscillator which is set at 170 Hz.
Describe and explain the variation of sound intensity heard by a man who walks at a slow speed of 0.50 m s^{-1} along the line joining the two speakers. [Use data.]

16.9 The diagram shows a way of representing stationary sound waves in an open tube, e.g. a recorder.

(a) Explain carefully the formation of the nodes N and antinodes A. [Remember that sound is a longitudinal wave motion.]

(b) If the length of the tube is 40 cm, calculate the frequencies of the two stationary waves shown. [Use data.]

(c) Draw two similar diagrams to represent stationary waves in the same tube when one end of the tube is closed.

(d) What are the now two lowest stationary wave frequencies?

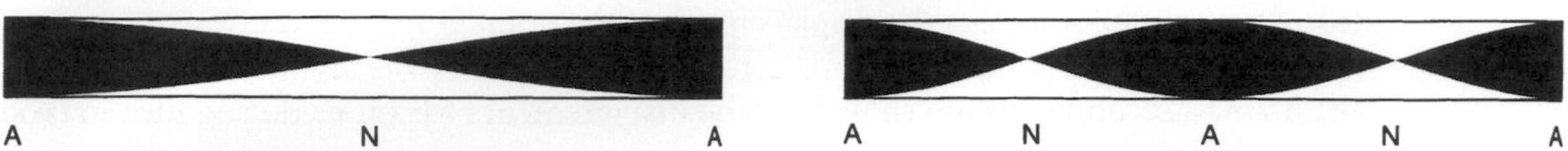

16.10 Draw successive positions in the vibration of a stretched string oscillating in its third harmonic, i.e. with nodes at each end and two other nodes. Hence explain why adjacent nodes in the stationary wave pattern are $\lambda/2$ apart.

16.11 A small speaker emitting a note at 600 Hz is held above a tall glass measuring cylinder which is 0.50 m tall. Sketch a graph of the sound intensity *I* heard by a nearby observer against the depth of water *d*, as water is poured down the side of the measuring cylinder until it is full. Mark a scale on the *d* axis. [Use data.]

16.12 Two radio transmitters T_1 and T_2 emit vertically polarised e-m waves at a frequency of 1.44 MHz. A stationary wave pattern is set up along the line T_1T_2.

(a) What is the distance between adjacent nodes in this pattern? [Use data.]

(b) A car is driving along the line between the transmitters (e.g. along the M4) and its radio is receiving the programme carried by the 1.44 MHz waves. Describe what the driver hears if the car is travelling at just over 30 m s^{-1}.

16.13 In the diagram T is a microwave transmitter and P a detecting probe. As P is moved towards the metal reflecting sheet, a series of maxima are found separated by 16 mm.

(a) Deduce, with a full explanation, the wavelength of the microwaves.

(b) As P approaches the metal sheet the intensity of the minima is found to become closer and closer to zero. Explain why this is so.

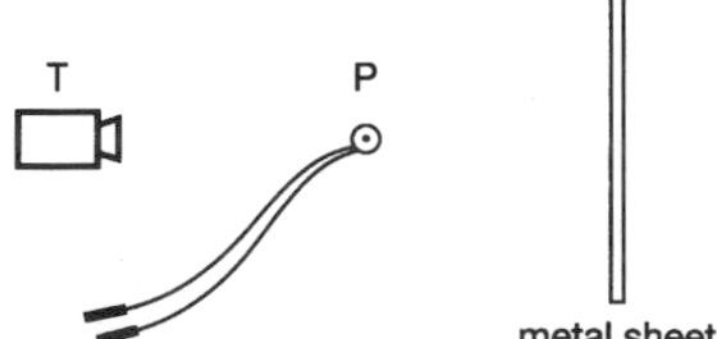

16.3 Two-source patterns

In this section you will need to

- remember that, for two sources which are in phase, the result of superposition at a point P is given by:
  $S_1P - S_2P = n\lambda$ (maximum or constructive interference)
  $S_1P - S_2P = n\lambda + \lambda/2$ (minimum or destructive interference)
- understand that energy is not lost in an interference pattern, it is only redistributed
- use the relationship for light from two adjacent slits:
  wavelength = (slit separation) × (fringe width) ÷ (slit to screen distance)
- understand how to produce two coherent optical sources.

16.14 The diagram shows an experimental arrangement for investigating the superposition of sound waves from two sources S_1 and S_2. The sources are in phase and produce sound of wavelength 80 mm.

(a) When S_1M is 800 mm the trace on the oscilloscope is a maximum. Suggest three possible values for S_2M. [Use data.]

Without moving any of the apparatus the leads to one of the speakers are reversed, i.e. the sources are now in antiphase.

(b) Explain the meanings of *in phase* and *in antiphase* and describe how the trace on the oscilloscope changes when the leads are reversed.

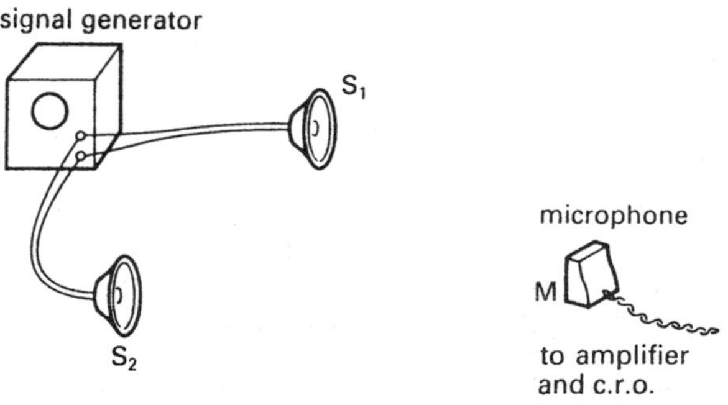

16.15 Two ripple tank dippers S_1 and S_2 are vibrating in phase.

(a) Draw (you will need a compass) a series of arcs to represent the wavefronts from S_1 and S_2, using a wavelength such that $S_1S_2 = 3.5\lambda$, e.g. make $\lambda = 1$ cm and $S_1S_2 = 3.5$ cm.

(b) Mark on your diagram the antinodal lines, that is lines joining places where the two series of arcs overlap.

(c) Describe, without further drawing, how your pattern would differ if the two sources were in antiphase.

16.16 The diagram on the next page shows the position of two in phase sources, S_1 and S_2, e.g. ripple tank dippers. The full lines are antinodal lines, i.e. places where the superposition of waves from the two sources produces a maximum. The dashed lines – lines of minima – are nodal lines.

(a) For point P on the line $n = 1$, use a ruler to show that $S_1P - S_2P = 5.5$ mm. Hence explain why the wavelength of these waves is 5.5 mm.
(b) Choose another point on the line $n = 1$ and confirm that the wavelength is 5.5 mm.
(c) Choose a point R on the line $n = 2$. Predict the value of $S_1R - S_2R$ and check your prediction using the ruler.
(d) Check that, when Q is on a nodal line, $S_1Q - S_2Q$ is equal to a whole number of wavelengths plus half a wavelength.

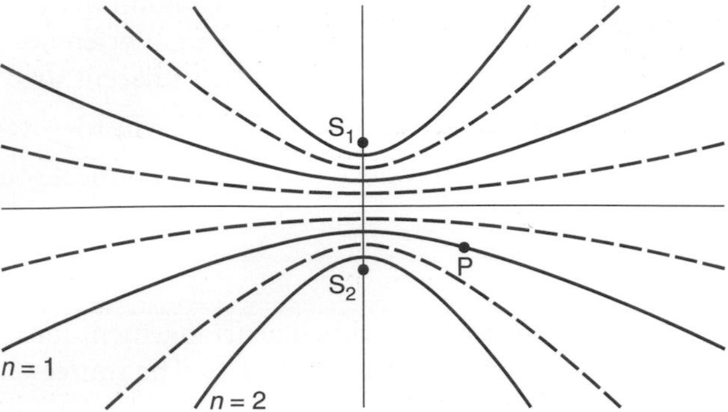

16.17 In another experiment with the apparatus described in question 16.14, the height of the trace on the oscilloscope was found to be a minimum when $S_1M = 0.80$ m and $S_2M = 1.00$ m.
(a) Why can the wavelength of the sound waves not be found from these measurements alone?
(b) What further observations are needed before the wavelength can be found?

16.18 The only practical way to produce visible two-source interference patterns with light is, in effect, to derive two sources from a single source. Explain why this is so and describe one method of achieving the two sources.

16.19 The photograph is of interference fringes formed on a screen in a Young's slit experiment. The slit separation was 0.48 mm and the distance from slits to screen was 0.95 m.
(a) By taking measurements from the photograph, which is full size, find a value for the wavelength of the light used and suggest its colour.
(b) In the photograph there is no light energy reaching the minima – the dark lines. Explain where the energy from the two slits has gone in this interference pattern.

16.20 The eye is more sensitive to wavelengths in the range 500 nm (green) to 600 nm (orange) than to colours in the rest of the visible spectrum.
(a) For a Young's slits arrangement with a slit separation of 0.68 mm, calculate the fringe separation for each of these wavelengths at a distance of 0.80 m from the slits.
(b) Show that the sixth green fringe from the centre of the interference pattern coincides with the fifth orange fringe at this distance.

16.21 This is a question about the geometry of the two-source Young's slits experiment. (You will need a calculator with at least an 8-digit display).

(a) In the diagram, which is *not* to scale, sources at S_1 and S_2 send light to a point P. If $a = 0.400$ mm, $D = 800.0$ mm and $x = 3.000$ mm, calculate

(i) the lengths of S_1P and S_2P using Pythagoras's theorem

(ii) the path difference of $S_2P - S_1P$ between light arriving at P from the two sources.

(b) If the wavelength of the light is known to be 0.60×10^{-6} m, explain whether P is at a bright (maximum) or a dark (minimum) place in the interference pattern.

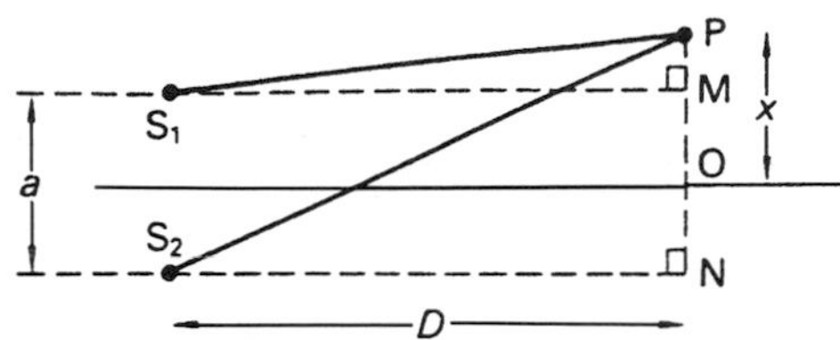

16.22 In a Young's slits type of experiment with microwaves of wavelength 30 mm, the distance between maxima detected at a distance of 42 m from the sources while moving across the line of symmetry was found to be 2.5 m. What was the separation of the microwave sources?

16.23 **(a)** A microwave transmitter T and receiver R are placed side by side facing two sheets of material M and N as shown in the diagram. It is found that a very small signal is registered by R; what can you deduce about the experimental set up?

(b) When M is moved towards N a series of maxima and minima is registered by R. Explain this and deduce the wavelength of the e-m waves emitted by T if the distance moved by M between the second and seventh minimum is 70 mm.

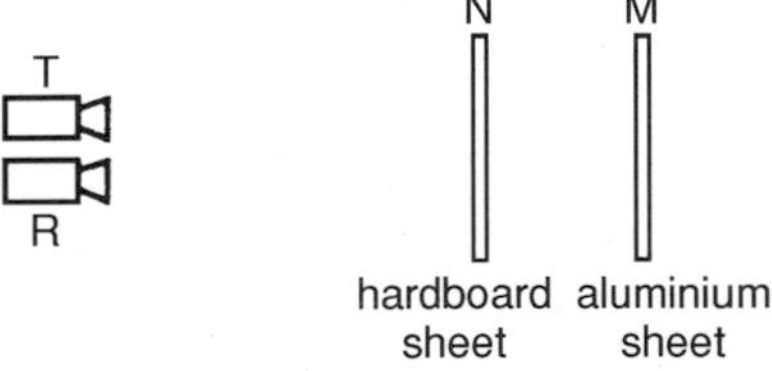

16.24 In order to cut out unwanted reflections and increase the percentage of light which is transmitted from a spectacle lens, a process called blooming is used. This involves coating the lens with a very thin film of calcium fluoride in which light travels at a speed v, less than the speed of light c.

(a) If $v = 0.69c$, what is the wavelength in the film of light which has a wavelength of 580 nm in air?

The diagram shows two reflected light rays, R_1 and R_2, which will superpose to the left of the film.

(b) What is the minimum thickness of film which will give rise to R_1 and R_2 being in antiphase?

(c) Suggest why, for this thickness of film, other wavelengths will not be completely cut out.

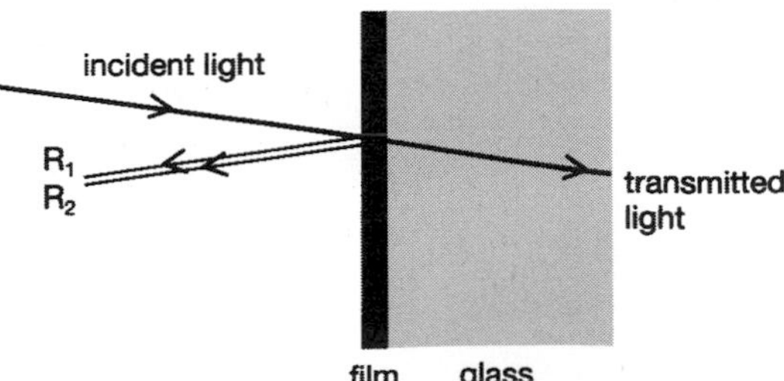

16.25* The diagram below shows a transmitter T of electromagnetic waves of wavelength $\lambda = 0.30$ m. A receiver R is placed 6.0 m away as shown and a plane reflecting surface M is held in such a way that the perpendicular distance from M to the line TR is 2.0 m.

(a) Express the two path lengths TR and TMR by which waves can reach R from T in terms of the wavelength λ.

(b) In reflecting from M the wave which follows the path TMR undergoes a phase change which is equivalent to it having travelled an extra distance of $\lambda/2$. Are the two waves reaching R in phase or in antiphase?

(c) Describe as fully as possible how the signal received at R varies as M is slowly moved towards the line TR until it almost lies on it.

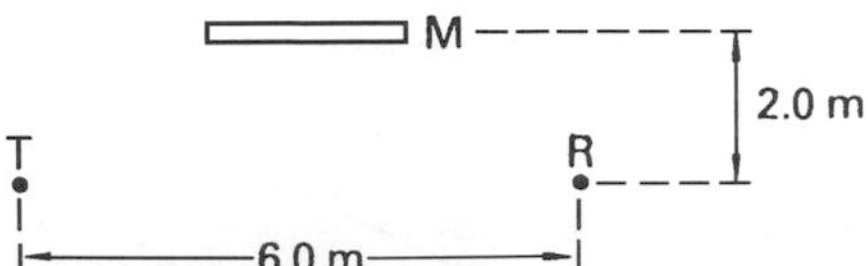

16.26* Two loudspeakers L_1 and L_2 driven from a common oscillator are arranged as shown in the diagram. A detector is placed at D. It is found that, as the *frequency* of the oscillator is gradually changed from 200 Hz to 1000 Hz the detected signal passes through a series of maxima and minima.

(a) Explain why this is so.

(b) Calculate the frequency at which the first minimum above 200 Hz is observed. [Use data.]

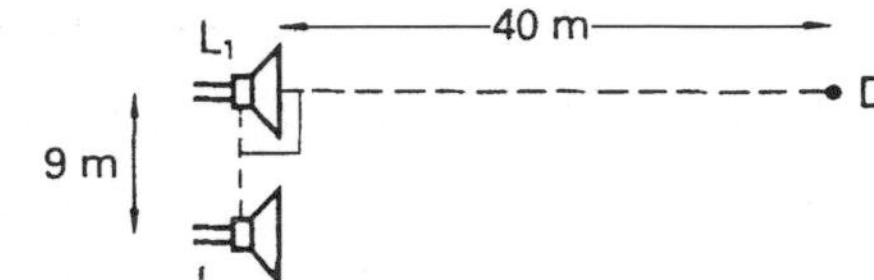

16.4 Diffraction patterns

In this section you will need to

- understand that waves diffract; that is, their energy spreads into the shadow area when part of a wavefront is blocked
- draw graphs of intensity against position for diffraction at a single slit
- use the equation $\sin\theta_m = \lambda/d$ for the first minimum in a single slit diffraction pattern
- understand that a diffraction grating produces a series of narrow maxima at angles θ_n given by the formula $n\lambda = s\sin\theta_n$
- explain how a diffraction grating can produce spectra.

16.27 **(a)** Sketch two diagrams to illustrate the behaviour of plane water waves of wavelength **(i)** 0.1 m **(ii)** 1.0 m as they arrive at a barrier in a children's paddling pool in which there is a gap 1.0 m wide.

(b) Describe what is happening to the wave energy in each case.

16.28 A loudspeaker system in a large hall consists of a number of rectangular speakers, each measuring 1.2 m by 0.15 m. The speakers are mounted about 3 m above the floor with the long side vertical.
Explain why the speakers are this shape and why they are mounted in this way.

16.29 The graph shows the diffraction pattern produced by a single slit illuminated by a parallel beam of light of wavelength λ. The width of the slit is 0.20 m. Calculate λ.

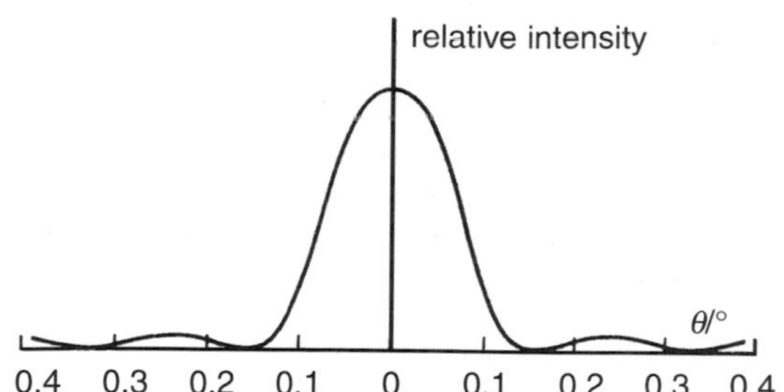

16.30 Draw a diffraction pattern graph, similar to that in the previous question, for the diffraction of infrared radiation of wavelength 1.4 μm through a slit of width 0.10 mm. Label the *x*-axis carefully.

16.31 A laser beam is used to illuminate a flag at night. The flag is 400 m from the laser and approximately 2 m square. The laser produces light of wavelength 6.6×10^{-7} m and the initial beam width is 2.0 mm. Will the laser light adequately illuminate the flag? [Assume that the light spreads out by diffraction through the end of the laser tube.]

16.32 Light diffracts when it enters the pupil of the eye. Take the average wavelength of visible light to be 550 nm and a typical pupil to have a diameter of 3.0 mm.

(a) Estimate the angle to the first minimum in the diffraction pattern formed on the retina when looking at a point source, e.g. a star. [It is an estimate because the formula giving the first minimum is for a slit and not for diffraction through a round hole.]

(b) For an eyeball which is a sphere of diameter 38 mm, what will be the width of the bright central patch formed by the diffraction pattern on the retina?

(c) Explain why two stars which are very close together in the night sky might not be registered as two separate images, but rather as a single blur.

16.33 The diagram shows a graph of intensity against distance on a screen placed 2.0 m away from an illuminated double slit in a typical 'Young's slits' experiment. The light was of wavelength 590 nm and the minima occur every 4.2 mm.

(a) Explain the general shape of the curve (the full line).

(b) Calculate the slit separation.

(c) Calculate the *width* of each slit.

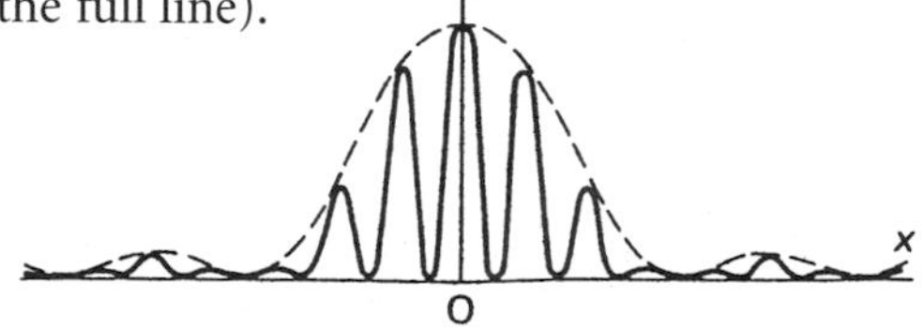

16.34 A diffraction grating was set up so that parallel light is perpendicularly incident on it. For light of wavelength 589 nm the first order spectral lines are observed in directions making an angle $\theta = 22.0°$ with the straight through position.

(a) Calculate **(i)** the value of the slit separation s **(ii)** the number of slits per millimetre in the grating.

(b) Calculate the wavelength of light which would give a first order spectral line at $\theta = 24.3°$.

16.35 How many spectral orders are produced within 15° of the incident direction when a parallel beam of yellow light of wavelength 590 nm falls on a coarse diffraction grating the lines of which are 18 μm apart?

16.36 A diffraction grating having 5.00×10^5 lines per metre is illuminated with a parallel beam of light which contains strong lines of wavelengths 750 nm and 400 nm.

(a) Calculate the angular dispersion between these two (red and violet) lines in the first order spectrum.

(b) Show that the second order of the red line will overlap with the third order of the violet line.

(c) Explain why this overlap will always occur, no matter what the line spacing in the diffraction grating.

16.37 When white light passes through a diffraction grating the grating *disperses* the light. Each slit in the grating *diffracts* the light and the resulting superposition of light leads to a pattern of peaks or maxima of light intensity at angles of *deviation* which vary from colour to colour. Write brief notes, including diagrams where appropriate, to explain in your own words each of the italicised words.

17 Oscillations

Data: gravitational field strength $g = 9.81\ \text{N kg}^{-1}$

17.1 Simple harmonic oscillators

In this section you will need to

- remember that the time for one complete oscillation is called the period T, the number of oscillations per unit time is called the frequency f and that, $f = 1/T$
- understand that, for a body moving with simple harmonic motion:
  - the period is independent of the amplitude
  - the displacement varies sinusoidally with time
  - the maximum velocity occurs when the displacement is zero
- use the equations $x = x_0\cos 2\pi ft$ and $v_0 = 2\pi f x_0$
- use the relationships $a \propto -(\text{constant})x$ and $a = -(2\pi f)^2 x$
- describe how to produce displacement–time graphs for oscillating objects
- understand the relation between sinusoidal x–t, v–t and a–t graphs.

17.1 Estimate the time periods of each of the following motions and hence calculate (to 1 significant figure) their frequencies:
(a) a child on a playground swing **(b)** a baby rocked in its mother's arms
(c) the free swing of your leg from the hip.

17.2 The graph shows two periods of a simple harmonic oscillation. On a copy of the graph mark with a
(a) **(i)** P any place where the speed is a maximum
(ii) Q any place where the speed is zero.
(b) Explain where the acceleration is a maximum and where it is a minimum.

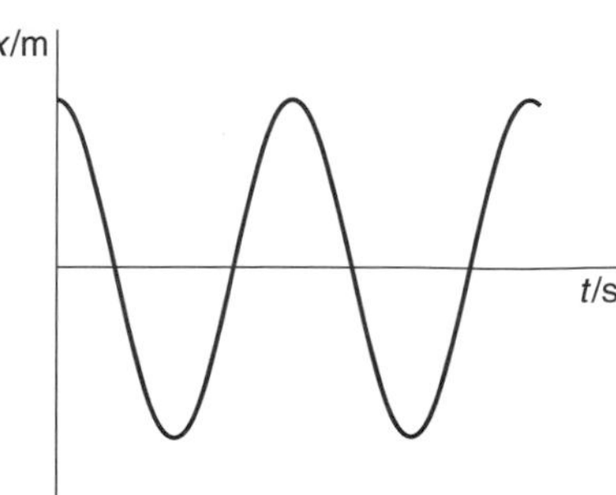

17.3 Write an explanation which you could give to a non-scientist of what is meant by simple harmonic motion. Use a situation with which he or she will be familiar to illustrate your explanation.

17.4 What is **(i)** the frequency **(ii)** the period of:
(a) the rise and fall of the sea
(b) the beat of a normal heart
(c) piano strings which oscillate when middle C is played?
Express your answers in hertz (frequency) and seconds (period).

17.5 The equation defining linear s.h.m. is

$$a = -(\text{constant})x$$

(a) What units must the constant have?
(b) Two s.h.m.s, A and B, are similar except that the constant in A is nine times the constant in B. Describe how these s.h.m.s differ.

17.6 For a s.h.m. of period 2.00 s and amplitude 16.0 cm calculate the displacement, remembering that $2\pi ft$ is in radians not degrees,
(a) at $t = 0$ **(b)** at $t = 1.00$ s **(c)** at $t = 0.50$ s **(d)** at $t = 0.25$ s.

17.7 A bored student holds one end of a flexible plastic ruler against the laboratory bench and flicks the other end, setting the ruler into oscillation. The end of the ruler moves a total distance of 8.0 cm as in the diagram and makes 28 complete oscillations in 10 s.
(a) What are the amplitude x_0 and frequency f of the motion of the end of the ruler?
(b) Use $x = x_0\cos 2\pi ft$ to produce a table of values of x and t for values:
t/s = 0, 0.04, 0.08, 0.12, 0.16, 0.20, 0.24, 0.28, 0.32, 0.36.
[Be sure that your calculator is switched to rad and not deg. Your value of $\cos 2\pi ft$ for $t = 0.04$ s should be 0.76, for example.]
(c) Draw a graph of x against t and use it to find the maximum speed at the end of the ruler.

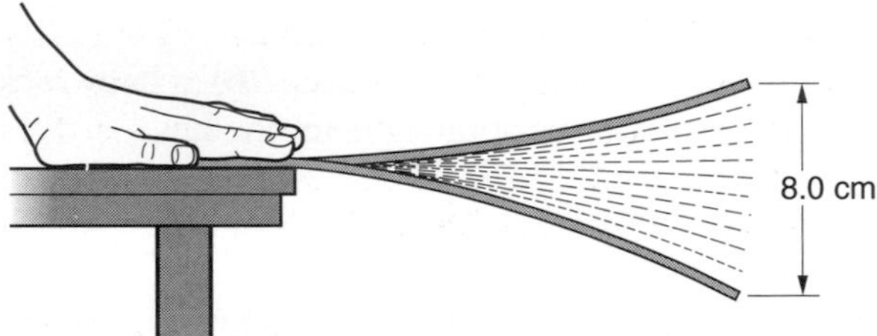

17.8 A body oscillates with s.h.m. described by the equation

$$x = (1.6\text{ m})\cos(3\pi\text{ s}^{-1})t$$

(a) What are **(i)** the amplitude **(ii)** the period of the motion?
(b) For $t = 1.5$ s, calculate **(i)** the displacement **(ii)** the velocity **(iii)** the acceleration of the body.

17.9 The diagram shows three sinusoidal (sine-shaped) graphs for displacement x, velocity v and acceleration a for s.h.m. The origin for time t is not given.
(a) Explain the relationship between
(i) the x–t and the v–t graphs
(ii) the v–t and the a–t graphs
(iii) the x–t and the a–t graphs.
(b) Illustrate your answer to **(iii)** by sketching a graph of x against a.

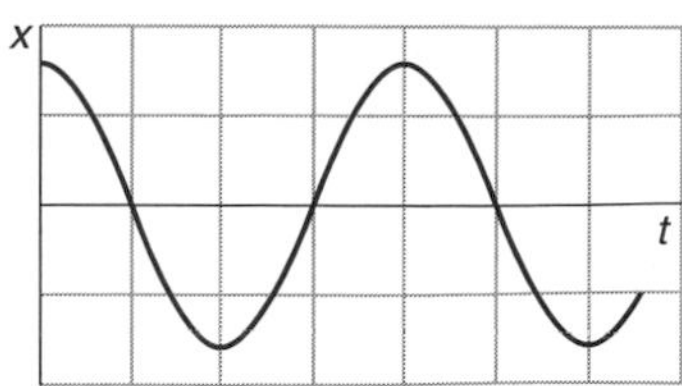

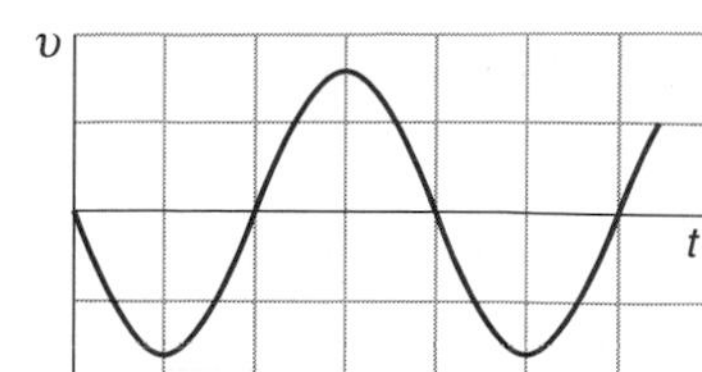

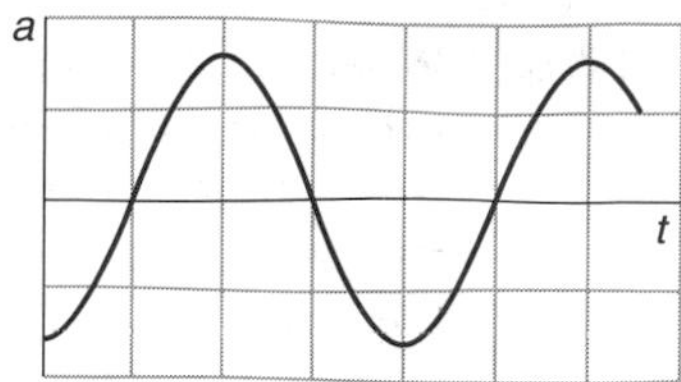

17.10 In order to test how well pilots can recognise objects when seated in a juddering helicopter they are subjected to vibrations of frequency from 0.1 Hz to 50 Hz.
If a pilot is being tested with vibrations of frequency 35 Hz and amplitude 0.60 mm, what is

(a) his maximum velocity

(b) his maximum acceleration during the test?

17.11 A loudspeaker produces musical sounds by oscillating a light diaphragm.
If the amplitude of oscillation is limited to 1.2×10^{-3} mm, what frequencies will result in the acceleration of the diaphragm exceeding 10 m s^{-2}, i.e. g?

17.12 The needle of a sewing machine moves up and down with simple harmonic motion. If the total vertical motion of the tip of the needle is 12 mm and it makes 30 stitches in 7.0 s, what is the maximum speed of the tip of the needle?

17.13 A potential difference which alternates sinusoidally is applied to the Y-plates of an oscilloscope. A stationary trace, with an amplitude of 4.0 div and a wavelength of 1.5 div, is obtained with the time base set at 1.0 ms div^{-1}. When the time base is switched off the trace becomes a vertical line. Calculate the maximum speed of the spot of light on the screen when producing the vertical line. Take 1.0 div to equal a length of 10 mm.

17.14* A dock has a tidal entrance at which the water is 10 m deep at 12 noon, when the tide is at its lowest. The water is 30 m deep when the tide is at its highest, which follows next at 6.15 pm. A tanker, needing a depth of 15 m, needs to enter the dock as soon as possible that afternoon. Calculate the earliest time it could just clear the dock entrance. [Assume that the water rises and falls with s.h.m.]

17.2 Pendulums and mass-spring oscillators

In this section you will need to

- use the equation $T = 2\pi\sqrt{(l/g)}$ for the period of a simple pendulum
- use the equation $T = 2\pi\sqrt{(m/k)}$ for the period of a mass-spring oscillator.

17.15 **(a)** What is the length of a pendulum which has a time period of 2.00 s?

(b) Show that the acceleration of the bob of a simple pendulum of length l is related to its displacement x from the centre of the oscillation for small oscillations by the equation $a = -gx/l$.

(c) For such a pendulum calculate the acceleration of the bob **(i)** when $x = 5$ cm **(ii)** when $x = 10$ cm.

17.16 The pendulum bob of a grandfather clock swings through an arc of length 196 mm from end to end. The period of the swing is 2.00 s.
(a) What is **(i)** the amplitude **(ii)** the frequency of the bob?
(b) With what speed does the bob pass through the centre of the swing?

17.17 Someone suggests that you could 'draw' a displacement–time graph for a body moving with s.h.m. as follows.
(1) Hang a bucket of fine dry sand on a rope slung in a doorway.
(2) Punch a small hole in the bottom of the bucket.
(3) Move a long strip of carpet at a steady speed beneath the bucket as it swings from side to side in the doorway.
Draw a sketch of the arrangement and discuss whether the experiment would succeed in its object.

17.18 Describe how you would use a small steel sphere and some thread, together with normal laboratory apparatus, to determine the acceleration of free fall g.

17.19 What is the period of a pendulum on Mars, where the free-fall acceleration is about 0.37 times that on Earth, if the pendulum has a period of 0.48 s on Earth?

17.20 A pendulum 650 mm long is hung from the ceiling of a hotel lift. When the lift is in operation the pendulum is found to have a period of 1.55 s. Describe the motion of the lift when this measurement is made. Be as quantitative as possible.

17.21 A mass hangs in equilibrium from a light vertical spring and when disturbed oscillates with simple harmonic motion. Would the period be increased or decreased if, separately,
(a) the mass was doubled
(b) a second identical spring was placed in parallel with the first
(c) the spring was cut in two, and one half used to replace the original spring
(d) the system was transferred from the Earth to the Moon?

17.22 **(a)** Two identical springs of stiffness k are connected **(i)** in series **(ii)** in parallel, and support a mass m. What is the ratio of their periods of vertical oscillation?
(b) What mass would be required in case **(ii)** to make their periods of oscillation equal?

17.23 To 'weigh' himself an astronaut ties himself into a harness which is held to the side of the space vehicle by strong springs of stiffness 1.1 kN m^{-1}. He then sets himself oscillating and times 10 oscillations as taking 15.8 s. Calculate his mass if the mass of the harness is 4.0 kg.

17.24 **(a)** Sketch two graphs of displacement against time up to 4 s for a mass-spring oscillator for which $m = 1.3$ kg and $k = 5.0$ N m^{-1} when the amplitude of the oscillation is **(i)** 0.40 m **(ii)** 0.20 m.
(b) **(i)** What are the maximum speeds in each case?
(ii) What property of the graphs shows these maximum speeds?

17.25 A man of mass 80 kg bounces on the seat of a motorcycle. He finds that he is at the bottom of his bounces every 0.50 s. Calculate the spring constant of the suspension. State any assumptions you make.

17.26 A fisherman's scale stretches 2.6 cm when a 1.9 kg fish is hung from it.

(a) What is the spring stiffness?

(b) Calculate the frequency with which the fish will vibrate if it is pulled down a little and released.

17.27 A 'baby bouncer' is a device for amusing babies before they can walk. It consists of a harness suspended from the lintel of a doorway.
Such a device has ropes 1.20 m long which stretched to 1.42 m when a baby of mass 8.5 kg was placed in the harness. The baby was then pulled down 8.0 cm and released. Calculate

(a) the spring constant for the baby bouncer

(b) the period of the baby's motion

(c) the baby's maximum speed.

17.3 Energy transfer and resonance in oscillators

In this section you will need to

- draw energy flow diagrams (Sankey diagrams) to illustrate energy transfer in simple harmonic oscillators
- describe the nature of damped harmonic oscillators
- understand that the energy stored in a Hooke's law spring is proportional to the square of the extension of the spring (see also section 9.2)
- use the equation $W = \frac{1}{2}kx^2$ for the energy stored in a spring
- understand the condition for resonance in mechanical oscillators
- describe an experiment to demonstrate the resonant condition of a mechanical system.

17.28 The diagram shows the energy transfer in an undamped simple harmonic oscillator. The p.e. might be elastic or gravitational potential energy.
Draw a similar diagram to illustrate the energy transfers in a damped harmonic oscillator and explain any features new to your diagram.

17.29 The expendable springs often used in school and college laboratories have a stiffness of 30 N m^{-1}.

(a) Calculate the energy stored in such a spring for extensions x/cm = 4, 8, 12, 16, 18, 20.

(b) Plot a graph of energy stored against extension for these springs.

17.30 A block of wood of mass 0.25 kg is attached to one end of a spring of stiffness 100 N m^{-1}. The block can oscillate horizontally on a frictionless surface, the other end of the spring being fixed. The graph overleaf shows how the elastic potential energy E_p of the system varies with displacement x for a horizontal oscillation of amplitude 0.20 m.

(a) Show that the graph is correctly drawn.

(b) Copy the graph and sketch a second curve to show how the kinetic energy of the mass varies with the displacement x.

(c) Calculate the maximum speed of the block and its speed when the displacement is 0.10 m.

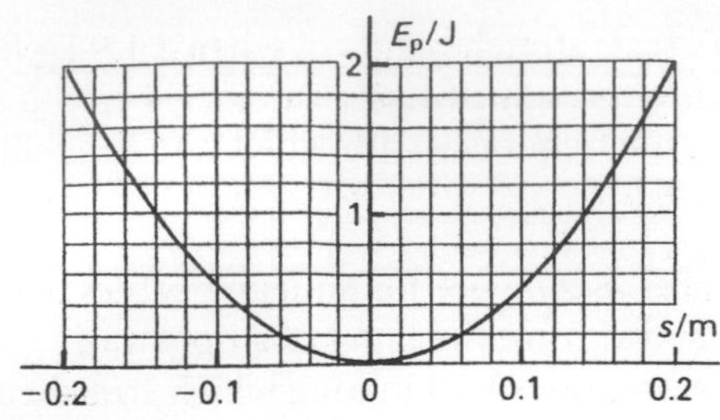

17.31 The graph shows how the kinetic energy of a heavy pendulum varies with time. During these swings the total mechanical energy can be taken to be constant.

(a) Make a rough copy of the graph and add a line showing the total mechanical energy of the pendulum.

(b) Add a dashed line showing how the gravitational potential energy of the pendulum varies with time.

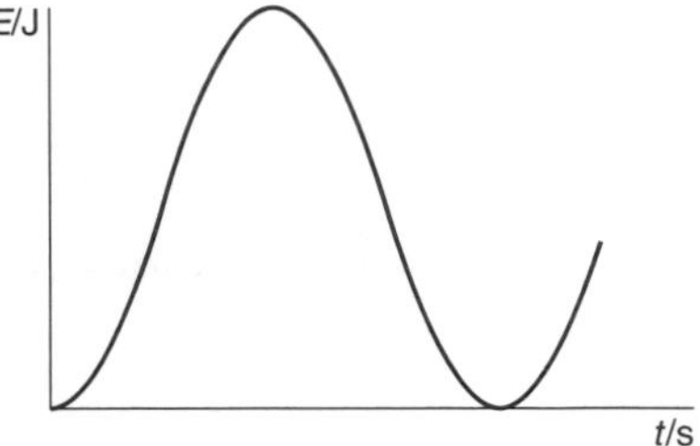

17.32* The relationship between the displacement x of a simple harmonic oscillator and the time t is

$$x = (1.2\ \text{m})\cos(4.0\ \text{s}^{-1})t$$

The variation of its elastic potential energy E_p with time is given by $E_p = \frac{1}{2}kx^2$, so that

$$E_p = (58\ \text{J})[\cos(4.0\ \text{s}^{-1})t]^2$$

when k, the restoring force per unit displacement of the oscillator, is 80 N m^{-1}.

(a) For values of t/s = 0, 0.2, 0.4, 0.6, 0.8, 1.0, 1.2, 1.4, 1.6 make a table of values for x and E_p.

(b) Plot a graph of x against t and, below it using the same scale on the same time axis, plot a graph of E_p against t. [Be careful about the shape of the $E_p - t$ graph near points where $E_p = 0$.]

17.33 Explain the meaning of the terms **(a)** damped oscillations **(b)** forced oscillations **(c)** resonance.

17.34 The diagram shows a system of light pendulums which oscillate perpendicular to the page as if they were suspended from points along the line AC.

(a) Describe the motion of the pendulums when the heavy metal sphere sets the cord ABC oscillating with s.h.m.

(b) Suggest how you could alter the damping of the small spheres and describe the result of making them more heavily damped.

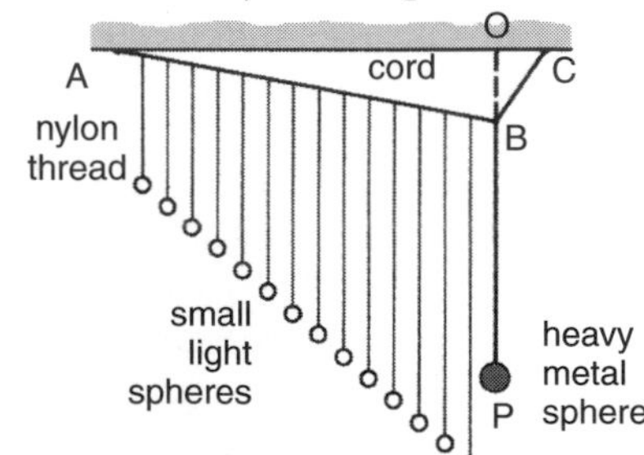

17.35 At certain engine speeds, parts of a car, such as a door panel, may vibrate strongly. Explain the physical reason for this strong vibration and suggest how it might be reduced by alterations to
(a) the door **(b)** the engine assembly.

17.36 The diagram shows a graph of the amplitude of oscillation x_0 of a mass-spring oscillator against the frequency f at which it is being forced to oscillate.
(a) Explain the significance of f_0.
(b) Copy the graph and add labelled lines to show the variation of x_0 with f for
(i) light damping
(ii) heavy damping.

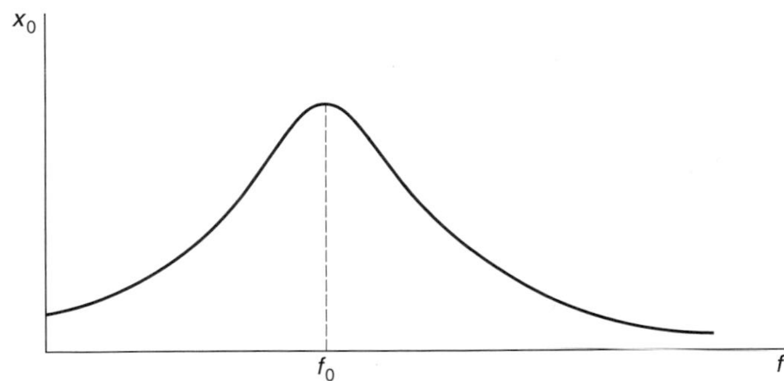

17.37 Our legs and the front legs of many four-legged animals swing like pendulums the period of which depends on the length of the leg. Suppose that a giraffe has a leg which is similar in shape to that of a horse, but that it is four times longer. Find the ratio of the following quantities for the giraffe and the horse:
(a) the period of the leg's free swing
(b) the frequency of the leg's free swing
(c) the length of a stride
(d) the speed of walking.

18 Capacitance

Data: $\varepsilon_0 = 8.85 \times 10^{-12}$ F m^{-1}

18.1 Charge and capacitance

In this section you will need to

- understand that when a capacitor stores charge q there is a charge $+q$ on one plate and $-q$ on the other
- use the equation $q = CV$ and remember that the farad, the unit of capacitance, is so large that capacitances are usually measured in μF or pF
- use equations for capacitors in parallel and in series

$$C_{par} = C_1 + C_2 + C_3 \text{ and } \frac{1}{C_{ser}} = \frac{1}{C_1} + \frac{1}{C_2} + \frac{1}{C_3}$$

- describe how to measure capacitance.

18.1 A capacitor of capacitance 10 μF is connected to a battery of e.m.f. 12 V. What are the charges on its plates?

18.2 What potential difference must be applied between the plates of a 100 μF capacitor for the charges on them to be + 0.025 μC and –0.025 μC?

18.3 Copy and complete the following table which gives the charge q on capacitors of capacitance C when there is a p.d. V across them. Express your answers using prefixes (not in standard form).

V	12.0 V		1.5 V		100 V	600 mV
C	2.2 μF	5000 μF	4.7 μF	220 μF		
q		10 mC		66 μC	1.0 nC	280 μC

18.4 **(a)** What charge is carried on each of the plates of a 470 mF capacitor when there is a potential difference between the plates of 30 V?
(b) If this capacitor is charged through a large resistance in 4.0 s, what is the average charging current?

18.5 A large (electrolytic) capacitor is used as shown in a camera flash unit. Calculate
(a) the charge it stores when the control system connects the capacitor to the 30 V power supply
(b) the average discharge current if, when the control system connects the capacitor to the discharge tube, it discharges in 0.20 ms.

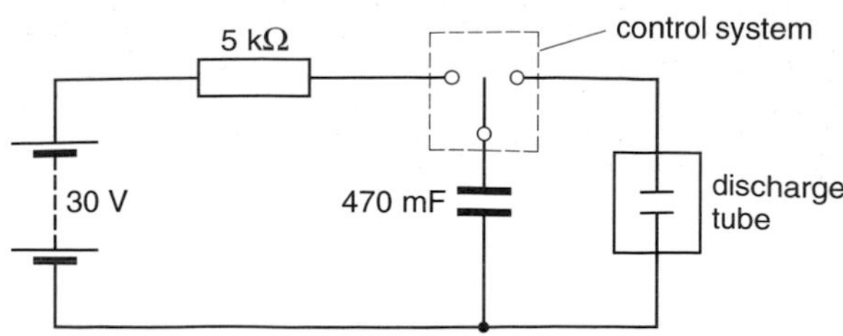

18.6 A 33 μF capacitor is charged and then insulated. During the next minute the potential difference between the plates falls by 4.0 V.
What is the average leakage current between the plates?

18.7 A steady charging current of 50 μA is supplied to the plates of a capacitor and causes the p.d. between its plates to rise from 0 V to 5.0 V in 20 s.
What is its capacitance?

18.8 Show that the unit of capacitance, the farad, is equivalent to $A^2\ s^4\ kg^{-1}\ m^{-2}$.

18.9 The pair of intersecting rings in the diagram is a symbol for a constant current source, which here produces a current of 0.48 mA. The switch in the circuit is closed for a short time and the digital (high resistance) voltmeter, connected across a 22 μF capacitor, then registers 74 mV. For how long was the switch closed?

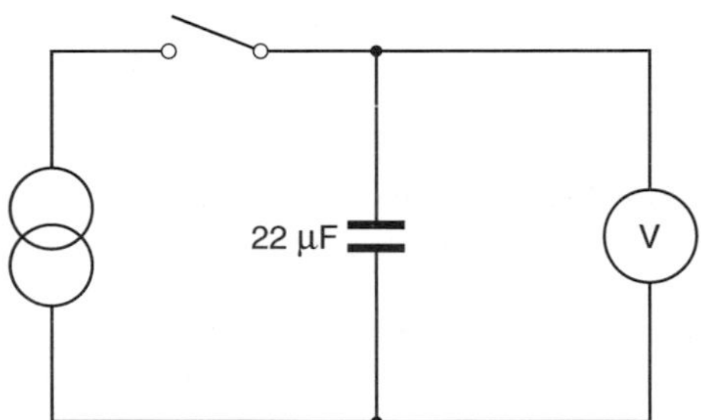

18.10 A capacitor of capacitance 0.47 μF is connected across the Y-input terminals of an oscilloscope. The horizontal (time-base) control is set at 100 ms div^{-1} and the Y-input control at 5.0 V div^{-1}. A battery in series with a resistor is then connected across the capacitor for a short time t and the oscilloscope trace shown in the diagram is observed.

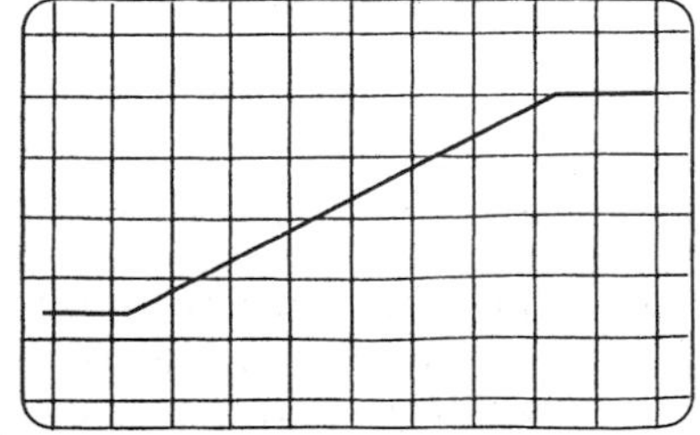

(a) Estimate the time t.
(b) Estimate the change of p.d. across the capacitor.
(c) What is the rate of change of p.d. ($\mathrm{d}V/\mathrm{d}t$)?
(d) What can you say about the charging current?

18.11 A capacitor is joined in series with a resistor, a low resistance microammeter, a switch and a 12 V battery as shown in the diagram on the next page.
(a) When the switch is first closed, what is **(i)** the p.d. across the capacitor, **(ii)** the p.d. across the resistor **(iii)** the current **(iv)** the charge on each of the capacitor plates?

(b) Calculate the same quantities as in **(a)** when the switch has been closed for some time so that steady conditions have been reached.

(c) If the capacitor had continued to charge at its initial rate, how long would it have taken to become fully charged?

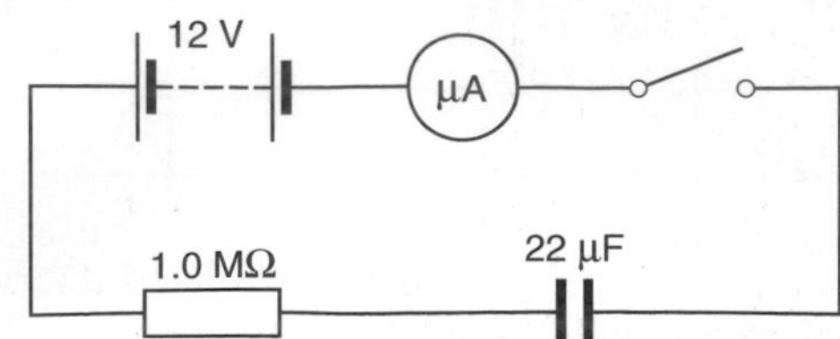

18.12 In the previous question, at a certain instant after the switch has been closed the microammeter reads 9.5 μA. Calculate, for this instant

(a) the p.d. across the resistor

(b) the p.d. across the capacitor

(c) the charge on each of the capacitor plates.

18.13 Suppose in the circuit diagram for question 18.11 the resistor was a variable resistor with a maximum value of 1.0 MΩ. Explain how you could use the circuit to charge the capacitor at a steady rate.

If the capacitance of the capacitor was not known, what measurements would you take and what addition(s) to the circuit would be required in order to measure the capacitance of the capacitor?

18.14 A vibrating reed is used to connect a capacitor alternately to a battery and to a meter as shown in the diagram. In this way the capacitor is fully charged by the battery and fully discharged through the meter 50 times per second.

If the e.m.f. of the battery is 12 V and the meter registers an average current of 2.4 mA, what is the capacitance of the capacitor?

reed vibrating at frequency of a.c. supply

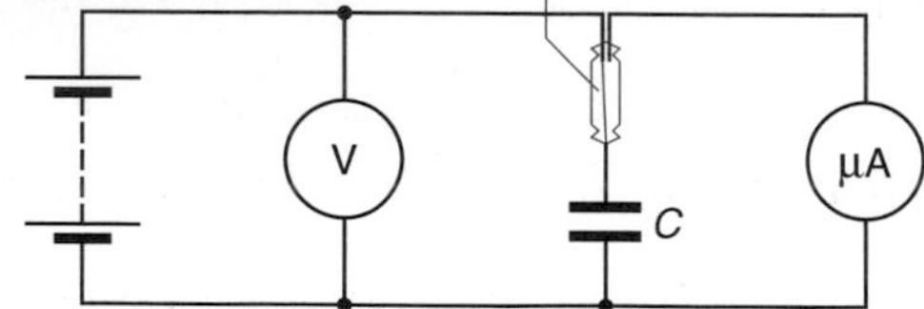

18.15 A capacitor of capacitance 180 pF is charged to a potential difference of 12 V and then discharged through a sensitive meter. This sequence of operations is repeated using a special switch 250 times per second.

What is the average current registered by the meter?

18.16 The leaf electrometer shown in the photograph has a capacitance of 12 pF. The graph shows how the deflection θ of the leaf varies with the p.d. V between the cap and the case.

(a) What is the p.d. indicated by a deflection **(i)** of 60° **(ii)** of 30°?

(b) If it takes 75 minutes for the deflection to fall from 60° to 30°, estimate the average leakage current during this time.

(c) Hence estimate the electrical resistance between the cap and the case of this leaf electrometer.

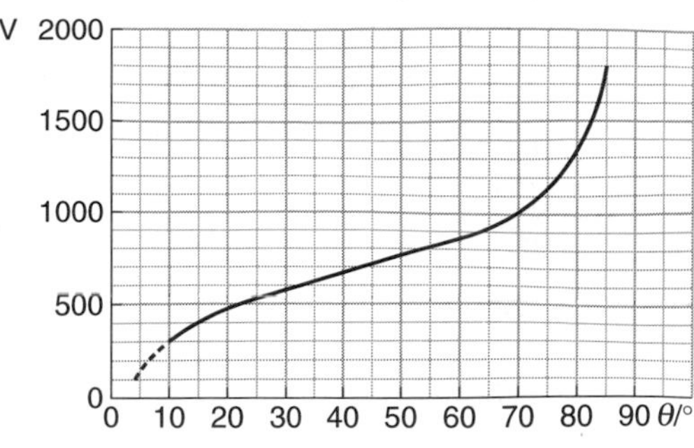

18.17 Three capacitors of capacitances 2.0 μF, 3.0 μF and 6.0 μF are joined **(a)** in parallel **(b)** in series.
What is the combined capacitance in each case?

18.18 Sketch a graph showing how the charge q in a 1.5 μF capacitor varies with the potential difference V across it for values of p.d. from 0 up to 30 V.

18.19 Two capacitors are arranged in series as in the diagram. A battery is connected between P and Q and it is found that the charge on plate w is +60 μF.

(a) Explain why the charge on plate z of the other capacitor is –60 μF.
[Hint: consider the current in the wires leading from the battery to the capacitors.]

(b) What are the charges on plates x and y?

(c) Calculate the p.d. across each capacitor.

(d) What is the p.d. between P and Q?

(e) Calculate the capacitance of a single capacitor placed between P and Q which would store 60 μF of charge when connected to the same battery.

30 μF 60 μF
P w x y z Q

18.20 A 22 μF capacitor is connected in series with a 47 μF capacitor to a battery. Say which of the following statements are true and for each of the true statements explain why it is true.

(a) Each capacitor has the same charge.

(b) The charge on the 22 μF capacitor is about half the charge on the 47 μF capacitor.

(c) The p.d. across the 22 μF capacitor is about half the p.d. across the 47 μF capacitor.

18.21 A steady p.d. of 200 V is maintained across a combination of two capacitors, A of capacitance 2.0 μF and B of capacitance 0.50 μF in series.
What is

(a) the combined capacitance

(b) the charge stored in each capacitor

(c) the potential difference across each capacitor?

18.22 If you have three capacitors of capacitance 3.0 μF, 6.0 μF and 8.0 μF, how could you produce a combination of capacitors of capacitance 10 μF?

18.23 In the arrangement shown

(a) work out the p.d. across each capacitor

(b) calculate the charge in each capacitor.

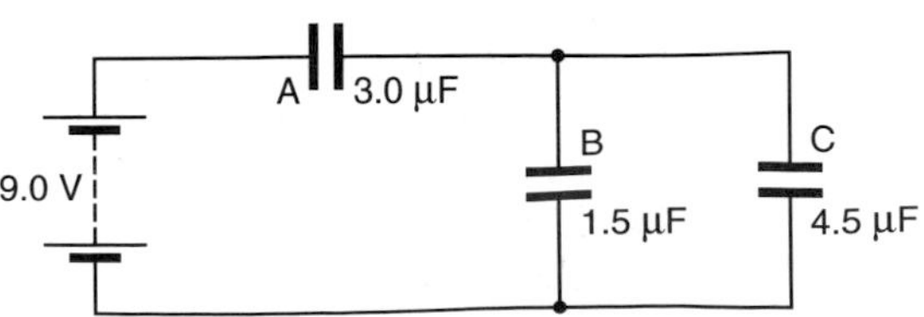

18.24 A 47μF capacitor C_1 is charged from a 6.0 V battery and disconnected. It is then connected across a 22 μF capacitor C_2.
(a) Calculate the initial charge on C_1.
(b) What is the total final charge on the two capacitors?
(c) What is the total capacitance of these two capacitors connected in parallel?
(d) Calculate the common voltage across C_1 and C_2.
(e) What are the final charges on C_1 and C_2?

18.25* A capacitor of unknown capacitance is charged by connecting it across a 30 V supply. It is then disconnected from the supply and connected to the input terminals of a digital voltmeter. The voltmeter has a capacitor of capacitance 0.47μF connected across its terminals. The voltmeter registers 0.25 V.
(a) Calculate the charge delivered to the 0.47 μF capacitor.
(b) Show that the unknown capacitance is 4.0 nF.
(c) Calculate the charge which remains on the unknown capacitor.
(d) Calculate the percentage of the charge on the unknown capacitor that remains on it after it has been connected across the digital voltmeter.
(e) Explain how your calculations enable a high resistance digital voltmeter to be adapted to measure charge, i.e. to be used as a coulombmeter.

18.2 Energy and capacitors

In this section you will need to

- understand that a capacitor can store energy
- use the equations for the energy stored in a capacitor

$$E = \tfrac{1}{2}qV = \tfrac{1}{2}CV^2$$

- understand that there is a useful analogy between a capacitor and a spring.

18.26 A capacitor of capacitance 22 μF is charged by connecting it to a 400 V supply, and is then discharged.
(a) Calculate the energy transferred during the discharge.
(b) If the discharge takes 10 μs, what is the average power of the discharge?

18.27 The graph shows how the p.d. across a capacitor changes as it is charged.
(a) What is the capacitance of the capacitor?
(b) **(i)** Calculate the energy stored in the capacitor when it has a charge of $q/\mu C$: 1, 2, 3, 4, 5.
(ii) Sketch a graph of energy stored against charge.
(iii) What shape would a graph of energy against p.d. have?

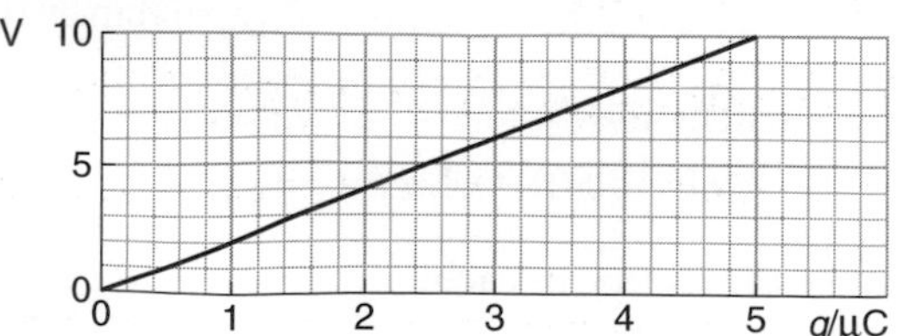

18.28 **(a)** A capacitor of capacitance 16 μF is connected across a 150 V d.c. supply. What is the charge on the capacitor plates, and what energy is stored?

(b) A second (initially uncharged) capacitor of capacitance 8.0 μF is now connected in parallel with the first. Calculate **(i)** the total capacitance **(ii)** the new potential difference across the capacitors **(iii)** the energy now stored in the system.

(c) How do you account for the change in the energy stored?

18.29 An 8.0 μF capacitor is charged by joining it to a 500 V supply through a resistor.

(a) What charge flows through the supply and the resistor?

(b) How much electrical energy is taken from the supply?

(c) How much electrical energy is stored in the capacitor?

(d) How do you account for the difference between these two amounts?

18.30 A capacitor of capacitance 22 μF is charged to a p.d. of 200 V and then isolated from the supply.

(a) What is the energy stored in it?

(b) If an identical capacitor, initially uncharged, is joined across it, what is the energy now stored in the pair of capacitors? Comment on your answer.

18.31 In the circuit shown, calculate the energy stored in the 4.0 μF capacitor

(a) with the switch S closed

(b) with the switch S open.

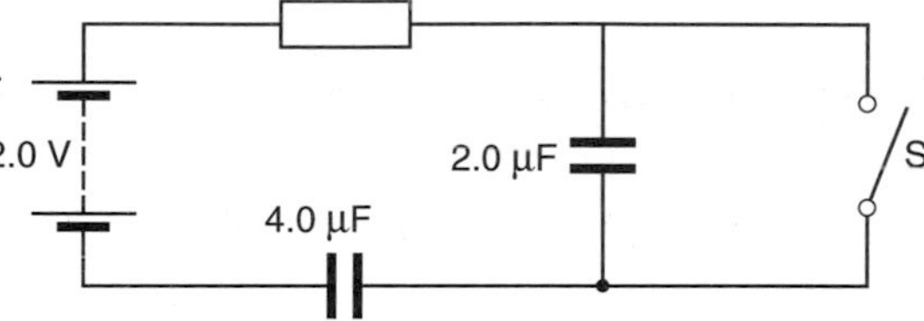

18.32 The diagram shows a water circuit which can be used as an analogy for an electrical circuit. The tank contains a rubber membrane through which the water cannot pass but which stretches further and further the harder the pump works.
Sketch the water circuit and add alongside it the analogous electrical circuit using standard symbols. Write a few words explaining the way in which each component of the water circuit acts like its partner in the electrical circuit.

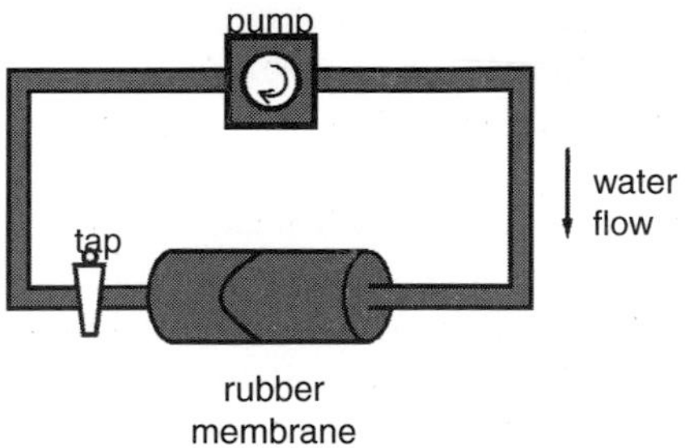

18.3 Capacitor design

In this section you will need to

- use the equations $C = \varepsilon_0 A/d$ and $C = \varepsilon_0 \varepsilon_r A/d$ for the capacitance of a parallel plate capacitor.

18.33 Sketch two graphs to show how the capacitance C of a parallel plate capacitor varies **(a)** with the area A of the plates **(b)** the separation d of the plates.

18.34 Calculate the capacitance of a parallel plate capacitor made from two circular metal plates of diameter 16 cm separated by an air gap of 0.30 mm. [Use data.]

18.35 The diagram shows a variable capacitor such as you will find connected to the tuning knob of a radio. The capacitance of this capacitor can be varied from 80 pF to 320 pF.

(a) Explain how turning the knob can vary the capacitance.

(b) Suggest how, using the same basic design, a capacitor with a wider variation of capacitance could be designed.

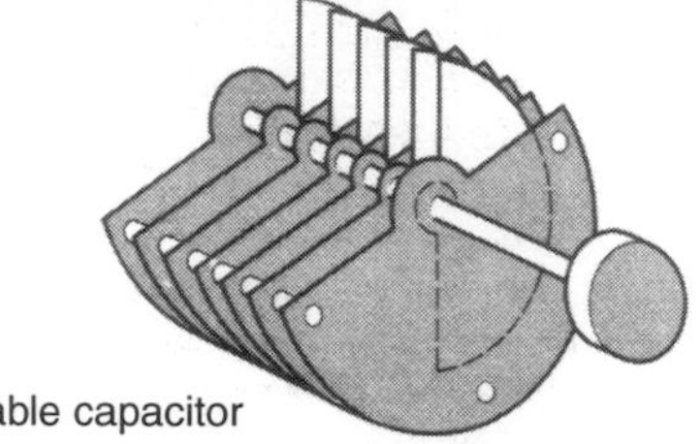

variable capacitor

18.36 The relative permittivity of polythene is 2.4. Calculate the dimensions of the plates of a parallel plate capacitor of capacitance 120 pF formed by two square plates separated by a film of polythene 0.15 mm thick.

18.37 The diagram shows a computer keyboard key. The metal plates, which measure 4.0 mm by 4.0 mm, are held apart by a squashy insulating material of relative permittivity 3.5 forming a parallel plate capacitor. When the key is depressed the separation of the plates decreases from 5.0 mm to 0.20 mm.
Calculate the change in capacitance of the key when it is depressed.

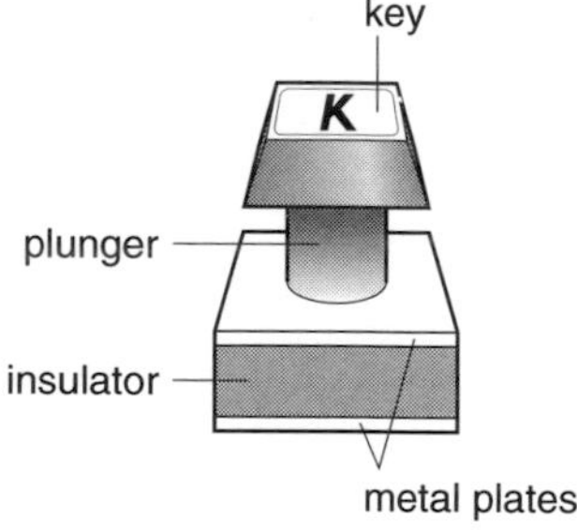

18.38 We live inside a giant capacitor! Its plates are the Earth's surface and the ionosphere, a conducting layer of atmosphere beginning at an altitude of about 60 km.

(a) What is the capacitance of this giant capacitor? Take the radius of the Earth to be 6.4×10^3 km. [Hint: you can treat it as a parallel plate capacitor.]

(b) When the p.d. between the Earth's surface and the ionosphere is 6.0 MV, how much energy is stored in this giant capacitor?

18.39 Two square metal plates (0.10 m by 0.10 m) are arranged horizontally one above the other 2.5 mm apart. The lower plate is earthed, and the upper one is raised to a potential of 0.50 kV and then insulated. Calculate

(a) the capacitance of the arrangement

(b) the charge carried on each plate

(c) the energy stored in the arrangement.

If the upper plate is now raised (keeping it insulated) until the gap between the plates is 5.0 mm, calculate

(d) the new potential of the upper plate

(e) the energy now stored.

By what means has the extra energy been supplied to the apparatus?

18.4 Capacitor discharge

In this section you will need to

- draw graphs showing the variation of charge and potential difference against time for the discharge of a capacitor through a resistor
- use the equation $I = \Delta q/\Delta t = C\Delta V/\Delta t$ for the rate of flow of charge from a capacitor
- understand the equation $dq/dt = -q/CR$ for the discharge of a capacitor through a resistor, the solution of which is called an exponential decay curve
- remember that the product CR is called the time constant of the circuit and that the time for the charge to halve is $CR\ln 2 \approx 0.7CR$
- use the equation

$$q = q_0 e^{-t/CR}$$

18.40 The graph shows the charge on a capacitor as it discharges through a resistor. What is the current in the resistor **(a)** at $t = 0$ **(b)** at $t = 15$ s? In each case show how you obtain your answer.

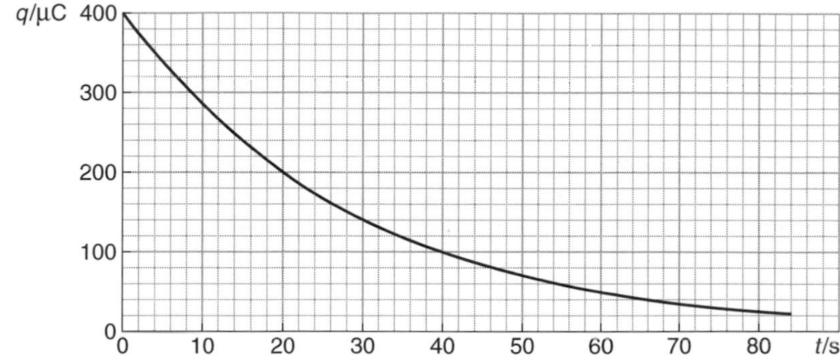

18.41 A charged capacitor is connected to a resistor. Sketch graphs to show the variation with time, as the capacitor discharges through the resistor, of
(a) the charge on the capacitor **(b)** the p.d. across the capacitor
(c) the p.d. across the resistor **(d)** the current in the resistor.

18.42 The graph used in question 18.40 is one showing exponential decay.
(a) Use the graph to find the time constant for this C-R circuit.
(b) Sketch a graph of current against time for this discharge.

18.43 Show that the unit of CR is the second.

18.44 A 47 μF capacitor is charged to 6.0 V and then discharged through a 1.0 MΩ resistor.
(a) What is the initial charge stored in the capacitor?
(b) Calculate the initial current and hence find the time t_d the capacitor would take to discharge fully if this discharging current were to remain constant.
(c) What is the product CR, the time constant, for this circuit?
(d) Show that, in general $CR = t_d$.

18.45 A poor-quality 100 μF capacitor, whose insulation is not perfect, is charged and connected to a digital voltmeter (of very high resistance). It is found that the p.d. across the capacitor falls to half its initial value in 300 s. What is
(a) the time constant of this discharge process
(b) the resistance of the insulation of the capacitor?

18.46 In one method for measuring the speed of a rifle bullet v, the bullet is made to break strips of conducting foil first at A and then at B, where the distance AB is d as shown in the diagram.
(a) Design a *CR* circuit in which opening a switch at A starts the discharge of a capacitor and opening a switch at B stops the discharge. Add a voltmeter which would enable you, if *R* and *C* are known, to find the time taken by the bullet to get from A to B.
(b) If v is about 300 m s^{-1}, suggest suitable values for d, *R* and *C* which would enable v to be measured. [Hint: you would wish *C* to lose about half its charge during the experiment.]

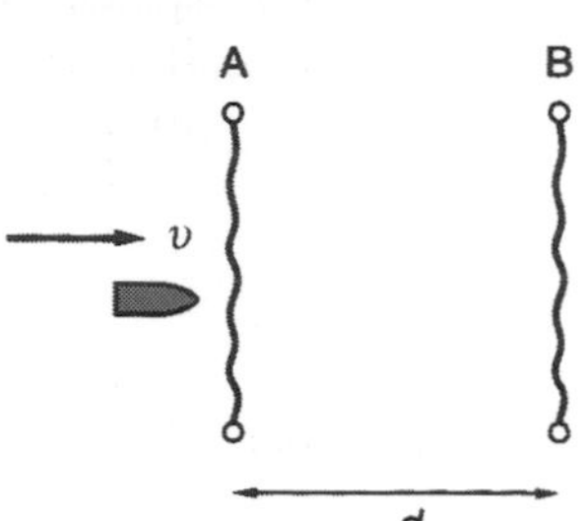

18.47 In one time constant the charge on a capacitor discharging through a resistor falls to 0.37 of its initial value.
How many time constants must elapse before a capacitor in a *CR* circuit discharges 99% of its initial charge?

18.48 The voltage across a capacitor of capacitance 220 μF is monitored as it discharges through a resistor. Use the graph to calculate
(a) (i) the initial rate of change of voltage across the capacitor
(ii) the initial rate of flow of charge, i.e. the current, through the resistor.
(b) What is the current in the resistor after 4.0 s?

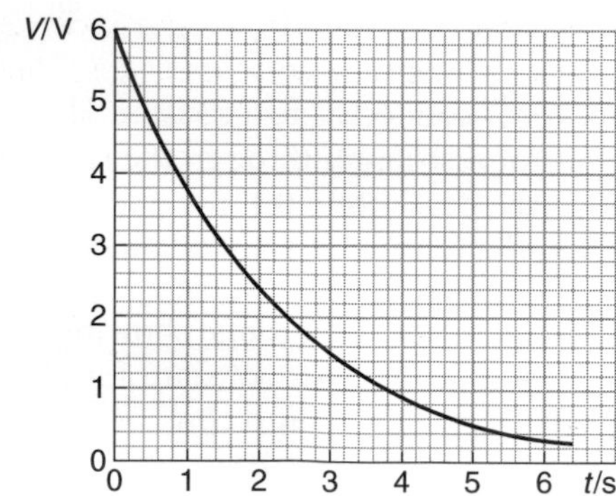

18.49 The graph used in the previous question is one showing exponential decay. What is **(a)** the half-time period **(b)** the time constant, for this circuit?

18.50 The diagram shows a capacitor and a resistor connected in series with a switch and a 1.5 V cell. At time $t = 0$ the switch is closed and the graph shows how the voltage across the resistor then varies with time. During this process the voltage across the resistor and the voltage across the capacitor are related by the equation $V_R + V_C = 1.5$ V.
Sketch the variation with time of **(a)** the voltage across the capacitor
(b) the charge on the capacitor **(c)** the current in the resistor.

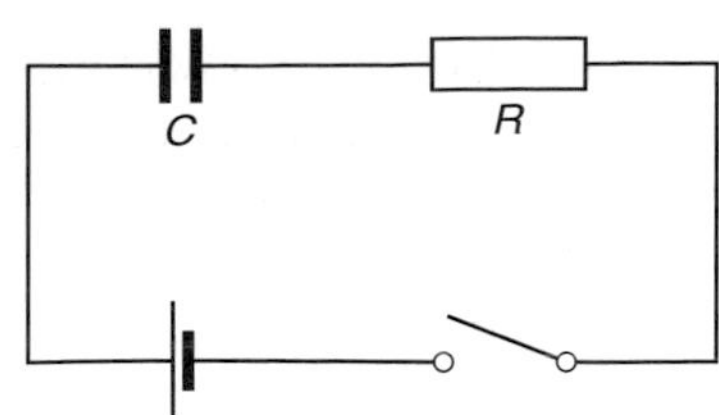

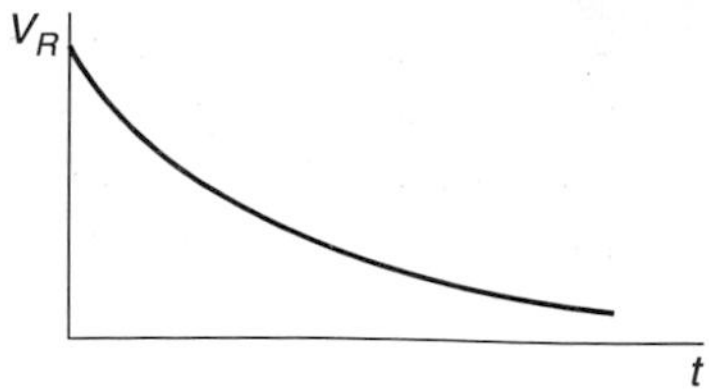

18.51 A capacitor of capacitance 22 μF is charged to 12 V and isolated. The charge leaks away through a resistance of 10 MΩ until the p.d. across the capacitor is 8.0 V.

(a) **(i)** Calculate the charge which leaks off the capacitor.

(ii) Assuming that the average p.d. driving the discharge is 10 V, what is the average leakage current?

(iii) Deduce how long the charge took to leak away.

(b) In reality the p.d. falls exponentially according to the equation $V = V_0 e^{-t/CR}$. Use this equation to calculate how long the charge took to leak away.

(c) What is the percentage difference between your answers to **(a)** and **(b)**?

18.52 A capacitor was joined across a digital voltmeter and charged to a potential difference of 1.00 V. The p.d. V was then measured at 20 s intervals and at time $t = 30$ s a resistor of resistance $R = 1.5$ MΩ was connected across the capacitor. The following results were obtained:

t/s	0	20	40	60	80	100	120
V/V	1.00	1.00	0.81	0.54	0.35	0.23	0.15

(a) What is the current in the resistor at $t = 30$ s?

(b) Plot a graph of V against t, and measure the rate of decrease of V immediately after $t = 30$ s.

(c) Hence calculate the capacitance of the capacitor.

(d) **(i)** Plot a graph of ln(V/V) against t to demonstrate the exponential fall of p.d.

(ii) deduce the time constant RC of the decay process.

(e) Use your value for the time constant to calculate a second value for the capacitance of the capacitor.

(f) Explain which method of finding C gives the more reliable value.

19 Electromagnetism

Data: electronic charge $e = 1.60 \times 10^{-19}$ C

magnetic field constant $\mu_0 = 4\pi \times 10^{-7}$ N A^{-2}

mass of electron $m_e = 9.11 \times 10^{-31}$ kg

19.1 Magnetic fields

In this section you will need to

- draw, in two dimensions, magnetic field (flux) patterns near a long straight wire, a flat coil and a long solenoid
- remember a rule for predicting the relation between the sense of a current and that of the field lines
- understand that magnetic flux density B is a vector quantity
- use the equation $B = \mu_0 I/2\pi r$ for a straight wire and $B = \mu_0 nI$ for a solenoid.

19.1 A student is plotting the magnetic field pattern around a magnet. The three parts of the diagram show features to be found in different parts of his diagram.
In **(a)** he has two lines of force crossing: explain why this cannot be correct.
In **(b)** he has two adjacent lines of force running in opposite directions: what can you say about the field between them?
In **(c)** he has two lines of force which are converging: what can you say about the field between them?

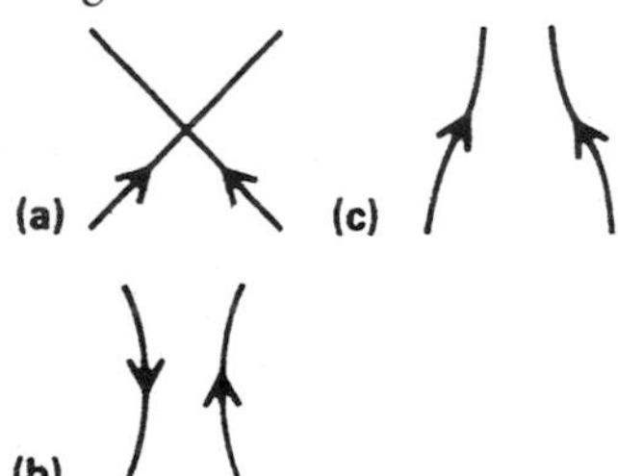

19.2 Two ceramic (black magnadur) magnets are placed on a table. They are made to stand with unlike poles facing each other. Sketch the magnetic field pattern between them, in the plane of the table, when they are about 40 mm apart. Indicate the region where the field is approximately uniform.

19.3 **(a)** Copy the diagram, which shows wires carrying currents in and out of the paper, and sketch the separate magnetic field patterns in the plane of the paper caused by these currents.

⊗ ⊙

(b) On a second copy of the diagram sketch the resultant magnetic field pattern. [As is usual, the lines should be close together where the field is stronger.]

(c) Sketch the magnetic field pattern in and around a current-carrying solenoid.

19.4 The external magnetic field of the Earth is similar to that which would be produced by a giant flat current-carrying coil placed inside the Earth. The radius of the coil would be about half that of the Earth and it would be tilted at about 11° to the equator. The current in the coil would be such as to produce a magnetic field emerging from somewhere near the south geographic pole. Draw a diagram of the Earth and its external magnetic field.

19.5 At a latitude of 55° the Earth's magnetic flux density is 50×10^{-6} T or 50 μT. In the northern hemisphere this field is directed downwards into the Earth at an angle of 65° to the horizontal. Calculate
(a) the horizontal component B_h **(b)** the vertical component B_v of this field.

19.6 The diagram shows how the magnetic flux density B varies along the axis of a solenoid.
(a) Describe how you would produce the experimental data needed to plot a graph of this kind.
(b) Express the magnetic flux density at the ends of the solenoid as a fraction of that at its centre.
(c) By considering the magnetic flux density at the join when two such solenoids are placed end-to-end, argue that your answer to **(b)** is theoretically predictable.

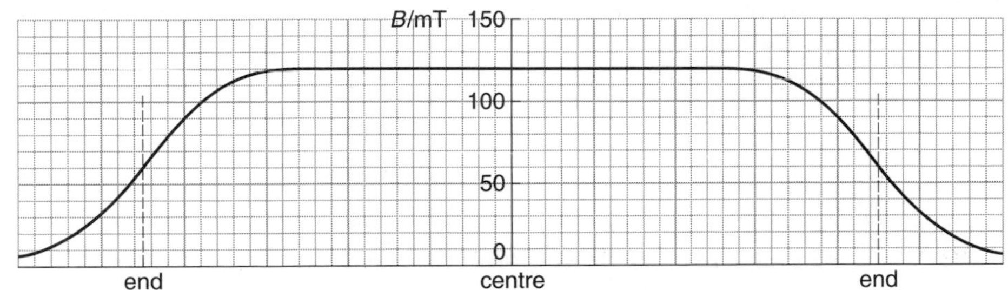

19.7 What is the magnetic flux density near the centre of a long solenoid which has 1200 turns, is 0.50 m long and carries a current of 3.0 A?

19.8 A solenoid used in magnetic resonance imaging (MRI) is wound from a niobium-titanium superconducting wire 2.0 mm in diameter, with adjacent turns separated by an insulating layer of negligible thickness. (The solenoid is 2.4 m long and 95 cm in diameter.) Calculate the current in the superconducting wires necessary to produce a magnetic field of flux density 1.5 T at the centre of the solenoid.

19.9 Show that the units of the equations $B = \mu_0 nI$ and $B = \mu_0 I/2\pi r$ are consistent.

19.10 The magnetic flux density at a distance of 10 mm from a long staight wire carrying a current of 10 A is 0.20 mT. What is the magnetic flux density
(a) at a distance of 6.0 mm, with the current still 10 A
(b) at a distance of 10 mm, if the current is 3.0 A?

19.11 A long straight wire carries a current of 10 A. Draw a graph with labelled axes to show how the magnetic flux density varies with distance r from the wire for values of r from 5 mm to 50 mm.

19.12 Two parallel straight wires A and B carry currents in opposite directions: the currents are respectively, 2.0 A and 3.0 A. The wires are 0.10 m apart.
At what distance from wire A will the resultant magnetic flux density be zero?

19.13 The diagram shows a flat coil in which there is a steady current. Sketch a graph to show the reading you would expect a Hall probe to record if you moved it, with its plane parallel to the plane of the coil, along the line AOA′.

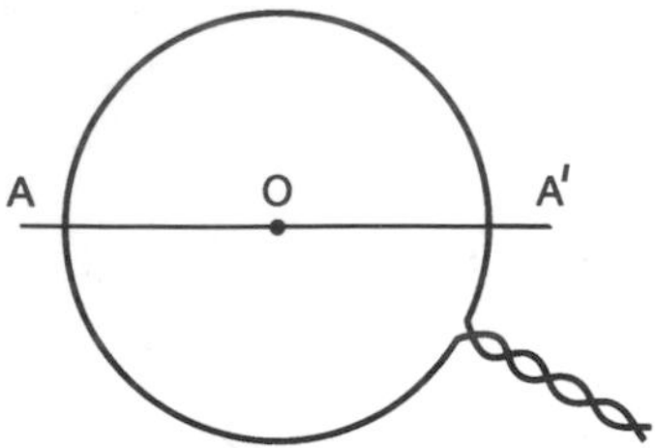

19.14 The cross-channel direct current cable between the United Kingdom and France may carry a maximum current of 14.8 kA. A typical depth below the surface of the sea is 50 m.
(a) What magnetic flux density is created at the surface of the sea by the maximum current?
(b) Your answer to **(a)** is a significant fraction of the strength of the Earth's magnetic field but in practice ships need make no allowance for it in navigating. Why not?

19.15* The photograph shows a coil fitted to an aircraft during the second world war. The purpose was to detonate magnetic mines. A change in the vertical magnetic field of 5 μT was necessary to detonate the mine.
For a single-turn coil of radius r, the magnetic flux density on the axis of the coil a distance d from its centre is given by $B = \mu_0 I\, r^2/2(d^2 + r^2)^{3/2}$.
Supposing that a mine is 15 m below the water surface and the aircraft is flying 20 m above the sea, what current would be needed in a coil of radius 9.0 m?

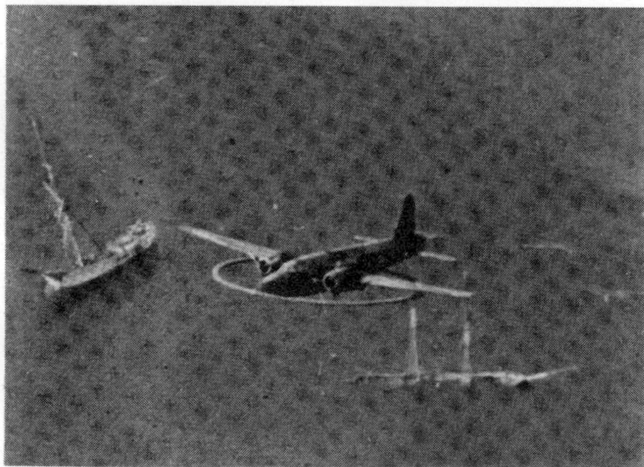

19.2 Magnetic forces

In this section you will need to

- remember the rule for predicting the direction of the magnetic force on a current-carrying wire placed in a magnetic field
- draw magnetic fields lines which show the direction of fields into the plane of a diagram as ×
- use the equation $F = B_{\perp}Il$ for the force on a current-carrying wire

- use the equation $F = B_{\perp}qv$ for the force on a charged particle moving in a magnetic field
- understand that a charged particle deflected by a magnetic field will move in a circular path
- use the expression v^2/r for centripetal acceleration.

19.16 What is the unit of magnetic flux density in terms of
(a) N, A and m **(b)** the base units of the SI?

19.17 A horizontal conductor of length 50 mm carrying a current of 3.0 A lies at right angles to a horizontal magnetic field of flux density 0.50 T.
(a) What is the size of the magnetic force on it?
(b) Draw a diagram to show the directions of the current, the field and the force.

19.18 A short horizontal length of wire carrying a current is held in the gap between a pair of ceramic magnets on an iron yoke placed on a top pan balance. When there is a current of 3.2 A in the wire the balance reading changes by 0.43 g. If the length of the wire between the magnets is 40 mm, calculate the magnetic flux density in the gap between the magnets.

19.19 A power cable crosses a gap of 80 m between two pylons and carries a current of 2500 A. The Earth's magnetic field in the region of the pylons is 48 μT at an angle of 20° to the vertical. If the cable is perpendicular to the Earth's field calculate
(a) the force on the cable **(b)** the vertical force on the cable.

19.20 There is a current of 5.0 A through each of the conductors OA, OB, OC and OD shown in the diagram. The conductors are in a magnetic field of flux density 0.15 T parallel to the plane of the diagram. What is the size and direction of the magnetic force on each conductor?

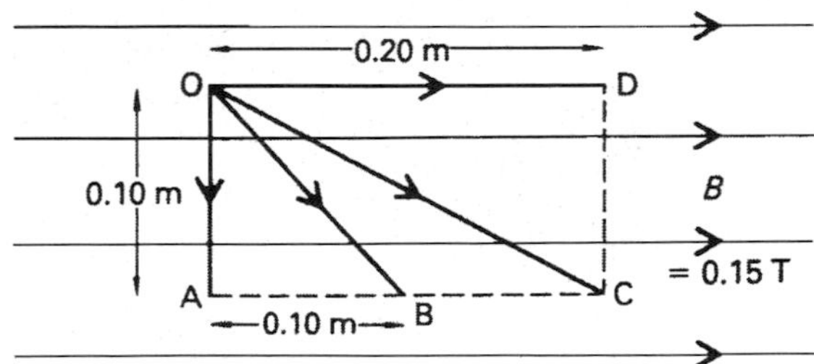

19.21 The diagram on the next page shows a rectangular 3-turn coil PQRS in which there is a steady current $I = 0.40$ A. The coil is free to rotate about an axis perpendicular to a magnetic field of flux density $B = 0.15$ T. On the right are diagrams showing, end-on, the coil in two positions as it rotates.
(a) Copy the two end-on diagrams and add the magnetic forces acting on the top and bottom of the coil in each case.
(b) If PQ = RS = 24 cm and PS = QR = 15 cm, calculate **(i)** the size of each of these magnetic forces **(ii)** the turning effect or torque which they produce on the coil in each case.

(c) Discuss the effect, if any, which the magnetic forces acting on the sides PS and QR have on the coil as it rotates.

(d) Explain why, if the coil is to continue to rotate, i.e. the system is to act as a d.c. motor, the current must be arranged always to enter the upper side of the coil.

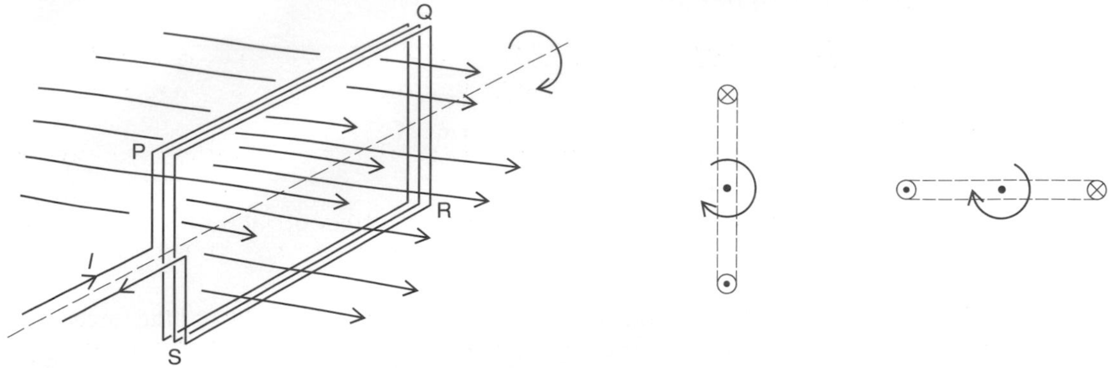

19.22 An ampere is defined as that current which, if flowing in each of two infinitely long parallel wires, causes one of them to exert a force of 2×10^{-7} N on a one-metre length of the other.

(a) Draw two parallel wires, a distance r apart, with a current I in the same direction in each.

(b) Add the magnetic field of *one* of the wires and write down the strength of this field at the second wire.

(c) Write an expression for the magnetic force on a length l of the second wire.

(d) Use the definition of the ampere to show that $\mu_0 = 4\pi \times 10^{-7}$ N A^{-2}.

19.23 When a current is switched on in a solenoid, do the coils attract or repel each other? Describe how, given any apparatus you might reasonably expect to find in your laboratory, you could investigate this effect quantitatively.

19.24 Use the equations $F = BIl$ and $I = nAqv$ to show that the force on a moving charge is given by $F = Bqv$. [Hint: the number of particles in a length l of wire of cross-sectional area A is nAl, when the number of particles per unit volume is n.]

19.25 **(a)** What is the size of the magnetic force on an electron moving with a velocity of 2.0×10^7 m s^{-1} at right angles to a uniform magnetic field of flux density 15 mT?

(b) Show on a diagram the directions of the field, the velocity and the force.

19.26 A stream of helium nuclei each carrying a charge of 3.2×10^{-19} C is travelling with a speed of 1.5×10^7 m s^{-1} at right angles to a magnetic field of 2.0 T.

(a) What is the magnetic force on each particle?

(b) What effect does this force have on the path followed by the nuclei?

(c) If the mass of each nucleus is 6.6×10^{-27} kg, calculate the acceleration of each nucleus.

19.27 An electric current of 5.0 A is passing through a copper wire of radius 0.50 mm which lies perpendicular to a magnetic field of 0.25 T. The density of free electrons in the wire is 1.0×10^{29} m^{-3}. Calculate

(a) the drift speed of the electrons [Use $I = nAqv$.]

(b) the magnetic force acting on each electron
(c) the number of free electrons there are in a 0.40 m length of the wire
(d) the force on the wire.
Compare your answer to **(d)** with the force F given by the expression $F = BIl$ for this length of wire.

19.28 The diagram shows a uniform magnetic field: its direction is into the paper and its strength is 0.25 T. An electron at P is fired in the direction shown with a speed of 2.0×10^6 m s^{-1}.
(a) On a copy of the diagram mark the direction of the magnetic force on the electron and calculate the size of the force.
(b) Explain why the electron will move in a circle at a constant speed.
(c) Taking the centripetal acceleration as v^2/r, calculate the radius of the electron's circular path.
(d) How long will the electron take to make one complete circular orbit?
(e) A second electron is fired from P in the same direction with twice the speed. Calculate the radius of its path.
(f) How long will it take the second electron to complete one circular orbit? Comment on your answer.

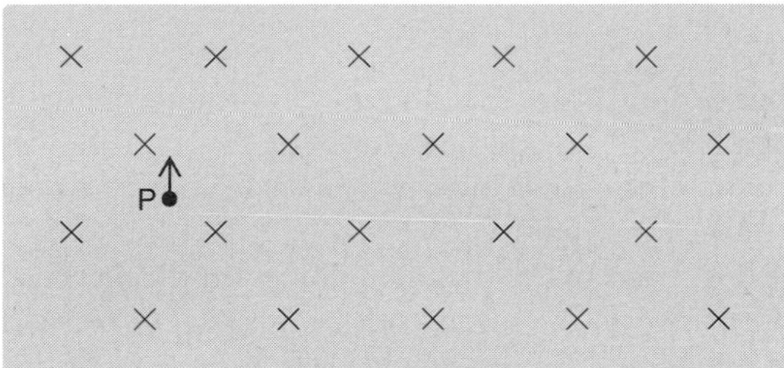

19.29 If a particle with a charge q and mass m is moving with speed v at right angles to a magnetic field of flux density B, and moves in a circular path of radius r, derive an expression
(a) for r in terms of q, m, v and B
(b) for the period T in terms of B, q and m.

19.30 In one type of mass spectrometer two beams of singly ionised neon atoms are fired perpendicular to a magnetic field B. The two beams have the same speed v but are of different isotopes of neon with masses $20m$ and $22m$.
What is the ratio of the radii in which the isotopes move? State which has the greater radius.

19.31* The charge-to-mass ratios, or specific charges, of electrons and α-particles are 1.76×10^{11} C kg^{-1} and 4.82×10^7 C kg^{-1} respectively.
Calculate the flux density of the magnetic field that will deflect into an arc of radius 0.10 m
(a) electrons moving at 2.0×10^7 m s^{-1}
(b) α-particles moving at 2.0×10^7 m s^{-1}.

19.3 Electromagnetic induction

In this section you will need to

- use the equation $\mathcal{E} = B_{\perp}lv$ for the e.m.f. produced across a conductor moving in a magnetic field
- use the equation $\Phi = B_{\perp}A$ which defines magnetic flux
- understand that electromagnetic induction involves the transfer of mechanical energy to electrical energy
- understand that e.m.f.s are induced when the magnetic flux linking a circuit, e.g. a loop of wire, is changing
- use Faraday's law of electromagnetic induction: $\mathcal{E} = N(\Delta\Phi/\Delta t)$, i.e. the induced e.m.f. is equal to the rate of change of flux linkage
- remember that the direction of an induced e.m.f. is such as to oppose the change causing it
- understand the principle of the transformer
- use the ideal transformer equation $V_P/V_S = N_P/N_S$ for primary and secondary voltages and turns.

19.32 A teacher demonstrates that an e.m.f. is produced when a wire is moving between the poles of a strong U-shaped magnet. The diagram shows the direction of the field and the way in which the wire is moved. The circle shows an enlargement of part of the wire. On a copy of the inset mark

(a) the directions of the magnetic forces on the positive ion and the electron

(b) the positive and negative ends of this part of the wire

(c) the direction a current would have if the ends of the wire were connected by a circuit none of which passed through the magnetic field.

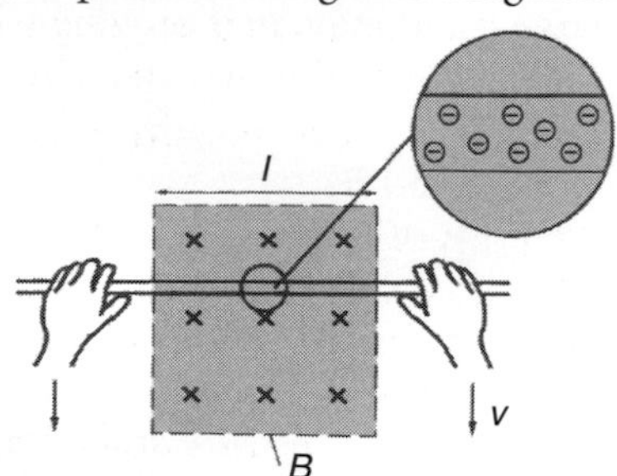

19.33 **(a)** If, in the previous question, there is an induced current in the wire, in which direction is the magnetic force produced by this current?

(b) Is this direction the same as, or opposite to, the direction in which the wire was being moved? Comment on your answer.

19.34 If, in question 19.32 the magnetic field is of flux density 0.12 T, the length of the wire in the field is 25 mm, and the teacher moves the wire at a maximum speed of 1.5 m s^{-1}, what is the maximum e.m.f. produced?

19.35 A large U-magnet is placed on a bench so that the magnetic field between its poles is horizontal. A long wire is connected to a datalogging device.
Draw sketch graphs of V against t as recorded by the datalogger when the wire is moved
(a) at the same steady speed vertically downwards and then vertically upwards between the poles of the magnet
(b) slowly downwards and then quickly upwards between the poles of the magnet
(c) at a steady speed horizontally from one pole of the magnet to the other.

19.36 Show that the units on the two sides of the equation $\mathcal{E} = Blv$ are consistent.

19.37 The wingspan of a Concorde aircraft is 25.6 m. The aircraft cruises at an altitude of 35 000 feet and travels at 550 m s^{-1}.
(a) Explain whether the size of the e.m.f. produced between its wing-tips depends on what direction the aircraft is flying at a particular place.
(b) Calculate the size of the e.m.f. at a place where the vertical component of the Earth's magnetic field is 50 μT.
(c) Where might Concorde be flying if the e.m.f. produced was zero?

19.38 A ceiling fan has four blades each 0.55 m long and rotates about a vertical axis at 96 revolutions per minute. An e.m.f. is induced between the axle of the fan and the tips of the blades as it cuts the vertical component of the Earth's magnetic field which is 48 μT.
(a) Calculate the average speed of a blade, i.e. the speed of the centre of the blade.
(b) Hence find the value of the induced e.m.f.

19.39 Calculate the magnetic flux through a football pitch which measures 110 m by 70 m at a place where the flux density of the Earth's magnetic field is 49 μT or 49 μWb m^{-2} in a direction making an angle of 22° with the vertical.

19.40 A large electromagnet has circular pole pieces of diameter 0.20 m. The total flux produced by the magnet is 0.050 Wb.
Calculate the average flux density between the pole pieces.

19.41 The diagram on the next page shows a wire loop ABCD whose centre is O. It is placed at right angles to a uniform magnetic field of flux density 0.50 T. The dimensions of the loop and the space covered by the field are shown on the diagram.
(a) What is the flux through the loop?
(b) What is the size of the change of flux through the loop when the loop
(i) is moved, in its own plane, 0.20 m to the right
(ii) is moved, in its own plane, 0.30 m to the right
(iii) is rotated, in its own plane, through 90°, about its centre O
(iv) is rotated, about the line BC, through 90°
(v) is rotated, about the line BC, through 180°
(vi) is raised 0.10 m, parallel to the field (i.e. out of the paper).

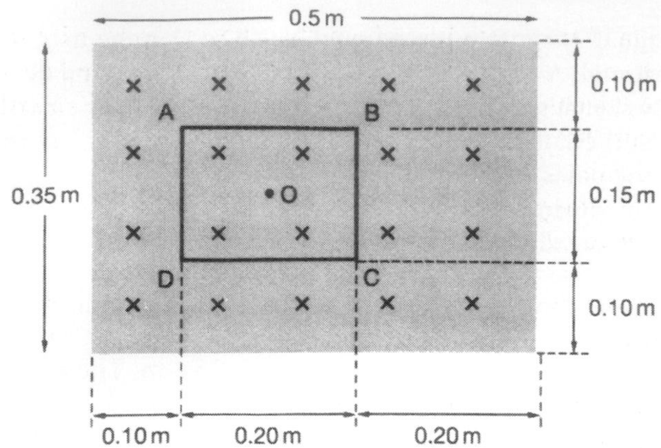

19.42 A bar magnet is moved towards a circular coil as shown in the diagram. An e.m.f. is induced in the coil.

(a) Explain why an e.m.f. is induced.

(b) By considering the direction of the induced current, deduce the direction of the induced e.m.f.

(c) What is the source of the energy which the induced current transfers to the coil?

(d) State the effect of **(i)** moving the magnet more quickly **(ii)** moving the magnet away from the coil.

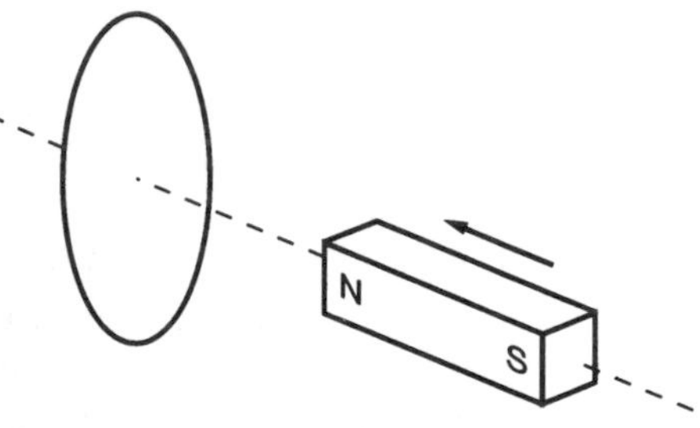

19.43 A circular coil is placed with its axis vertical and a bar magnet, with its axis aligned with the axis of the coil, is held above the coil and then dropped. A datalogger connected to the coil records the e.m.f. induced in the coil at short time intervals and later draws a graph to show how the e.m.f. varies with time.

(a) The diagram shows the graph obtained as the magnet falls through the coil. Explain the shape of the graph.

(b) Give two arguments, one based on forces and one based on energy, to explain why the magnet would take longer to fall if the datalogger were removed and, instead, the ends of the coil were connected together.

(c) Copy the graph, and on the same axes sketch the graphs which would have been obtained if, separately (with the datalogger again connected)

(i) the coil had been replaced with one with twice the number of turns

(ii) the magnet had been dropped from about twice the height.

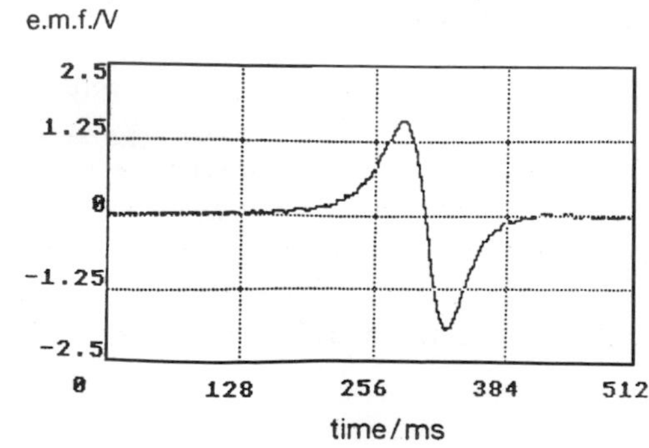

19.44 A metal framed window which measures 0.80 m by 1.2 m, pivots about a vertical side and faces south. When closed its plane is vertical and at right angles to the Earth's magnetic field which has a local horizontal flux density of 18 μT. The window is opened through 90° in 1.5 s. Calculate

(a) the initial flux through the window

(b) the change of flux through it in 1.5 s

(c) the e.m.f. induced in the window frame.

19.45 A ring of copper wire of area 100 cm^2 and electrical resistance 0.012 Ω is placed with its plane at right angles to a uniform magnetic field. If the field is changing at the rate of 2.0 T s^{-1}, calculate

(a) the e.m.f. induced in the ring

(b) the induced current in the ring.

19.46 A circular coil and a copper ring are placed, as in the diagram, flat on a horizontal table.

(a) What is the direction of the current in the ring if the current in the coil is

(i) clockwise and increasing **(ii)** clockwise and decreasing

(ii) anticlockwise and increasing **(iv)** anticlockwise and decreasing?

(b) The current in the coil is increased from zero to 4.0 A in 2.0 s. How will the current in the ring be affected if the current in the coil is increased from zero **(i)** to 2.0 A in 2.0 s **(ii)** to 2.0 A in 1.0 s **(iii)** to 4.0 A in 1.0 s?

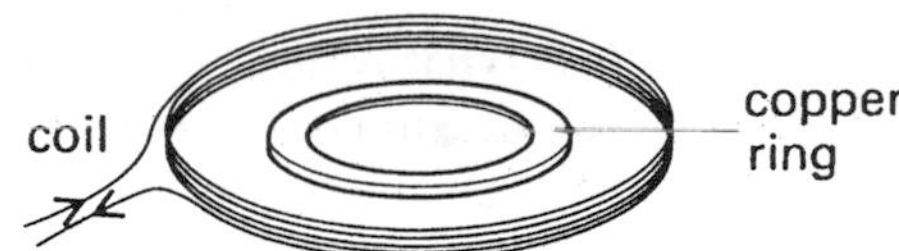

(c) How will the energy converted in the ring be affected if the three processes given in **(b)** occur?

19.47 A circular coil of radius 35 cm contains 90 loops of thin copper wire and is positioned perpendicular to a magnetic field of flux density 0.45 T. The coil is quickly stretched as shown in the diagram until the magnetic flux through the coil is a quarter of its initial value. If this stretching took 0.40 s, calculate the e.m.f. induced in the coil.

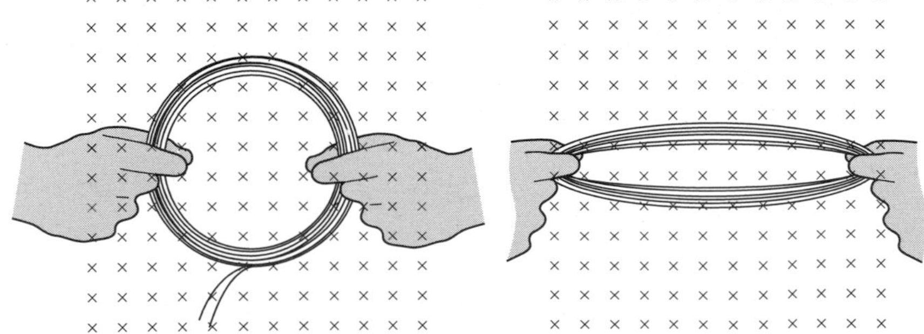

19.48 A coil with 5000 turns of mean area 1.0 cm^2 – a search coil – is placed in a magnetic field. The flux density of the field varies as shown in the graph overleaf.

(a) Explain at what time or times the e.m.f. induced in the coil will be a maximum.

(b) What is the maximum rate of change of the magnetic field in T s^{-1}?

(c) Estimate the maximum e.m.f. induced in the coil.

(d) Assuming that the induced e.m.f. is sinusoidal, sketch its variation with time and comment on the relationship between your sketch and the graph in the diagram.

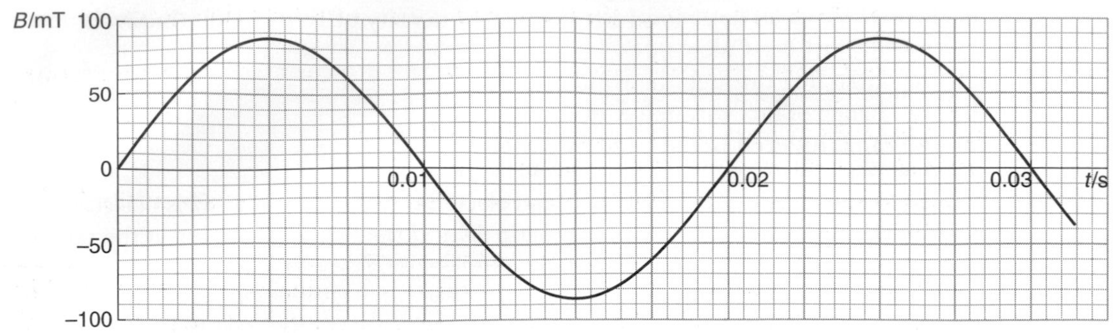

19.49 The diagram shows two straight wires AB and DC, each of length l, linked by fine flexible wires. The wires lie in a plane which is at right angles to a magnetic field of flux density B which passes into the page.
Suppose the wire AB is fixed in position, and you move the wire DC away from it at a speed v. A current will be induced in the wire: call this I.

(a) What is the magnetic force on the wire DC?

(b) What is the force F which you must use to pull the wire DC at a steady speed v.

(c) Calculate your mechanical rate of working P_m.

The current I is produced as a result of the e.m.f. $\mathcal{E}$ induced in the circuit ABCD as you move the wire.

(d) What is the electrical power P_e produced in the circuit?

(e) Why must your mechanical rate of working be equal to the electrical power produced?

(f) Hence derive an expression for $\mathcal{E}$ in terms of B, l and v.

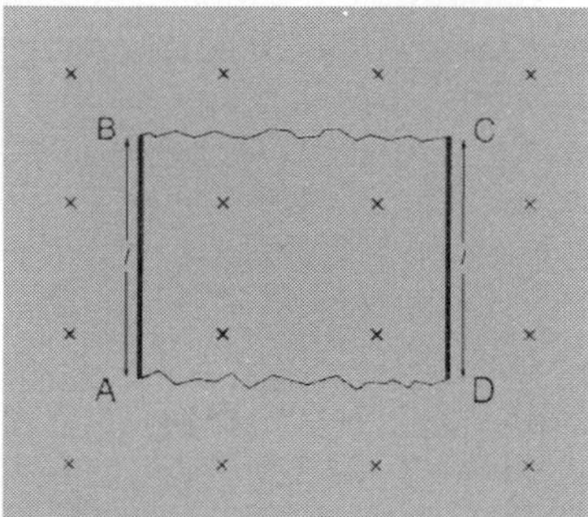

19.50 The diagram gives information about an ideal transformer. Explain as fully as possible, supporting your explanation with calculations, why this transformer is said to be ideal.

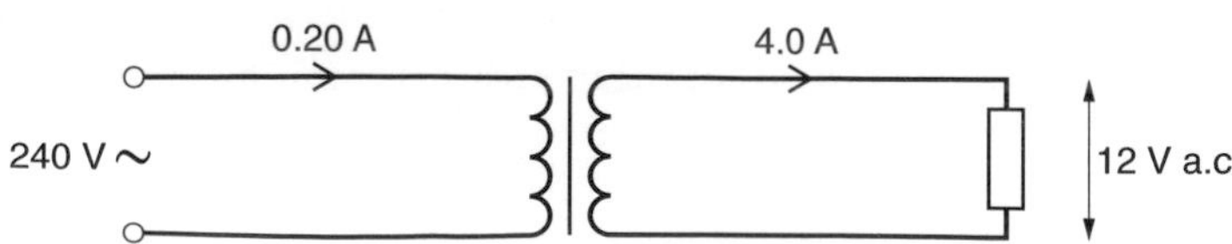

19.51 An ideal transformer has 4800 primary turns and 120 secondary turns. It is used to light a 6.0 V 24 W lamp.
What is **(a)** the primary voltage **(b)** the primary current **(c)** the primary power?

19.52 An overhead projector lamp is labelled '24 V 250 W'. It is connected via a transformer to the 240 V mains supply.
(a) What must be the ratio of turns in the transformer?
(b) Calculate the current in the projector lamp.
(c) Explain which of the following fuses you would use in the primary circuit: 13 A, 10 A, 2 A, 1 A.

19.53 A 12 V alternating supply is connected to the primary terminal of a transformer which steps down the p.d. to 6.0 V. The secondary terminals are connected to a rheostat which is set to a resistance of 10 Ω.
(a) What is the power in the rheostat?
(b) If the resistance of the rheostat is reduced to 5.0 Ω, what does the power in the rheostat become?
(c) Where has the additional power come from?

19.54 A building site has a transformer which steps down the p.d. from 11 kV to 415 V. Assume that the transformers in this question are 100% efficient.
(a) If the number of turns on the primary is 3000, what is the approximate number of turns on the secondary?
(b) The p.d. of 415 V is used to supply a crane which has a maximum power of 60 kW. Calculate the current drawn from the 11 kV supply when this crane is working at maximum power.
The site has a second transformer which, for safety reasons, steps down the p.d. of 415 V to 25 V for some 100 W hand-held lamps.
(c) When five of these hand-held lamps are in use, what current is drawn from the 11 kV supply?

19.55 A laboratory power supply contains a transformer so that the mains p.d. of 240 V may be stepped down to low p.d.s increasing in steps of 2 V from zero to 12 V.
The primary coil has 1200 turns. Explain how the secondary p.d.s may be obtained.

20 Inverse square law fields

Data: gravitational field strength $g = 9.81$ N kg^{-1}

universal gravitational constant $G = 6.67 \times 10^{-11}$ N m^2 kg^{-2}

mass of Earth = 5.97×10^{24} kg

radius of Earth = 6.37×10^6 m

electric field constant $k = 8.99 \times 10^9$ N m^2 C^{-2}

permittivity of free space $\varepsilon_0 = 8.85 \times 10^{-12}$ F m^{-1}

electronic charge $e = 1.60 \times 10^{-19}$ C

20.1 Gravitational and electrical forces

In this section you will need to

- use Newton's law of gravitation for two masses m_1 and m_2 separated by a distance r

$$F = Gm_1m_2/r^2$$

- understand that gravitational forces are only noticeable when one of the attracting bodies is at least of planetary size
- use Coulomb's law for two charges q_1 and q_2 separated by a distance r

$$F = kq_1q_2/r^2$$

- remember that the laws for both gravitational and electrical forces apply to point masses or charges and also to spherical masses and charged spheres.

20.1 **(a)** Calculate the size of the gravitational pull of a sphere of mass 10 kg on a sphere of mass 2.0 kg, when their centres are 200 mm apart.
(b) What is the gravitational pull of the 2.0 kg sphere on the 10 kg sphere?

20.2 A woman of mass 60 kg stands on the Earth's surface.
(a) Calculate, to two significant figures, the pull of the Earth on her
(i) using Newton's law of gravitation
(ii) using the value of g given in the data above.
(b) Sketch a graph to show how the pull of the Earth on her would vary if she were **(i)** 6.4×10^6 m **(ii)** 12.8×10^6 m **(iii)** 19.2×10^6 m above the Earth's surface.

20.3 Calculate **(i)** the gravitational pull of the Earth on **(ii)** the acceleration of
(a) a satellite near the Earth, mass 80 kg, distance from centre of Earth 8.0×10^6 m
(b) a geosynchronous satellite, mass 100 kg, distance from centre of Earth 4.2×10^7 m
(c) the Moon, mass 7.3×10^{22} kg, distance from centre of Earth 3.8×10^8m. [Use data.]

20.4 Show that the unit for G, the gravitational constant, can be expressed as $m^3\ s^{-2}\ kg^{-1}$.

20.5 Calculate the gravitational force of attraction between
(a) two touching lead spheres of radius 50 mm
(b) two touching lead spheres of radius 100 mm.
The density of lead is $1.13 \times 10^4\ kg\ m^{-3}$.

20.6 An astronaut stands on the surface of the Moon. He is carrying a life-support pack of mass 62 kg. The mass of the Moon is 7.3×10^{22} kg and its radius is 1.6×10^6 m. Calculate the push of the pack on him.

20.7 The diagram shows a space vehicle S at a point between the Earth and the Moon at which the pull of the Earth on it is equal in size to the pull of the Moon on it. By equating the forces using Newton's law of gravitation express the mass of the Earth m_E as a multiple of the mass of the Moon m_M.

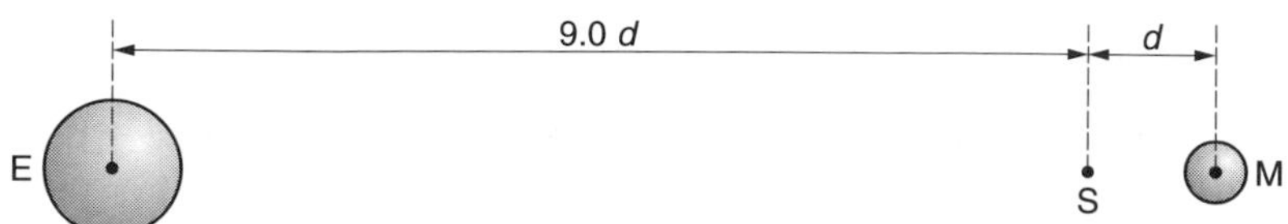

20.8 The weight of a mass of 5.00 kg at the Earth's surface is measured to be 49.1 N.
(a) Use $F = Gm_1m_2/r^2$ with the values of G and the radius of the Earth from the data to calculate a value for the mass of the Earth.
(b) Deduce the mean density of the Earth.
When scientists first measured G they said that they were 'weighing the Earth'.
(c) Explain what they meant.

20.9 Two raindrops falling side by side carry electric charges of +4.0 pC and –5.0 pC, respectively.
(a) What electrical force does one drop exert on the other when they are 12 cm apart?
(b) How many times bigger than your answer to **(a)** is the pull of the Earth on each raindrop, if each has a mass of 15 mg?

20.10 A small sphere carrying a charge of +1.0 nC is situated a distance of 180 mm from another small sphere carrying a charge of +4.0 nC.
(a) What is the size of the electrical force F between the two spheres?
(b) What would be the size of this force if they were moved until they were separated by distances of r/mm = 90, 60, 30, 20, 10?
(c) Sketch a graph to show the variation of F with r in this case.

20.11 A uranium nucleus can be regarded as a sphere of radius 8.0×10^{-15} m containing 92 uniformly distributed protons. What is the force on an α-particle just 'touching' the surface of the nucleus?

20.12 The metal top of a small Van de Graaff machine may be regarded as a sphere of radius 0.15 m. It carries a charge of –4.0 nC.
A small polystyrene sphere supported by a thin insulating thread is held near the metal sphere so that their centres are in the same horizontal plane and a distance 0.20 m apart.

The pull of the Earth on the polystyrene sphere, its weight, is 0.25 mN and the thread hangs at an angle of 45° to the vertical.

(a) Sketch the situation. What is the electrical force on the polystyrene sphere?
(b) Calculate the charge on the polystyrene sphere.

20.13* Two small conducting spheres P and Q, each of mass 1.5×10^{-5} kg, are suspended from the same point O by insulating threads 0.10 m long. When the spheres are charged (with equal charges) they come to rest with both threads inclined at 30° to the vertical, as shown in the diagram.

(a) Show that the force of repulsion between the spheres is 85 μN.
(b) Calculate the distance between the spheres.
(c) What is the charge on each sphere?
(d) When half the charge has leaked off one sphere, but none has leaked off the other, it is found that *both* threads are then inclined at 16° to the vertical. Explain why the angles must be equal.

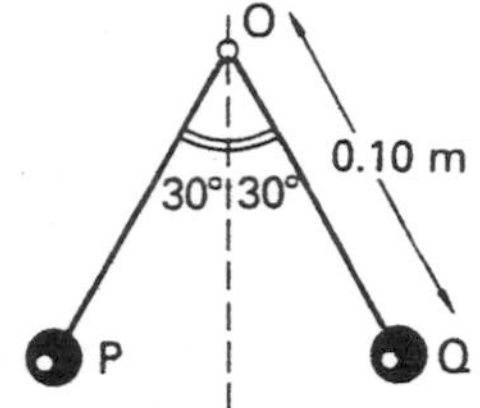

20.14 **(a)** Calculate **(i)** the gravitational force **(ii)** the electrical force between two protons which are 2.0×10^{-10} m apart. Take the mass of a proton to be 1.7×10^{-27} kg. [Use data.]
(b) What is the ratio of the electrical to the gravitational force?

20.2 Uniform *g*-fields and *E*-fields

In this section you will need to

- use the equation $g = F/m$ for the gravitational field strength at a point where F is the gravitational force on a mass m placed at the point
- use the equation $E = F/q$ for the electric field strength at a point where F is the electrical force on a charge q placed at the point
- remember that the difference in gravitational potential energy (g.p.e.) for a body of mass m between two points separated by a vertical distance Δh in a uniform g-field is $mg\Delta h$
- remember that the difference in electrical potential energy for charge q between two points in a uniform E-field is $qE\Delta x$ where Δx is measured parallel to the field
- understand that the gravitational potential difference between two points is independent of the mass of the body moving between the points
- understand the relationship between electrical potential difference (p.d.) and electric field strength and use the equation $E = V/d$
- draw diagrams to show field lines and equipotential surfaces for both gravitational and electric fields.

20.15 The values of g on the surface of the planets Mars, Saturn and Jupiter are: 3.7 N kg^{-1}, 8.9 N kg^{-1} and 23 N kg^{-1} respectively. A mass of 20 kg is lifted 1.6 m from the surface of each planet. Calculate for each planet

(a) the force needed to lift the mass
(b) the gain in gravitational potential energy of the mass.

20.16 In the previous exercise what is the gravitational potential difference, i.e. the change in g.p.e. per unit mass, between the surface of each planet and

(a) a position 1.6 m above its surface

(b) a position 3.2 m above its surface?

20.17 A multi-storey car park has six levels each 3.00 m above the other. Sketch the car park and label the entry level 1.

(a) What is the change in g.p.e. of a 1400 kg car as it moves **(i)** from level 1 to level 4 **(ii)** from level 6 to level 4?

(b) Label the levels on your diagram with values for the gravitational potential, giving level 1 a potential of 0 J kg^{-1}.

20.18 The diagram shows part of the gravitational field near the surface of the Earth. On this scale the field is essentially uniform.

(a) What is the distance apart of the equipotential surfaces?

(b) What is the gravitational force on a mass of 2.0 kg **(i)** at A **(ii)** at B?

(c) What is the gravitational potential energy of a mass of 2.0 kg **(i)** at A **(ii)** at B?

(d) How much work must be done to move the mass from B to A?

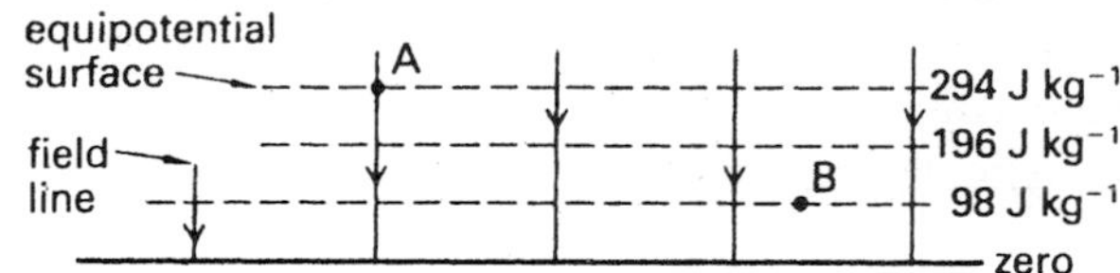

20.19 **(a)** Copy the diagram used in the previous question and draw two possible paths along which you could throw a heavy stone to pass through both A and B.

(b) By how much does the value of the kinetic energy per unit mass of the stone at A and B differ for each of your paths?

20.20 Draw and label with values of gravitational potential a set of equipotential surfaces separated by 10 m at the Moon's surface. Take the gravitational field strength at the surface of the Moon to be 1.7 N kg^{-1}.

20.21 If the Earth were flat, of uniform thickness d and density ρ, and 'went on' for ever, the uniform gravitational field above its surface would be given by

$$g = \tfrac{1}{2} G\rho d$$

(a) Show that the units of both sides of this expression are the same.

(b) Calculate a value for the thickness d needed to give a value of g of 9.8 N kg^{-1} assuming a value for ρ of 5500 kg m^{-3}.

20.22 Copy and complete the following table showing values of charge q, electric field E and electrical force F:

q/C	E/N C^{-1}	F/N
6.0×10^{-9}	3000	
3.2×10^{-19}		6.4×10^{-17}
	5.0×10^{6}	8.0×10^{-13}

20.23 A charged polystyrene ball of mass 0.14 g is suspended by a nylon thread from a fine glass spring. In the absence of any electric field the spring extends by 30 mm. The polystyrene ball is then placed in an electric field that acts vertically upwards, of strength 2.0×10^5 V m^{-1}, and the spring extends by a further 6.0 mm. What is the electric charge on the polystyrene ball?

20.24 A smoke particle weighing 3.0×10^{-15} N carries a charge of 8.0×10^{-19} C in an electric field of 5.0 kV m^{-1}.

(a) What electrical force acts on the smoke particle?

(b) Calculate the acceleration of the smoke particle if this force acts **(i)** vertically upwards **(ii)** horizontally.

20.25 By considering the equations defining the units involved, show that 1 V m^{-1} = 1 N C^{-1}.

20.26 Two parallel plates are fixed 20 mm apart and are maintained at a potential difference of 1.0 kV.

(a) What is the electric field strength between the plates?

(b) What force would act on a particle carrying a charge of 2.0×10^{-11} C when its distance from the negative plate is **(i)** 10 mm **(ii)** 2.0 mm?

20.27 A vacuum tube contains two parallel electrodes 7.5 mm apart. A p.d. of 150 V is maintained between them.

(a) What is the electric field strength in the gap?

(b) What is the force on an electron in the gap?

(c) How much electric potential energy does an electron gain in crossing the gap?

(d) How much kinetic energy is gained by such an electron if it travels from one side of the gap to the other?

20.28 The diagram shows some equipotential surfaces between two charged parallel plates which are 0.12 m apart.

(a) Copy the diagram and add values for the electric potential at each of the lines.

(b) Add some electric field lines and calculate a value for the *E*-field between the plates.

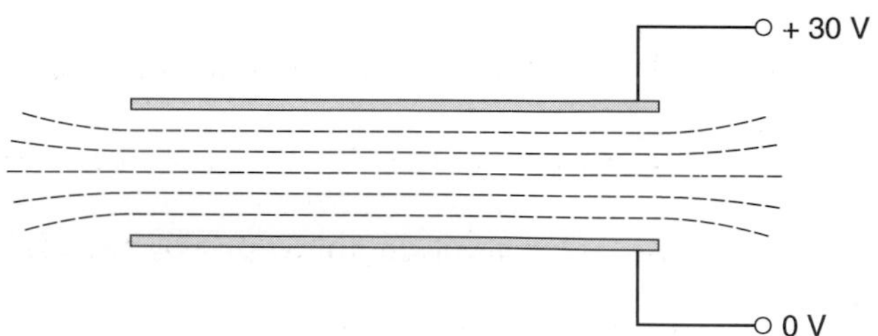

20.29 A charged oil drop of mass 2.0×10^{-15} kg is observed to remain stationary in the space between two horizontal metal plates when the potential difference between them is 245 V and their separation is 8.0 mm. Calculate the charge on the drop and comment on your answer.

20.30 What potential difference would you need to maintain between two horizontal metal plates 6.00 mm apart so that a particle of mass 4.00×10^{-15} kg with three surplus electrons attached to it would remain in equilibrium between them? Which plate would be the positive one? [Use data.]

20.31 Refer to the previous question.

(a) When the potential difference between the plates is 480 V, the particle moves slowly downwards in the space between them. When it has fallen 4.00 mm, what is the work done

(i) by the gravitational force acting on it

(ii) by the electrical force acting on it?

(b) Describe the energy transfers as the particle moves downwards.

20.32* A capacitor consists of two parallel plates 12 mm apart in air, each of area 0.040 m^2.

(a) When the potential difference between the plates is 4.0 kV calculate

(i) the capacitance of the capacitor C [Use data.]

(ii) the charge on the capacitor plates q

(iii) the charge density (charge per unit area) on the capacitor plates σ

(iv) the electric field between the capacitor plates E.

(b) **(i)** Show that the value of $\varepsilon_0 E$ is equal to σ. [Use data.]

(ii) Prove that $\varepsilon_0 E = \sigma$ algebraically.

20.3 Radial *g*-fields and *E*-fields

In this section you will need to

- understand that the Earth's gravitational field is radial and of size $g = Gm_E/r^2$, where m_E is the mass of the Earth
- use the equation $E = kq/r^2$ for the electric field strength around a charge q
- understand that g and E are vector quantities
- use the equation for the change of gravitational potential energy in the Earth's field between places at distances r_1 and r_2 from the centre of the Earth

$$\Delta(\text{g.p.e.}) = Gmm_E(1/r_1 - 1/r_2)$$

- use the equation $V = kq/r$ for the electric potential near a point or spherical charge.

20.33 The gravitational field strength on the surface of the Earth, a distance r_E from its centre is $g_0 = 9.8$ N kg^{-1}.

(a) What is the gravitational field strength at distances **(i)** $2r$ **(ii)** $3r_E$ **(iii)** $10r_E$ from the Earth's centre?

(b) What is the gravitational field strength **(i)** $r_E/2$ **(ii)** $r_E/10$ **(iii)** $r_E/100$ above the Earth's surface?

(c) How far above the Earth's surface will the gravitational field strength be **(i)** $g_0/2$ **(ii)** $g_0/3$ **(iii)** $g_0/10$? [Use data.]

20.34 **(a)** Calculate, using $g = Gm_E/r^2$, the Earth's gravitational field at its surface and 30 km above its surface.

(b) Show that the change is about 1% and predict at what height it will reduce by approximately a further 1%. Explain how you made your prediction.

20.35 **(a)** Express the mass of the Earth in terms of its radius and density ρ. Hence show that G and g_0, the value of g at the Earth's surface, are related by the equation

$$g_0 = \tfrac{4}{3}\pi G\rho r_E$$

(b) Check that the units of the right hand side of this relationship are m s^{-2}.

20.36 Suppose that the Sun shrank to become a neutron star of radius 1.5 km but retained its present mass. Calculate the gravitational field strength at its surface.
(In fact, the Sun would undergo changes before it became a neutron star which would involve it losing a considerable proportion of its mass.)

20.37 Refer to question 20.3. Write down the values given in the answers for the acceleration of the three bodies at their stated distances from the centre of the Earth.
(a) Produce three sets of values for $\lg(g/\text{N kg}^{-1})$ and $\lg(r/\text{m})$ and plot them on a graph grid of $\lg(g/\text{N kg}^{-1})$ against $\lg(r/\text{m})$.
(b) Explain why the points should lie on a straight line.
(c) Deduce the acceleration of a space probe which is **(i)** 1.0×10^8 m **(ii)** 2.0×10^9 m from the Earth.

20.38 **(a)** Calculate the gravitational pull F of the Earth on a body of mass 10 kg at distance r from the centre of the Earth given by $r/10^6$ m = 8, 10, 15, 20, 30 and plot a graph of F against r.
(b) The area between the graph and the r-axis represents the work done by the force F. Estimate the change of g.p.e. of the body when it moves from $r = 15 \times 10^6$ m to $r = 25 \times 10^6$ m.
(c) Check your estimate by calculating the change of g.p.e.

20.39 The diagram shows a region near the surface of the Earth.
(a) What is the gain in g.p.e. of a space vehicle of mass 1500 kg when it moves **(i)** from A to B **(ii)** from C to B?
(b) The space vehicle later returns through B at a speed of 4.0 km s^{-1}.
(i) What is then its kinetic energy?
(ii) What will be its kinetic energy when it reaches the Earth?
(iii) How fast will it be moving when it reaches the Earth?

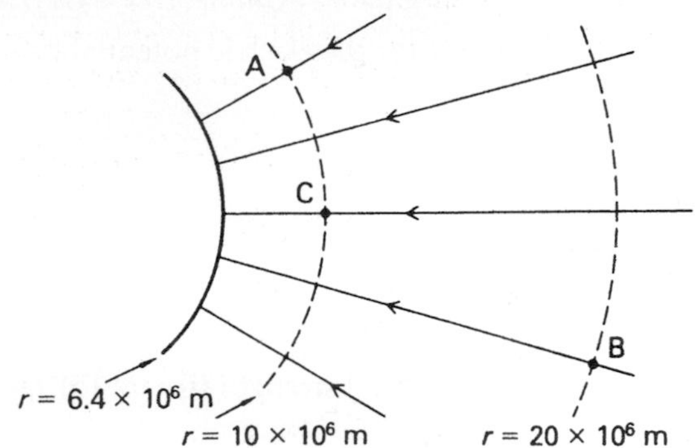

20.40 Show that the gravitational potential energy which must be given to a body of mass m at the Earth's surface if it is to escape the Earth's gravitational field, i.e. reach an infinite distance from the Earth, is equal to $mm_E G/r_E$.
(a) Hence derive an expression for the speed v_e at which a body must be projected from the Earth's surface (ignoring the atmosphere) if it is to escape from the Earth's g-field.
(b) Calculate a value for v_e and comment on the fact that it does not depend on the mass of the body.

20.41 **(a)** Calculate an escape speed for the Moon. Take the mass of the Moon to be 7.34×10^{22} kg and its radius to be 1.64×10^{6} m.

(b) The temperature of the Moon's surface rises to about 400 K when sunlight falls on it. At this temperature the average speed of oxygen molecules is over 500 m s^{-1}. Suggest why the Moon has no atmosphere.

20.42 Here are some data about the electric field near the charged sphere of a Van de Graaff generator:

distance from centre of sphere r/m	0.20	0.25	0.30	0.40	1.00
electric field strength E/kN C^{-1}	75	48	33	19	3.0

(a) Show that this E-field follows an inverse square law by

(i) plotting E against $1/r^2$

(ii) calculating values for Er^2.

Explain how each process tells you that $E \propto 1/r^2$.

(b) Calculate the charge on the surface of the sphere.

20.43 In a hydrogen atom the average distance apart of the proton and the electron is 5.3×10^{-11} m. Calculate

(a) the electric field at this distance from the proton

(b) the electrical force on the electron

(c) the acceleration such a force would produce on an electron. Take the mass of an electron to be 9.11×10^{-31} kg.

20.44 The graph shows how the electric potential V varies with distance r from the proton at the centre of a hydrogen atom. The proton charge is 1.6×10^{-19} C.

(a) Take readings from the graph to show that V is inversely proportional to r.

(b) Calculate a value for the electric field constant k.

(c) Calculate the energy which would be needed to move an electron from 10.5×10^{-11} m to a great distance from the proton.

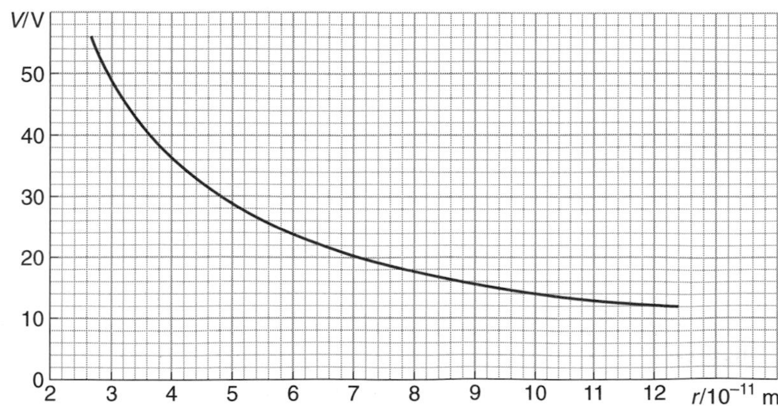

20.45 Two charges of size q, but of opposite sign, are placed on the line LM as shown in the diagram on the next page.

(a) Describe how the direction of the E-field produced by the two charges varies **(i)** along the line LM **(ii)** along the line AB. Explain your reasoning in each case.

(b) Sketch the shape of the E-field in the region around the charges.

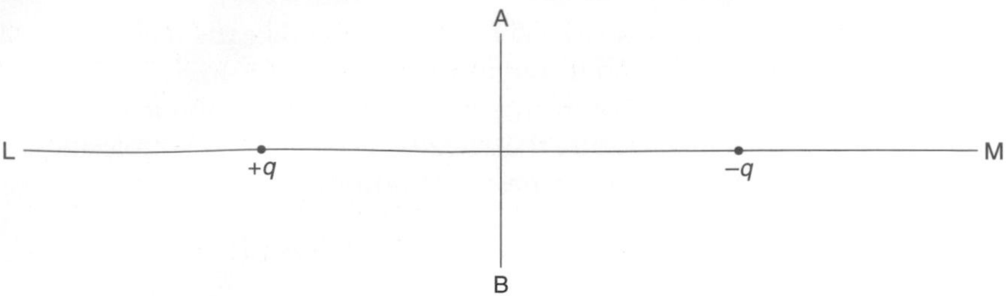

20.46 Two charges each of +3.0 μC, separated by a distance of 0.80 m, produce the electric field shown in the diagram.

(a) Calculate the electric potential at A, a distance of 0.50 m from each charge.

(b) Copy the diagram and add dashed lines to show the shapes of the equipotential surfaces in the plane of the paper.

***(c)** Calculate the resultant electric field at A.

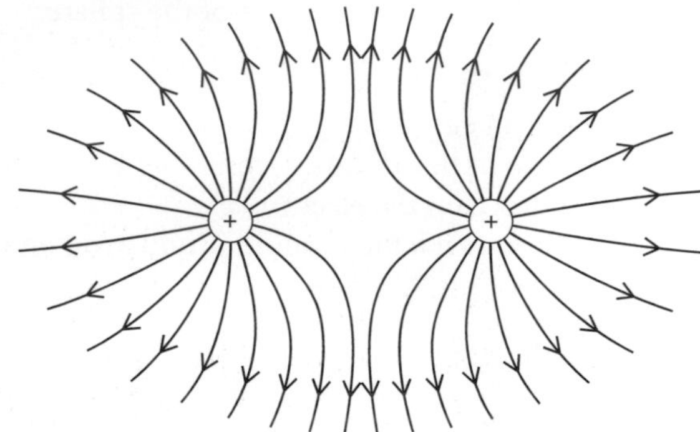

20.47 In a small Van de Graaff machine the metal sphere reaches a potential of 2.4×10^5 V before it discharges by sparking across an air gap.

(a) If the sphere is of radius 0.15 m, calculate the charge on the sphere just before discharge.

(b) What is the electric field at the surface of the sphere just before the discharge takes place?

20.48 Because of thunderclouds the upper atmosphere carries a permanent positive charge. The result is that in normal fair weather conditions there is a downward electric field at the Earth's surface of about 200 V m^{-1} or 200 N C^{-1}.

(a) Use $E = kq/r^2$ to show that the total charge on the surface is 9.0×10^5 C.

(b) Calculate the surface density of this charge in C m^{-2}. [Use data.]

20.4 Satellite motion

In this section you will need to

- understand that the centripetal acceleration, v^2/r, of bodies in circular orbits in a gravitational field is equal to the g-field at that orbit
- use Newton's second law combined with his law of gravitation to solve satellite problems
- explain that a person feels weightless when he or she is in a state of free fall.

20.49 A geosynchronous satellite moves in a circle of radius 4.2×10^7 m every 24 hours.

(a) Use this data to calculate **(i)** the speed **(ii)** the centripetal acceleration of the satellite.

(b) Write down the size of the Earth's *g*-field at the satellite's orbit.

(c) What is the centripetal acceleration of a satellite moving in an orbit of radius 2.1×10^7 m, i.e. half that of the geosynchronous satellite?

(d) Calculate the speed of this inner satellite.

(e) Hence deduce the orbital period of this inner satellite.

20.50 The photograph shows an astronaut who is moving in a circular orbit 350 km above the Earth's surface. (There is a space shuttle close by!)

(a) What is **(i)** the Earth's *g*-field at this height **(ii)** the acceleration of the astronaut towards the centre of the Earth?

(b) Explain **(i)** why the astronaut does not crash to the ground **(ii)** why the astronaut feels 'weightless'?

(c) Calculate **(i)** the speed of the astronaut in this orbit **(ii)** how long it takes the astronaut to circle the Earth.

20.51 The *Apollo 11* space capsule was placed in a parking orbit (around the Earth) of radius 6.56×10^6 m before moving onwards to the Moon.

(a) Calculate the values of the gravitational field strength at this radius.

(b) What was the acceleration of the capsule then?

(c) Use $W = mg$ to find the weight of a 70 kg astronaut in the capsule.

(d) What was his acceleration?

(e) Use $ma = F_{res}$ to find the resultant force on him.

(f) Did any force other than his own weight act on him?

(g) What would a weighing machine in the capsule have recorded if he had stood on it?

20.52 Isaac Newton knew that the mean Earth–Moon distance was 3.8×10^8 m and that the period of the Moon's revolution was 2.4×10^6 s. He also knew that the radius of the Earth was 6.4×10^6 m and that the acceleration of an apple at the Earth's surface was $9.8\ \text{m s}^{-2}$.

(a) Calculate the Moon's (centripetal) acceleration g_M towards the Earth.

(b) Express g_M as a fraction of g_0, the value of g at the Earth's surface.

(c) Express the Earth's radius as a fraction of the Earth–Moon distance.

(d) Use your answers to **(b)** and **(c)** to show how Newton could suggest that the gravitational field of the Earth varies inversely as the square of the distance from its centre.

20.53 The *Apollo 11* spacecraft was orbiting the Moon at a speed of 1.65 km s^{-1} before the first lunar module landed. The mass of the Moon is 7.34×10^{22} kg.

(a) Calculate the radius of the spacecraft's orbit around the Moon.

(b) If the Moon has a mean radius of 1.64×10^6 m, how far above the Moon's surface was the spacecraft orbiting?

20.54 The relationship between the period of revolution T of a body in a circular orbit and its distance r from the gravitationally attracting body at the centre of the orbit is

$$4\pi^2 r^3 = GMT^2$$

where M is the mass of the central body.

(a) Show that the units of both sides of this expression are the same.

(b) Use it to calculate the mass of the Sun, given that the mean Earth–Sun distance is 1.5×10^8 km.

(c) Jupiter has lots of moons, four of which can be seen using a good pair of binoculars or a small telescope. Explain how observations of the motion of the Moons of a planet enable the mass of the planet to be calculated.

20.55 The period T of a satellite around its gravitationally attracting central mass (a star or planet) is related to its mean distance r from the centre of this mass by the relationship $T^2 \propto r^3$ (Kepler's third law).

(a) Use the following data about four of the moons of the planet Uranus to plot a suitable graph to test Kepler's law:

T/hours	60.5	99.5	209	323
$r/10^3$ km	192	266	436	582

(b) If a further moon of Uranus were discovered with a period of 170 hours, what would be its mean distance from the centre of Uranus?

(c) Write an equation connecting the gravitational force on a moon of mass m with its mass and centripetal acceleration v^2/r. Use $v = 2\pi r/T$ to derive Kepler's law for circular orbits.

(d) Hence calculate the mass of Uranus.

20.56 **(a)** Calculate the orbital radius of a synchronous communications satellite, i.e. one which has a period of 24 hours so that it appears to remain stationary above one point on the Earth's surface. Express this radius as a multiple of the Earth's radius to the nearest whole number. [Use data.]

(b) Why must the satellite's orbit lie in the plane of the equator?

(c) Draw a scale diagram to estimate the angle above the horizon that a receiving aerial at a latitude of 45° must point in order to receive signals from the satellite.

21 Practising calculations

21.1 Numbers

In this section you will need to

- understand what is meant by expressing a number to a certain number of significant figures
- understand that the number of significant figures in a result should not be more than the smallest number of significant figures in the data
- understand what is meant by expressing a number in standard form
- use a calculator to add, subtract, multiply and divide numbers expressed in standard form
- use a calculator to raise numbers to powers, and to take square and cube roots of numbers
- remember the meaning of the following prefixes: p, n, μ, m, k, M, G, and know what they stand for: i.e., pica-, nano-, micro-, milli-, kilo-, mega-, giga-.

21.1 Express the following numbers in standard form, to two significant figures:
(a) 0.0342 **(b)** 0.005 291 **(c)** 0.145 **(d)** 153.2 **(e)** 674.

21.2 The numbers in these calculations have been given to different numbers of significant figures. In each case what should be the number of significant figures in the answer?
(a) $5.72 \times 11.1 \times 1.1$ **(b)** $6.35 \times (2.35)^2 \times 0.1152$ **(c)** $67 \times 2.46 \times 0.010$.

21.3 *Without using a calculator*, find the value of **(a)** $10^3 \times 10^6$ **(b)** $10^2 \times 10^4$ **(c)** $10^6 \times 10^{3}$ **(d)** $10^{-3} \times 10^{-6}$ **(e)** $10^6 \div 10^3$ **(f)** $10^4 \div 10^{-2}$ **(g)** $(10^6 \times 10^{-3})/10^{-4}$ **(h)** $1 \div 10^3$ **(i)** $1 \div 10^{-6}$.

21.4 *Using a calculator*, find the value of **(a)** $10^3 \times 10^{-5}$ **(b)** $10^2 \div 10^6$ **(c)** $10^{-3} \times 10^5$ **(d)** $1 \div 10^{-3}$ **(e)** $10^{-3} \div 10^5$.

21.5 Giving your answers to two significant figures, multiply together each of the following pairs of numbers:
(a) 1.34×10^4, 1.05×10^2 **(b)** 4.78×10^5, 1.34×10^{-2} **(c)** 9.11×10^{-31}, 7.432×10^{-12}.

21.6 Refer to the previous question. Giving your answers to two significant figures, divide the first number in each pair by the second number.

21.7 Do the following calculations, leaving your answers in standard form, to the appropriate number of significant figures:
(a) $\pi(2.35)^2$ **(b)** $4\pi^2(1.63 \times 10^{-3})^2$ **(c)** $\sqrt{(25/\pi)}$.

21.8 Express the following percentages as fractions to two significant figures (e.g. 50% = 0.50): **(a)** 20% **(b)** 2.0% **(c)** 0.040% **(d)** 300%.

21.9 What is the percentage increase in a quantity if it is **(a)** doubled **(b)** trebled?

21.10 If $x = 60$, what does x become if there is a percentage increase in it of **(a)** 50% **(b)** 100% **(c)** 120% **(d)** 500%?

21.11 Express the following fractions as percentages: **(a)** 0.10 **(b)** 0.87 **(c)** 2.3 **(d)** 0.0020.

21.12 **(a)** If 30% of x is 37.2, what is x? **(b)** If 18% of x is 43.9, what is x? **(c)** If 75% of x is 0.972, what is x?

21.13 Measurements of the speeds of 10 cars on a motorway gave the following results: $v/\text{m s}^{-1}$ = 30.1, 25.6, 28.9, 34.5, 32.7, 33.1, 29.6, 29.9, 38.2, 23.1. Which of these were within 5% of the 31.3 m s^{-1} speed limit?

21.14 Giving your answers to two significant figures, do the following calculations: **(a)** $\sqrt{(0.242)}$ **(b)** $\sqrt{(0.00476)}$ **(c)** the reciprocal of 43.98 **(d)** the reciprocal of 0.0824 **(e)** $\sqrt[3]{2.45}$ **(f)** $(9.73)^3$ **(g)** $(0.431)^{1/2}$ **(h)** $\pi(0.142)^2$ **(i)** $4.96^{3/2}$ **(j)** $(1\text{–}0.15^2)^{1/2}$.

21.2 Units and equations

In this section you will need to

- remember that the base units in the Système International (SI) are the metre, kilogram, second, ampère and kelvin
- understand that quantities need to be expressed in base units when used in equations for calculations
- rearrange equations so that a different quantity is the subject of the equation
- use equations to show how a derived unit is related to the base units
- know how to check that an equation is homogeneous with regard to units.

21.15 The base units in the m.k.s. system are the metre, the kilogram and the second. Express the following quantities in the appropriate base units, in standard form to two significant figures: **(a)** 6.34 cm **(b)** 12 mm **(c)** 832 km **(d)** 546 nm **(e)** 53.4 g **(f)** 500 t **(g)** 123 mg **(h)** 2.3 μg **(i)** 30 minutes **(j)** 23 ms **(k)** 24 hours **(l)** 45 litres.

21.16 Express the following areas and volumes in the appropriate base units: **(a)** 1.6 cm^2 **(b)** 5.3 mm^2 **(c)** 0.017 cm^2 **(d)** 7.8 cm^3 **(e)** 34 mm^3.

21.17 Calculate the areas of circles with these diameters, giving your answer in m^2 and to two significant figures: **(a)** 2.5 m **(b)** 54 cm **(c)** 2.6 mm.

21.18 Write down the following quantities as numbers in standard form (to two significant figures) together with the appropriate unit without any prefix (such as p or M): **(a)** 470 pF **(b)** 1.5 kV **(c)** 50 MW **(d)** 40 ns.

21.19 The force F exerted on an area A by a pressure p is given by the equation $F = pA$. Calculate F, giving your answer to the appropriate number of significant figures, when $p = 1.01 \times 10^5$ N m^{-2} and $A = 1.25$ m^2.

21.20 The volume V of a cylinder is given by the equation $V = \pi r^2 h$. Calculate V, giving your answer in the base unit and to the appropriate number of significant figures, when $r = 3.54$ cm and $h = 0.25$ m.

21.21 Rearrange these equations to make the quantity given in the bottom row the subject of the formula:

(a)	**(b)**	**(c)**	**(d)**	**(e)**	**(f)**	**(g)**
$x = vt$	$V = bdh$	$A = \pi r^2$	$V = \pi r^2 h.$	$\rho = m/V$	$v = u + at$	$v^2 = u^2 + 2ax$
v	h	r	h	V	a	a

(h)	**(i)**	**(j)**	**(k)**	**(l)**	**(m)**	**(n)**
$\eta = 1 - (T_2/T_1)$	$P = I^2 R$	$P = V^2/R$	$T = 1/f$	$T = 2\pi\sqrt{(l/g)}$	$V = E - Ir$	$V = \frac{4}{3}\pi r^3$
T_1	I	R	f	l	r	r

21.22 Use the equation $c = f\lambda$ to calculate f when $c = 3.00 \times 10^8$ m s^{-1} and $\lambda = 546$ nm.

21.23 Use the equation $A = \pi r^2$ to calculate r when $A = 12$ cm^2.

21.24 Use the equation $V = \frac{4}{3}\pi r^3$ to calculate r when $V = 3.2 \times 10^{-3}$ m^3.

21.25 Use the equation $E = Fl/eA$ to calculate e when $F = 12$ N, $l = 1.5$ m, $A = 0.010$ mm^2 and $E = 200$ GN m^{-2}.

21.26 Use the equation $GM = 4\pi^2 r^3/T^2$ to calculate T when $G = 6.67 \times 10^{-11}$ N m^2 kg^{-2}, $M = 5.97 \times 10^{24}$ kg and $r = 6.37 \times 10^6$ m.

21.27 What are the base units, in the SI system of units, of **(a)** length **(b)** mass **(c)** time **(d)** electric current **(e)** temperature?

21.28 The unit of force F is the newton (N). Use the equation $F = ma$, where m is mass and a is acceleration, to express the newton in terms of the base units.

21.29 The unit of pressure p is the pascal (Pa). Use the equation $p = F/A$, where F is force and A is area, to express the pascal in terms of the base units.

21.30 The unit of energy W is the joule (J). Use the equation $W = Fs$, where F is force and s is distance, to express the joule in terms of the base units.

21.31 The unit of power P is the watt (W). Use the equation $P = W/t$, where W is energy and t is time, to express the watt in terms of the base units.

21.32 The unit of electric charge Q is the coulomb (C). Use the equation $I = Q/t$, where I is electric current and t is time, to express the coulomb in terms of the base units.

21.33 The unit of potential difference V is the volt (V). Use the equation $V = W/Q$, where W is energy transferred and Q is electric charge, to express the volt in terms of the base units.

21.34 The unit of frequency f is the hertz (Hz). Use the equation $T = 1/f$, where T is the time period, to express the hertz in terms of the base units.

21.35 The unit of electrical capacitance C is the farad (F). Use the equation $C = Q/V$, where Q is electric charge and V is potential difference, to express the farad in terms of the base units.

21.36 Simplify the following expressions, performing the calculation on the units as well as the numbers. What type of quantity is each expression? (In answering parts of the question it may help you to know that N m can be written as J, and N m^{-2} as Pa.)
(a) $20 \text{ m s}^{-1} + (5 \text{ m s}^{-2})(2.0 \text{ s})$ **(b)** $\frac{1}{2}(20 \text{ N m}^{-1})(0.01 \text{ m})^2$
(c) $(1.29 \text{ kg m}^{-3})(9.81 \text{ N kg}^{-1})(2.3 \text{ m})$ **(d)** $\frac{4}{3}\pi(2.63 \text{ m})^3$

21.37 Calculate the following and express the results by means of numbers in standard form with a single unit:
(a) 470 pF × 1.5 kV given that 1 F = 1 C V^{-1}
(b) 50 MW × 40 ns given that 1 W = 1 J s^{-1}
(c) your answer to **(b)** divided by your answer to **(a)** given that 1 V = 1 J C^{-1}.

21.38 Check the homogeneity of the following equations with respect to units:
(a) $W = mgh$ where W = change of gravitational potential energy, m = mass, h = change of height
(b) $W = \frac{1}{2}\sigma\varepsilon$ where W = change of elastic potential energy per unit volume, σ = stress and ε = strain
(c) $p = \rho gh$ where p = change of pressure, ρ = density of fluid and h = change of height
(d) $T = 2\pi\sqrt{(g/l)}$ where T is the time period of a pendulum and l is its length
(e) $\Delta E = c^2\Delta m$ where ΔE is the energy equivalent of a mass Δm and c is the speed of light.

21.39 In an old scientific textbook the intensity of the solar radiation arriving at the Earth (before absorption, etc. by the atmosphere) is quoted as 2.0 cal per square centimetre per minute. Express this quantity in the appropriate m.k.s. unit, given that 1.0 cal = 4.2 J.

21.40 Tyre pressures for road vehicles are usually still given in British units, e.g. 30 lb/in^2. Express this quantity in the appropriate m.k.s. unit, given that 1.00 lb = 4.45 N and 1.00 inch = 2.54 cm.

21.41 Land areas are often still given in acres. One acre is an area which has a width of 1 chain and a length of 1 furlong. 1 chain = 22 yards; 1 furlong = 220 yards. The metric unit of land area is the hectare (ha): one hectare is an area which is 100 m square. Given that 1 yard = 0.914 metre, express 1 hectare in acres.

21.42 The cost of electrical energy is quoted as 6.51p per kilowatt-hour. Work out the cost of cooking a large meal, if this requires the use of three cooking rings for 2.5 hours; the rings are each rated at 2.8 kW.

21.3 Graphs and relationships

In this section you will need to

- understand what is meant by saying that one quantity is proportional to another
- understand what is meant by saying that there is a linear relationship between two quantities
- understand that an equation of the form $y = mx + c$ represents a linear relationship and that when y is plotted against x the result is a straight line

- understand that when an equation is not of the form $y = mx + c$ it may still be possible to obtain a linear graph by a suitable choice of quantities to plot
- measure or calculate the gradient of a straight-line graph
- measure or calculate the gradient at a point on a non-linear graph by drawing a tangent to the curve
- understand that the area between a graph and the horizontal axis may represent a quantity
- calculate the quantity represented by the area between a graph and the horizontal axis.

21.43 State the meaning of: **(a)** $t > 10$ s **(b)** 20 s $< t \leq$ 40 s **(c)** $(m_1 \sim m_2) = 0.35$ g **(d)** $A \propto r^2$ **(e)** $\Delta x = 0.35$ m.

21.44 Write these statement as algebraic relationships using only mathematical symbols:
(a) The resistance R of a conductor is directly proportional to its length l, and inversely proportional to its cross-sectional area A.
(b) The centripetal force F on a body moving in a circular path with constant speed is directly proportional to the square of its speed v, and inversely proportional to the radius r of its path.

21.45 What is **(a)** Δv if v changes from 5.3 m s^{-1} to 6.7 m s^{-1} **(b)** Δt if t increases from 2.34 s to 3.98 s **(c)** ΔV if V changes from 65 cm^3 to 18 cm^3?

21.46 Write the following statement out in full without mathematical symbols, where V refers to the potential difference across a thermistor, I refers to the current in it, and T is its temperature: $\Delta V = -2.0$ V, when $\Delta T = +10$ K, if $\Delta I = 0$.

21.47 Answer yes or no to the following:

	relationship	constant quantity	are these quantities proportional?
(a)	$F = pA$	p	F, A
(b)	$F = pA$	A	F, p
(c)	$W = Fs$	F	W, s
(d)	$W = Fs$	s	W, F
(e)	$P = W/t$	t	P, W
(f)	$P = W/t$	W	P, t
(g)	$P = W/t$	W	P, $1/t$
(h)	$E = \frac{1}{2}Fx^2$	x	E, F
(i)	$E = \frac{1}{2}Fx^2$	F	E, x
(j)	$E = \frac{1}{2}Fx^2$	F	E, x^2

21.48 In the diagram overleaf, which shows three graphs for varying quantities,
(a) which of the graphs show a linear relationship between y and x?
(b) which of the graphs show that y is proportional to x?

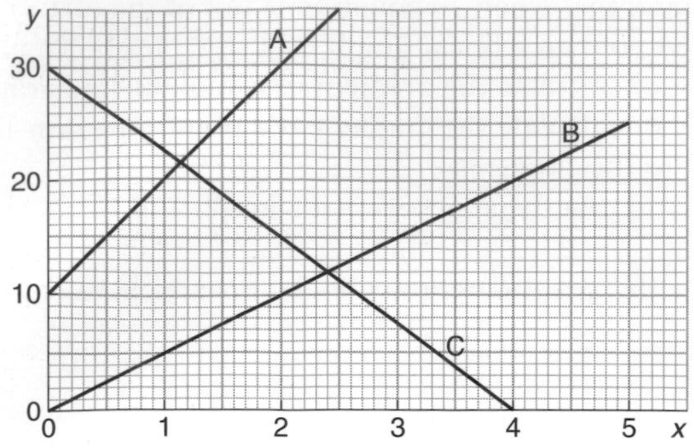

21.49 In the diagram, which shows three graphs for varying quantities,
(a) which of the graphs show a linear relationship between *y* and *x*?
(b) which of the graphs show that *y* is proportional to *x*?

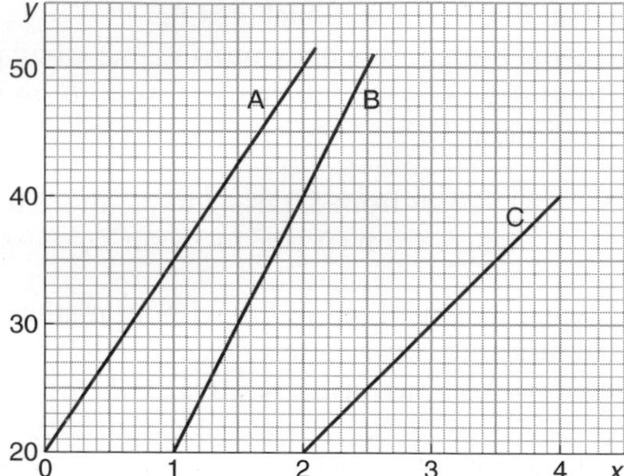

21.50 Calculate the gradients of these graphs. If you can, simplify the unit to express it in its simplest form.

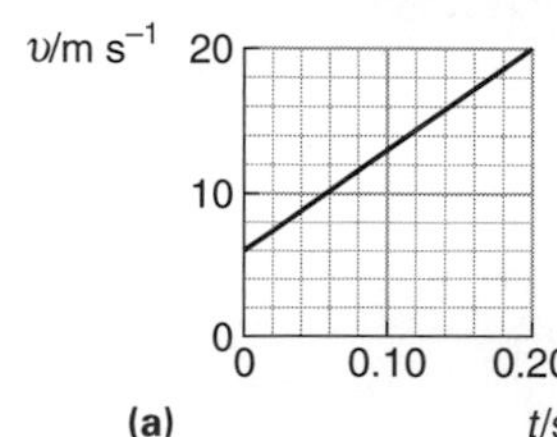

(a)

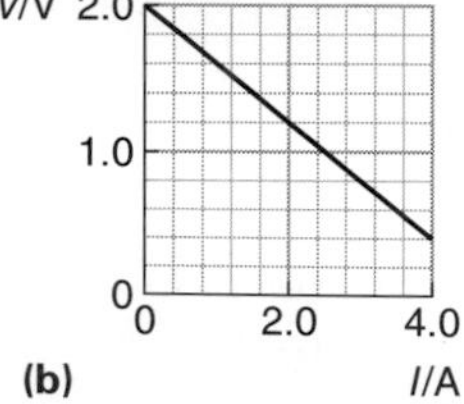

(b)

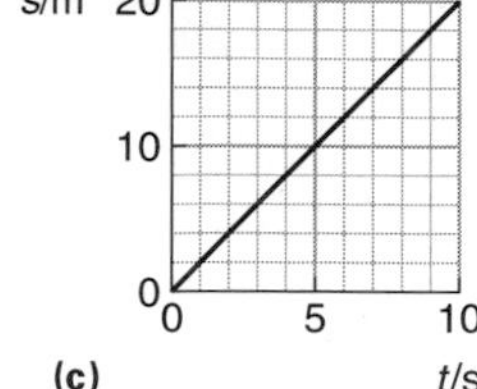

(c)

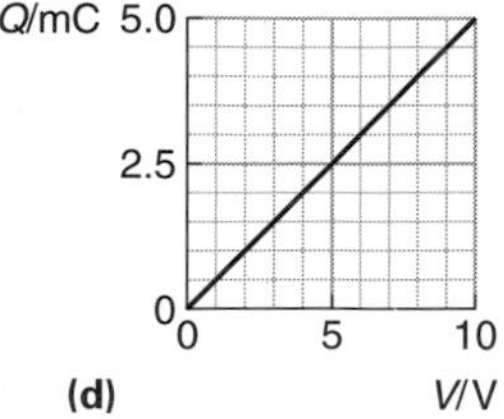

(d)

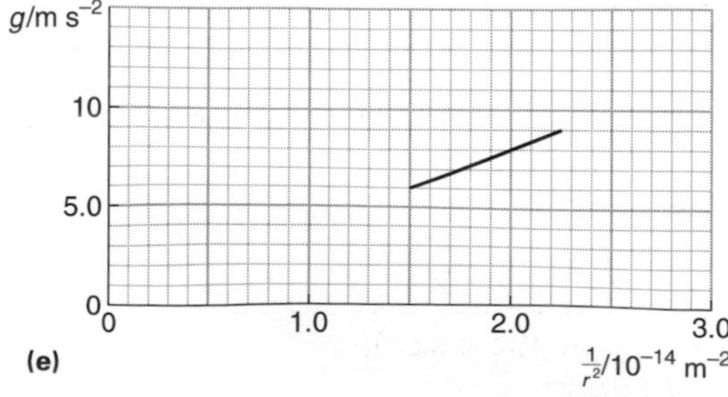

(e)

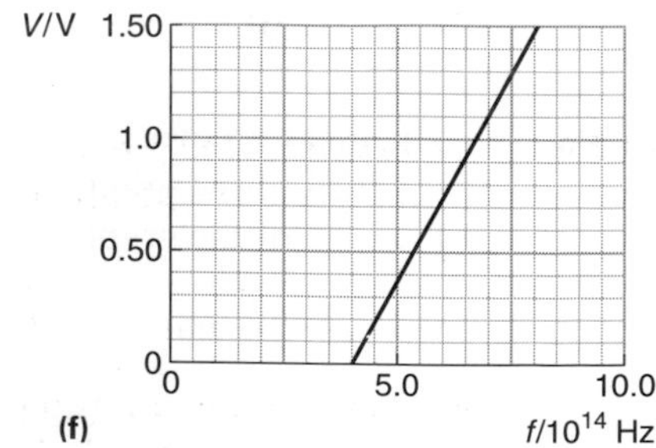

(f)

21.51 Calculate the quantity represented by the areas between these graphs and the horizontal axis. If you can, simplify the unit to express it in its simplest form.

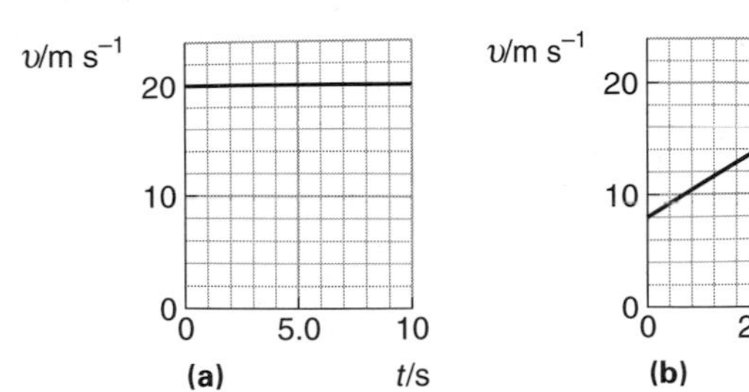

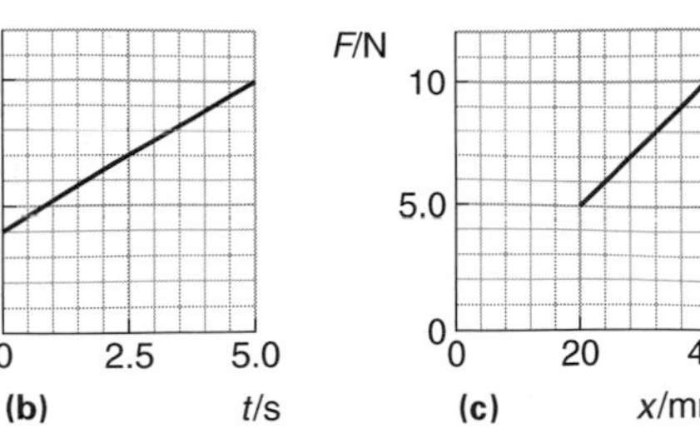

21.52 Calculate the quantity represented by the areas between these graphs and the x-axis. [Hint: first find the quantity represented by the small square and then estimate the number of these between the graph and the horizontal axis.] If you can, simplify the unit to express it in its simplest form.

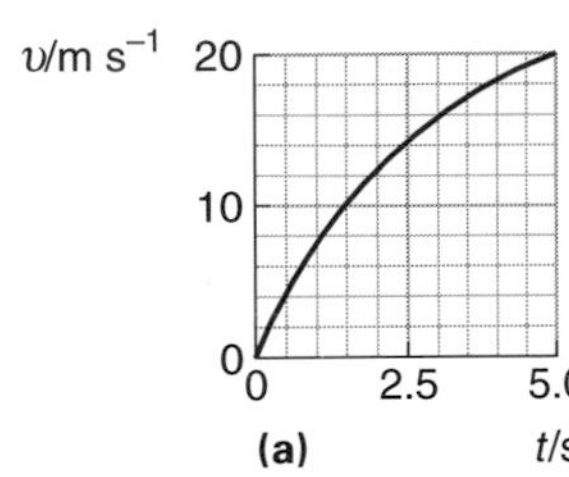

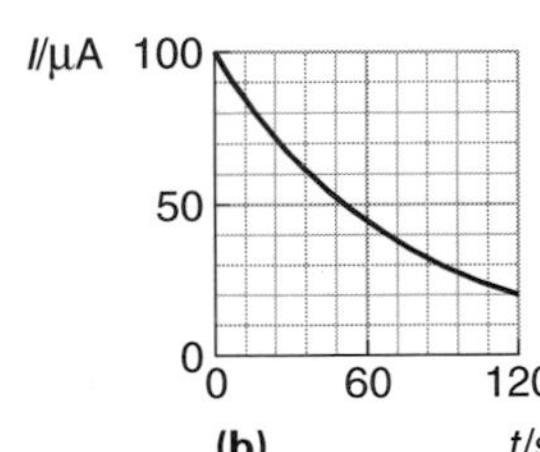

21.53 The table gives some relationships between quantities: in the next columns the two varying quantities are shown. All the other quantities are constant. Copy the table and in each case complete the columns to show what quantities you would plot on the y-axis and x-axis if you wanted to obtain a straight-line graph, and what would be the gradient.

	relationship	varying quantities	plot on x-axis	plot on y-axis	gradient
(a)	$W = \frac{1}{2}kx^2$	W, x			
(b)	$E = V/d$	E, d			
(c)	$F = Gm_1m_2/r^2$	F, r			
(d)	$c = f\lambda$	f, λ			
(e)	$C = \varepsilon_r\varepsilon_0 A/d$	C, A			
(f)	$C = \varepsilon_r\varepsilon_0 A/d$	C, d			
(g)	$T = 2\pi\sqrt{(l/g)}$	T, l			
(h)	$T = 2\pi\sqrt{(m/k)}$	T, k			
(i)	$eV_s = hf - \varphi$	f, V_s			

21.54 The extensions x of a copper wire are measured for different values of force F, and the following results are obtained:

F/N	10	20	30	40	50
x/mm	1.5	3.1	4.6	6.2	7.7

(a) Plot a graph of F (on the y-axis) against x (on the x-axis).
(b) Is there a linear relationship between F and x?

(c) Is F proportional to x?
(d) Calculate the gradient of the graph, remembering to give the units.
(e) If the gradient is equal to EA/l, where A, the cross-sectional area of the wire, is 0.10 mm^2, and l, the length of the wire, is 2.0 m, calculate the value of E, the Young modulus of copper.

21.55 The potential difference V across an electric cell is measured for different values of the current I taken from it, and the following results are obtained:

I/A	0	0.20	0.40	0.60	0.80	1.00
V/V	1.52	1.40	1.28	1.16	1.04	0.92

(a) Plot a graph of V (on the y-axis) against I.
(b) Is there a linear relationship between V and I?
(c) Is V proportional to I?
(d) What is the gradient of the graph?
(e) The equation connecting V and I for this cell is in the form $V = mI + c$. What are the values of the quantities m and c? (Give their units as well as the numbers.)

21.56 The following table gives a series of values of the wavelength λ of sound in air measured for different values of the frequency f.

f/kHz	0.20	0.40	0.60	1.00	1.50	2.00
λ/m	1.71	0.86	0.57	0.34	0.23	0.17

Plot graphs of **(a)** f against λ **(b)** f against $1/\lambda$, with f on the vertical axis in each case. What can you deduce from these graphs about the relationship between f and λ? Measure the gradient of whichever graph is a straight line. [Remember that 1 Hz = 1 s^{-1}.]

21.4 Sines, cosines and tangents

In this section you will need to

- understand that the ratios of the sides of right-angled triangles are called sine, cosine and tangent
- remember which ratio is referred to by the terms sine, cosine and tangent
- remember that the sines and cosines of angles such as 0°, 30°, 60°, 90°, 180° have simple values
- understand how to use, in a right-angled triangle, an angle and a side to calculate the other sides
- understand how to find the resolved part of a vector quantity
- understand how Pythagoras's theorem can be used to calculate the third side of a right-angled triangle.

21.57 If $\theta = 25.3°$, what, to three significant figures, is **(a)** sin θ **(b)** cos θ **(c)** tan θ?

21.58 If $\theta = 0°$, what is **(a)** sin θ **(b)** cos θ **(c)** tan θ?

21.59 If $\theta = 90°$, what is **(a)** sin θ **(b)** cos θ?

21.60 Why, if you use your calculator to find tan 90°, does the display read 'E' or 'Error'?

21.61 **(a)** In this triangle what is $\sin\theta$?
(b) Use the $\sin^{-1}$ function on your calculator to find θ to four significant figures.

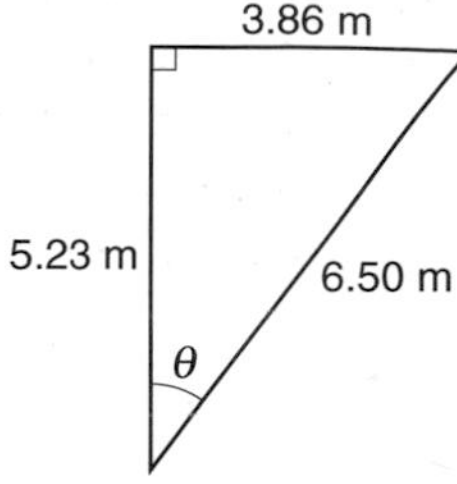

21.62 Refer to the previous question. Also find θ by first calculating
(a) $\cos\theta$ and using the $\cos^{-1}$ function on your calculator
(b) $\tan\theta$ and using the $\tan^{-1}$ function on your calculator.

21.63 In the triangle shown, what are the lengths of
(a) AB
(b) BC?

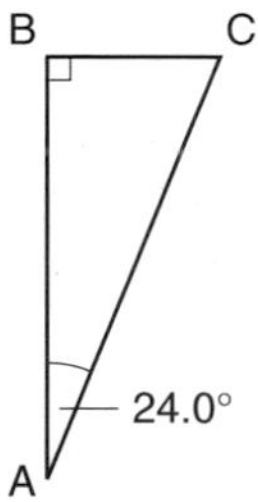

21.64 In the diagram what is the perpendicular distance from O to the line AB?

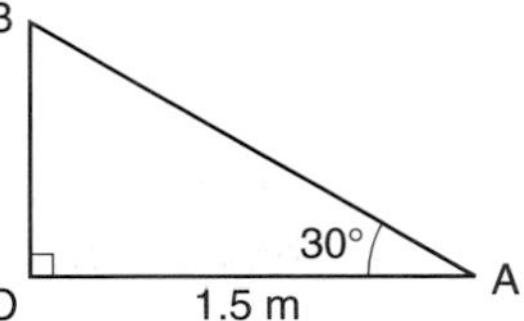

21.65 The diagram shows various vectors and two perpendicular directions x and y into which the vectors are to be resolved.
(a) In each case what is the other angle which makes up the right angle?
(b) Write down the two resolved parts in each case, using only the angle between the vector and the direction in which you are resolving. Leave your answer in the form $X\cos\theta$. You are not asked to do the calculations.

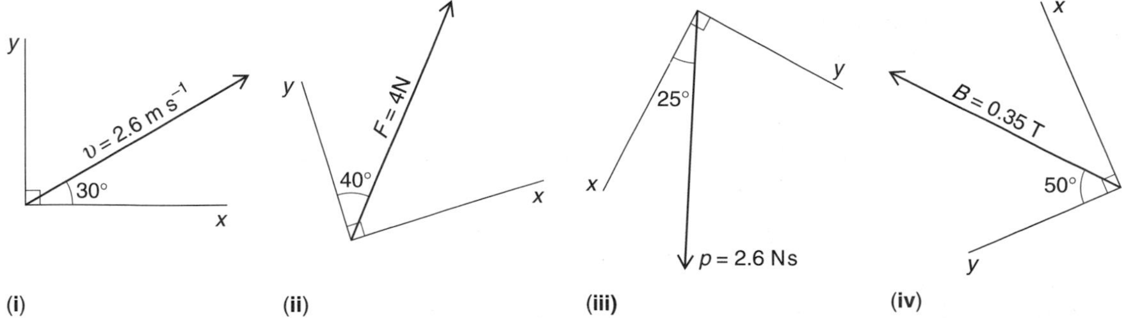

21.66 Use Pythagoras's theorem to check whether these could be the sides of right-angled triangles:
(a) 3 m, 4 m, 5 m **(b)** 5 m, 12 m, 13 m **(c)** 11 m, 22 m, 25 m **(d)** 9 m, 40 m, 41 m
(e) 15 m, 36 m, 39 m.

21.67 Find the third side of a right-angled triangle whose two short sides are **(a)** 6.32 m and 4.17 m **(b)** 5.21 m and 1.87 m.

21.68 Find the third side of a right-angled triangle if **(a)** its shortest side is 3.46 cm and its longest side is 7.21 cm **(b)** its shortest side is 0.542 cm and its longest side is 0.825 cm.

21.69 The diagram shows two forces of 24 N and 7.0 N acting at a point. Their resultant is given by the diagonal of the rectangle. Find **(a)** the size of the resultant of these forces **(b)** the angle which it makes with the 24 N force.

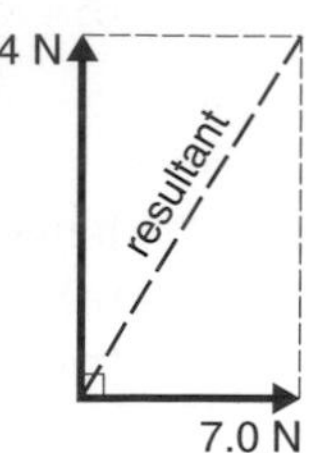

21.70 The diagram shows an air liner flying at a speed of 350 m s^{-1} relative to the air, in a cross wind of 50.0 m s^{-1}. Its velocity relative to the ground is given by the diagonal of the rectangle. Find the size and direction of this velocity.

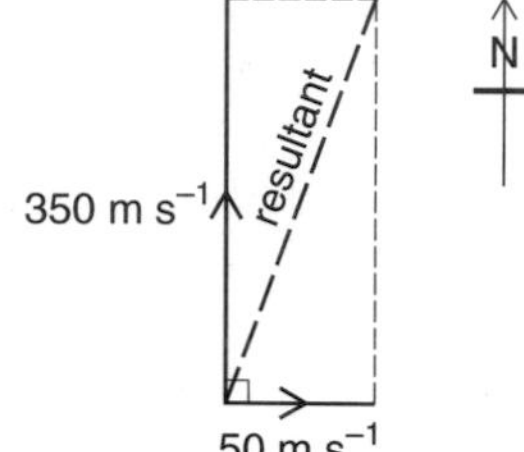

21.5 Angles

In this section you will need to

- understand that angles may be measured in radians as well as in degrees
- remember that π radians = 180°
- understand the meaning of the word 'subtend' in a phrase like 'the arc subtends an angle of 30° at the centre of the circle'
- understand that some calculations are simpler when the radian measure of angle is used
- understand that it is possible to make the approximation that $\sin\theta$ and $\tan\theta$ are equal to the angle θ measured in radians if the angle is small and answers are required to three or fewer significant figures.

21.71 Express the following angles in degrees, to three significant figures: **(a)** 0.10 rad **(b)** 1.00 rad **(c)** 2.00 rad **(d)** 2π rad.

21.72 Express the following angles in radians, as fractions or multiples of π rad: **(a)** 90° **(b)** 60° **(c)** 120°.

21.73 The diagram shows a segment of a circle of radius 1.50 m. The angle at the centre is 1.00 rad.

(a) What is the length of the arc?

(b) If the angle at the centre is increased to 2.00 rad, what is the new length of the arc?

(c) Write down an equation relating *s*, the arc length, to the radius *r* and the angle θ measured in radians.

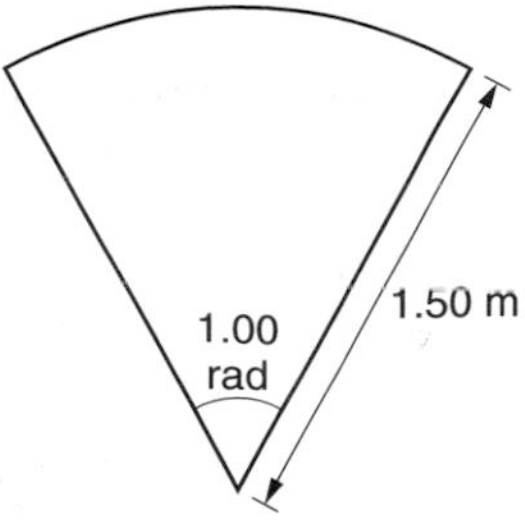

21.74 An arc of a circle of radius 2.50 m subtends an angle of 60.0° at the centre. What is **(a)** 60.0° in radians **(b)** the length of the arc?

21.75 A motor car tyre of radius 0.270 m is punctured in close succession by two nails on the road. The angle subtended at the centre of the wheel by the part of its circumference between the two tacks is 140°. What was the shortest possible distance on the road between the nails?

21.76 A pole of height 2.156 m is placed 550.0 m away on a distant hillside. A man looks at the pole; his eyes are level with the foot of the pole.

(a) Calculate, by these two methods, the angle subtended at his eye:

(i) Use trigonometry to work out the tangent of the angle, and hence the angle in degrees. Convert this angle to radians, giving your answer to four significant figures.

(ii) Assume that the pole is the arc of a circle of radius 550.0 m, and use the equation $\theta = s/r$ to calculate the angle.

(b) The second method is approximate – but are the answers the same to four significant figures?

(c) Which method is easier?

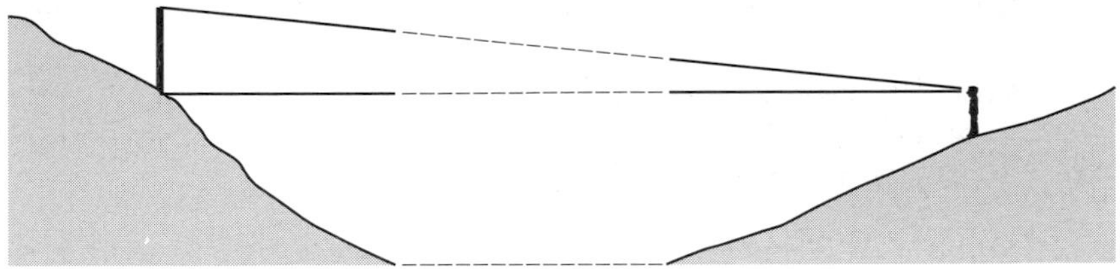

21.77 The top of a tree is seen at an elevation of 9.0° above its base when observed from a point at a distance of 120 m from it. Estimate the height of the tree, without using sine, cosine or tangent functions on your calculator.

21.78 The visible diameter of the planet Venus is 1.22×10^7 m, and its closest distance of approach to the Earth is 4.10×10^{11} m. Calculate its maximum angular size as seen from the Earth **(a)** in radians **(b)** in minutes of angle.

21.79 **(a)** A unit of distance used in astronomy is the parsec. This is defined as the distance from the solar system at which the radius of the Earth's orbit round the Sun (1.50×10^{11} m) subtends an angle of 1 second. (60 seconds = 1 minute; 60 minutes = 1 degree.) Calculate the size of the parsec in metres.

(b) The nearest star to the solar system is α-Centauri at a distance of 1.32 parsec. What angle does the radius of the Earth's orbit subtend at α-Centauri?

21.80 In order to measure the angular width of the scene photographed through a camera lens a boy takes a picture of a horizontal metre scale set at right-angles to the lens axis and at a distance of 1.2 m from it. The developed photograph shows 0.92 m of the scale. Calculate the angular width of the scene

(a) assuming that 0.92 m is the length of the circular arc between the extreme edges of the scene

(b) taking this distance more exactly as the shortest distance between these edges.

(c) What is the percentage error in using the approximate method of calculation?

21.6 Sinusoidal oscillations

In this section you will need to

- remember the shape of the sine and cosine functions for an angle θ as θ increases from 0° to 360° (or from 0 to 2π radians), and understand that they repeat indefinitely for larger angles
- remember what is meant by the amplitude, frequency and period of a sinusoidally varying quantity
- understand that an expression of the form $x = x_0\cos(2\pi ft)$ represents a quantity x which varies sinusoidally with time; its amplitude is x_0 and frequency f
- remember that the maximum rate of change of x is given by $2\pi fx_0$, and that when x is a displacement, this will give the maximum velocity
- remember that in a similar way the maximum rate of change of velocity, i.e. the maximum acceleration, is given by $2\pi f(2\pi fx_0)$ or $(2\pi f)^2x_0$.

21.81 **(a)** Sketch a graph to show how $\sin\theta$ varies with θ for angles from 0° to 360°.

(b) Also label the angle-axis with the same angles in radians.

(c) What are the maximum and minimum values of $\sin\theta$?

(d) What are the angles when $\sin\theta$ varies most rapidly?

21.82 Repeat the previous question for $\cos\theta$.

21.83 In the expression (1.5 m)cos($2\pi ft$) the frequency f is measured in Hz (which is the same as the unit s^{-1}) and t is measured in seconds. The angle is expressed in radians. Calculate the value of the expression where $f = 50$ Hz and $t = 12$ ms. [Hint: adjust your calculator so that it deals with angles in radians and not degrees, before finding the cosine.]

21.84 Repeat the previous question for $f = 100$ Hz and $t = 5.0$ ms.

21.85 **(a)** Plot a graph to show how x varies with t, where $x = x_0\cos(2\pi ft)$, $x_0 = 1.2$ m and $f = 0.25$ Hz, using the following values of t in s: 0, 0.25, 0.50, 1.00, 1.50, 1.75, 2.00. [Hint: the angle $2\pi ft$ will be in radians, so set your calculator to work in radians before calculating the cosines.]

(b) What is the period of this sinusoidal variation?

(c) If x represents displacement from the equilibrium position, what does the gradient of the graph represent?
(d) At what time(s) does the graph have a maximum gradient?
(e) Measure the gradient of the graph at $t = 1.00$ s, and verify that its size is equal to $2\pi f x_0$.
(f) What would be the maximum speed for the motion for which **(i)** $x_0 = 1.2$ m and $f = 0.50$ Hz **(ii)** $x_0 = 0.60$ m and $f = 0.25$ Hz?

21.86 When a tuning fork is set in vibration the displacement x of the end of one of the prongs from its equilibrium position at time t is given by $x = x_0 \cos 2\pi ft$.
(a) The frequency f of a particular fork is 440 Hz. Calculate the maximum speed v_0 of the ends of its prongs, if the maximum displacement x_0 is 0.50 mm.
(b) If the speed v of the ends of the prongs also varies sinusoidally, calculate the maximum acceleration of the prongs.

21.87 At a certain point in a radio set tuned to the BBC long-wave transmitter there is a sinusoidal alternating potential V of peak value 2.5 V and frequency 200 kHz. Calculate
(a) the maximum rate of change of V **(b)** the period of the oscillation
(c) the value of V at a time 0.20 μs after it changes direction.

21.88 The tides of the sea move up and down a certain beach at a frequency of approximately twice per day.
(a) What is this frequency in Hz? [Note: 1 Hz = 1 s^{-1}.]
(b) The amplitude of movement of the tide is 15 m (measured along the surface of the beach) on either side of the mean tide-line. The beach can be assumed to have a constant slope. What is the maximum rate at which the tide moves up the beach, assuming that the variation is sinusoidal?

21.7 Exponentially varying quantities

In this section you will need to

- understand what is meant by the logarithm of a number, and remember that the base of logarithms may be 10 ($\log_{10}x$ or lg x) or e ($\log_e x$ or ln x)
- understand that an exponentially varying quantity is one whose rate of increase is proportional to the quantity itself
- understand that exponential increases and decreases occur naturally and in many man-made situations
- understand that the general form of an exponential change is given by the equation $X = X_0 e^{-kt}$
- 'take logarithms' of both sides of an equation such as $X = X_0 e^{-kt}$ to give $\ln X = -kt + \ln X_0$
- understand that ln is the symbol for the natural logarithm of a number.

21.89 Use a calculator to find, to three significant figures **(a)** lg 25 **(b)** lg 250 **(c)** lg 2500.

21.90 Use a calculator to find, to three significant figures **(a)** ln 25 **(b)** ln 250 **(c)** ln 2500.

21.91 Look at your answers to the last two questions. What pattern do you see in the answers?

21.92 What is, to three significant figures **(a)** lg 10 **(b)** lg 100 **(c)** lg 1000?

21.93 What is, to three significant figures **(a)** ln e **(b)** ln e^2 **(c)** ln e^3?

21.94 What is the logarithm to base 10, to three significant figures, of x if **(a)** $x = 10^2$ **(b)** $x = 10^3$ **(c)** $x = 10^{2.5}$?

21.95 What is the logarithm to base e, to three significant figures, of x if **(a)** $x = e^2$ **(b)** $x = e^3$ **(c)** $x = e^{2.5}$?

21.96 Calculate the values of $e^{-\lambda t}$ where **(a)** $\lambda = 0.675\ s^{-1}$ and $t = 4.0$ s **(b)** $\lambda = 2.45\ min^{-1}$ and $t = 50$ s **(c)** $\lambda = 6.54 \times 10^{-5}\ s^{-1}$ and $t = 2.0$ hours.

21.97 It is thought that the current I in a circuit component varies with the applied p.d. V according to the equation $I = kV^n$, where k and n are both unknown quantities, at least for some values of V. The following measurements of V and I are made.

V/V	0	25	50	75	100	125	150	175	200
I/mA	0	3.2	9.1	16.9	25.6	30.1	32.0	32.3	32.3

(a) Take logarithms to base e of both sides of the equation and plot $\ln I$ on the y-axis against $\ln V$ (on the x-axis).
(b) Find the value of n from the straight part of the graph.
(c) Verify that $\log_e x = 2.30 \times \log_{10} x$, where x is any number. What difference would there have been in your graph, or the value for n, if you had taken logarithms to base 10?

21.98 The number of people in a certain group is 1000, and is increasing at a rate of 20 each year. If the increase is exponential, what is the rate of increase when the number of people in the group is **(a)** 2000 **(b)** 5000?

21.99 Why might you expect the rate of growth of the following to be exponential initially? In each case, why would the increase stop being exponential after a time?
(a) the number of bacteria in some food
(b) the number of people ill with flu
(c) the number of owners of mobile phones?

21.100 An exponentially decreasing quantity has the value 1000 at 12 noon. Its value at 1 p.m. is 800.
(a) What will be its value at **(i)** 2 p.m. **(ii)** 3 p.m.?
(b) What was its value at 11 a.m.?

21.101 A quantity X decreases according to the equation $X = X_0 e^{-at}$, where $X_0 = 400$ and $a = 0.500\ s^{-1}$.
(a) What is X when **(i)** $t = 2.00$ s **(ii)** $t = 3.00$ s **(iii)** $t = 4.00$ s?
(b) After what time will $X = 200$?
(c) After what time will X be 1% of its initial value?

21.102 A specimen of radioactive material contains 2.0×10^9 undecayed atoms at $t = 0$. This number N decreases exponentially. When $t = 10$ s, $N = 1.6 \times 10^9$.

(a) Use this information to draw a graph to show how N decreases with time t for $t = 0$ to $t = 40$ s.

(b) Call the initial number N_0. At what time does $N = \frac{1}{2}N_0$? What name is given to this time?

(c) At what time does $N = \frac{1}{4}N_0$?

(d) At what time does $N = \frac{1}{e}N_0$?

21.103 **(a)** Calculate the values of N, where $N = 10.0e^{-0.200t}$ for $t = 0$, 1.00, 2.00, 3.00, 4.00, 5.00.

(b) Plot a graph of $\ln N$ against t.

(c) How would you describe the relationship between $\ln N$ and t?

21.104 The current in one kind of circuit containing a resistor and a capacitor varies with time t according to the equation $I = I_0 e^{-t/RC}$, where R and C are the resistance and the capacitance. The initial current $I_0 = 100$ mA.

(a) What is the unit of the quantity RC? (You do not need to know the units of R and C to be able to answer this.)

(b) Suppose $R = 150$ kΩ and $C = 2.2$ μF. Calculate RC.

(c) After what time will the current in the circuit have fallen to **(i)** $1/e$ **(ii)** $1/e^2$ **(iii)** $1/e^3$ of its initial value?

(d) Find the time taken for the current to become 50 mA.

(e) How long will it take for the current to become **(i)** 25 mA **(ii)** 12.5 mA?

21.105 The equation $N = N_0 e^{-\lambda t}$ describes how the number N of undecayed nuclei varies with time t. N_0 is the initial number of undecayed nuclei and λ is the decay constant.

(a) What feature of the equation tells you that the number N decreases with time?

(b) What feature of the equation tells you how rapidly N decreases?

(c) In a particular case the following readings of N were taken for different values of t:

t/h	0	1.0	2.0	3.0	4.0
N	8532	5548	3601	2348	1523

(c) Take logarithms, to base e, of both sides of the equation and plot a graph of $\ln N$ against t.

(d) Use your graph to calculate the value of λ.

(e) The half-life $T_{\frac{1}{2}}$ is related to λ by the equation $\lambda T_{\frac{1}{2}} = \ln 2$. Calculate $T_{\frac{1}{2}}$ and use your graph to verify this result.

(f) Take *any* two times which are 1.6 h apart (e.g. $t = 1.0$ h and $t = 2.6$ h) and find the ratio of the final activity to the initial activity.

21.106 **(a)** Explain why if radioactive decay is a random process and each nucleus has the same chance of decaying in a certain time, the number of undecayed nuclei must decrease exponentially.

(b) If the number of undecayed nuclei decreases exponentially, why must the *rate* of decrease also decrease exponentially?

21.107 The activity of a radioactive source is 4200 s^{-1} when it is first measured. The half-life for this source is 2.8 h. After 2.8 h its activity will be 2100 s^{-1}.

(a) What is its activity after two half-lives?

(b) Continue this process to find the number of half-lives which will need to elapse for the activity to fall to less than 1% of its initial value.

(c) Will this be true for all radioactive sources?
(d) Why would there be no point in trying to continue the process to find out, for example, how many half-lives would elapse for the activity to fall to 0.1% of its initial value?

21.108 A certain bacterial population number N grows exponentially in the right environment according to the relation $N = N_0 e^{bt}$ where N_0 is the number of bacteria when $t = 0$. In an environment of tinned meat at room temperature the population doubles in 5.0 hours.
(a) What is the value of b for this population?
(b) If a tin of meat is polluted with a single live bacterium, what will the bacterial population become after one week?

21.109 The pressure in the atmosphere decreases with height above the Earth's surface according to the equation $p = p_0 e^{-kh}$, where $k = g_0\rho_0/p_0$.
(a) Calculate k, if $g_0 = 9.81$ m s^{-2}, $\rho_0 = 1.29$ kg m^{-3}, $p = 101$ kPa, giving the units, if any, of k.
(b) Why would you expect the unit of k to be what you have found?
(c) Calculate the value of p at heights of **(i)** 10 km **(ii)** 20 km.
(d) What do you notice about the sizes of p_0 and your answers to **(c)**?

21.110 Plot the exponential curve $z = z_0 e^{bt}$ for values of t at 2 s intervals from 0 up to 10 s, taking $b = 0.20$ s^{-1} and $z_0 = 2.0$.
Measure the gradient of this curve at a number of points by drawing tangents at these points; and plot another graph of the gradient against t. Test whether the value of b deduced from this graph agrees with that used above.

21.111 The excess pressure p in a leaky gas cylinder is measured at 9 a.m. every morning at various times t (in days) with the following results:

t/d	0	1	2	3	4	5
p/MPa	6.0	3.8	2.4	1.5	0.96	0.63

(a) Plot a graph of $\ln(p/\text{MPa})$ against t.
(b) Do you consider that the measurements justify stating that the pressure excess in the cylinder is decaying exponentially?
(c) If so, the relationship is of the form $p = p_0 e^{-at}$, where a is the time constant of the decay process. Calculate the time constant.
(d) After what time would you expect the pressure to have fallen to half its initial value?

21.112 In a particular electrical circuit the p.d. V across a component increases with time t according to the equation $V = V_0(1 - e^{-t/RC})$.
(a) The quantity RC has the numerical value of 2.00. What is its unit?
(b) If V_0 is 5.00 V what is V after a time **(i)** $t = 1.00$ s **(ii)** $t = 2.00$ s?
(c) After what time is $V = 3.50$ V?

22 Synoptic questions

The numbers in the margin (e.g. 3.2 4.1 9.1) give the sections in the other chapters to which these questions relate. The data needed for each question is provided before the question.

3.2 4.1 9.1

Data: gravitational field strength $g = 9.81\ \text{N kg}^{-1}$

22.1 Suppose a girl of mass 55 kg jumps off a wall, which is 3.0 m above the ground.

(a) What is her speed on reaching the ground?

She jumps on to a pressure pad, of the type used in intruder alarm systems. The resistance of the material inside the pad decreases when a force is applied to the pad. The pressure pad is connected to a data logger which records the current through the pressure pad.

(b) Sketch a graph to show how the current might vary from the time before she lands until the time when she is stationary on the pad.

(c) The data logger shows that she takes 25 ms to stop when she reaches the ground. What is the average force exerted on her by the ground?

(d) Estimate the maximum force exerted on her by the ground.

(e) The main bone in her leg is the tibia. If the minimum cross-sectional area of one of her tibia is $2.4 \times 10^{-4}\ \text{m}^2$, what is the maximum compressive stress in each of her tibia during the landing, assuming that the force is distributed equally between the two legs?

(f) This value is quite a large fraction of the maximum compressive stress of bone (about 16×10^7 Pa). What would you advise her to do when she reaches the ground?

2.1 21.7

Data: gravitational field strength $g = 9.81\ \text{N kg}^{-1}$

22.2 The diagram shows a cross-sectional view of a steel rod. A mass of 1.00 kg is supported by a cord which passes over the rod and is pulled as shown with a force *P*. The cord is in contact with the rod over a length which subtends an angle θ at the centre of the rod. A student is asked to investigate how the force *P* required to stop the mass falling depends on the angle θ. He obtains the following results:

$\theta/°$	45	90	180	270	360
P/N	8.25	6.94	4.91	3.35	2.46

He is told that theory shows that $P = We^{-k\theta}$, where *W* is the weight of the hanging mass and *k* is a constant which depends on the roughness of the rope and the rod.

(a) Explain why, to verify this relationship, he should plot a graph of $\ln(P/\text{N})$ on the *y*-axis against θ on the *x*-axis.

(b) What would be the significance of the gradient of the graph, and the intercept on the *y*-axis?

(c) Plot $\ln(P/\mathrm{N})$ against θ and deduce the value of k.
(d) Why would you expect the intercept on the y-axis to be 2.28?
(e) How many times, to the nearest quarter-turn, would the rope need to be wrapped round the rod if P is to be one-hundredth of the weight of the mass?
(f) It might be thought that the diameter of the rod would affect the size of P, since for a particular angle the length of contact of the rope, and hence the frictional force, would be greater. But the relationship between P and θ shows that P is independent of the diameter. Explain why, if the diameter increases, and the length in contact therefore increases, the friction force nevertheless remains the same.
(g) If, instead of P being required to stop the mass falling, P is required to raise the mass, the relationship between P and θ is $P = We^{k\theta}$. Assuming that the constant k has the same value in this situation, calculate the size of P when $\theta =$ **(i)** 90° **(ii)** 180°.
(h) To stop a large boat from moving away from the side of the dock while it is being moored, a rope may be wound round a bollard on the side of the dock. Why is it possible for one man by himself to do this, but impossible for him to pull the boat nearer to the dock?

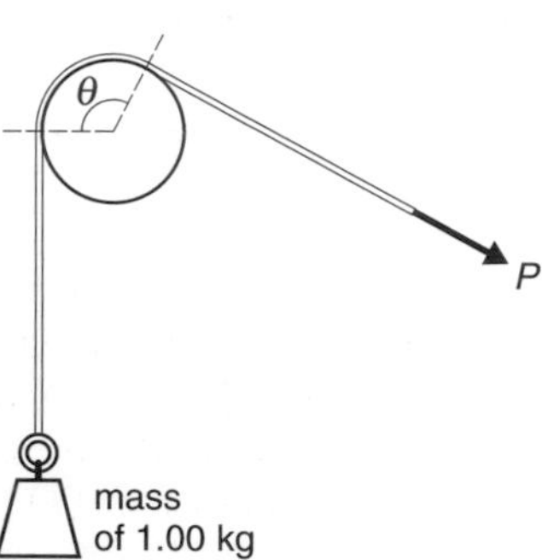

7.2 9.1

22.3 The diagram **(i)** shows a *strain gauge.* It consists of very thin strips of copper nickel alloy deposited on a plastic backing and doubled back on itself, as shown in **(ii)**, so that there are six parallel strips of alloy, each of length 5.0 mm. It has a resistance of 120 Ω. The strips are parallel to the length of the gauge. It can be fixed to a piece of material (e.g. a steel girder in a building) whose strain needs to be measured. If the girder extends, so does the wire in the strain gauge and its resistance increases. The increase in resistance can be detected remotely.

(a) If the resistivity of the copper nickel alloy is 50×10^{-8} Ω m, calculate the cross-sectional area, in mm^2, of the strips of material.

A resistance bridge network is constructed from four strain gauges connected as a 'rosette' as shown in part **(iii)** of the figure; the electrical connections are as shown in **(iv)**. The rosette is glued to a steel girder whose axis is parallel to the direction of the wires in P and S.

(b) How does including the gauges Q and R compensate for any changes of temperature?
(c) How does having both P and S present in the circuit double the effect of a change in strain?
(d) If the Young modulus of the steel is 340 GPa, and a tensile stress of 150 MPa is applied to the girder, what is the strain?
(e) What will be the strain in the wire of the strain gauge?
(f) Give two reasons why the resistances of P and S will increase.
(g) In which direction will current flow in the leads to the amplifier?

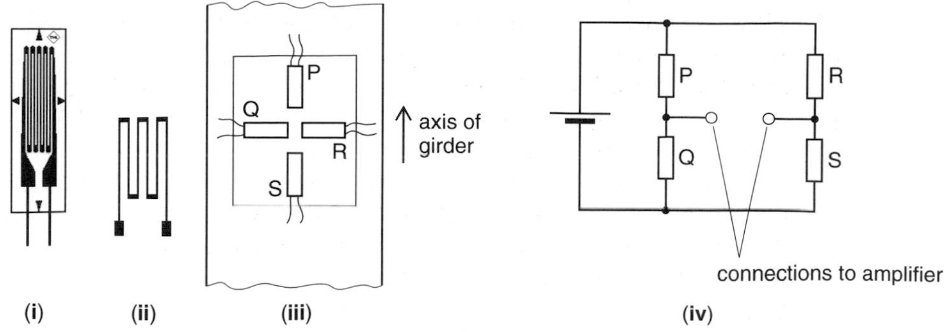

9.1 15.1 16.2

22.4 A steel rod of length 0.75 m is clamped at its centre in a horizontal position and struck axially with a hammer. Stationary waves are set up in the rod, so that musical notes are produced, but only the fundamental frequency has an amplitude large enough to be detected. This frequency can be measured by a microphone placed near one end. The microphone is connected to a data logger which counts the pulses of sound arriving at it in one-tenth of a second. This figure can be used to calculate the frequency.

(a) Draw a diagram to show the nodes and antinodes of the stationary wave produced in the rod which corresponds to the fundamental frequency.
(b) Is the wave transverse or longitudinal?
(c) What is the wavelength of the wave?
(d) If the data logger counts 344 pulses, what is the speed of elastic waves in the rod?
(e) The speed c of elastic waves in a rod is given by the equation $c = \sqrt{(E/\rho)}$ where E is the Young modulus of the material and ρ is its density. Show that the units of both sides of this equation are the same.
(f) The density of steel is 7860 kg m^{-3}. Calculate its Young modulus.
(g) The density of aluminium is 2710 kg m^{-3} but the speed of elastic waves in aluminium is the same as it is in steel. Deduce the Young modulus of aluminium.
(h) Why would you expect the speed of elastic waves to be large if the Young modulus is large and the density is small?
(i) When the hammer strikes the rod it sends a compression pulse along the rod. At the far end the pulse is reflected, and because it is a free end, it travels back as a rarefaction pulse. When this reaches the struck end of the rod it pulls the rod away from the hammer. If wires are connected to the rod and the hammer and a timer so that the time of contact can be measured, what time will be recorded?

1.1 1.3 11.3

Data $g = 9.81$ N kg^{-1}, A_r for bismuth = 209, 1 u = 1.66×10^{-27} kg, the Boltzmann constant $k = 1.38 \times 10^{-23}$ J K^{-1}

22.5 An early method of investigating the distribution of molecular speeds used apparatus like that shown in the figure. Atoms or molecules emerged from the oven and passed in a fine beam through the slit S and into a rotating drum. They crossed the drum and were deposited on a glass plate. The intensity of the deposition was measured optically. The slit was opened when the glass plate was in the position shown, and closed when the end B was opposite the slit.

(a) Draw a diagram of the glass plate, labelled AB as in the figure, and shade it to show what the distribution of atoms or molecules would be like.

(b) At which end were the fastest molecules? Explain.

(c) What is the r.m.s. speed of a bismuth atom, at 1100 K?

(d) If the diameter of the drum is 200 mm, how fast must it be rotated if atoms with this r.m.s. speed are to strike the glass plate 45° round the circumference from A?

(e) How far vertically did atoms with the r.m.s. speed fall in their flight between the slit and the drum, if this distance is 250 mm?

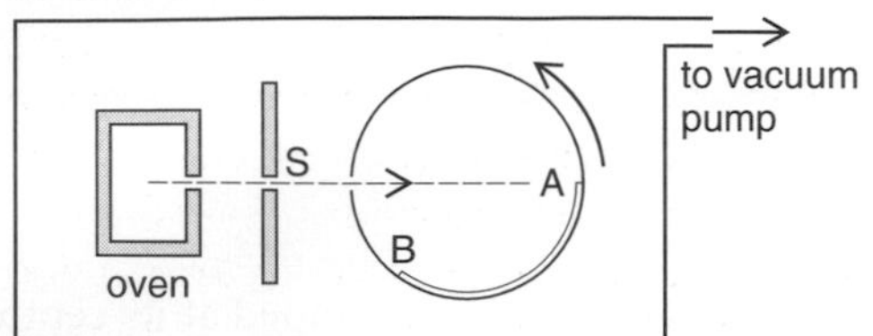

6.2 8.1 10.3 18.2 18.4

22.6 A capacitor of capacitance 10 mF is charged to a potential difference of 20 V and then discharged through a bundle of wire. The wire is an alloy of copper and nickel whose density is 8.9×10^3 kg m^{-3}. Its length is 2.0 m and its diameter is 0.38 mm.

(a) What is the mass of the wire?

(b) The resistivity of this alloy is 4.7×10^{-7} Ω m. What is the time constant for this *RC* circuit?

(c) The s.h.c. of the alloy is 420 J kg^{-1} K^{-1}. What is the temperature rise in the wire?

(d) To measure the temperature rise, the bundle of wire could be wrapped round the bulb of a mercury thermometer, or round one junction of a thermocouple. Which would give a better indication of the rise in temperature?

(e) If, instead, a p.d. of 10 V had been used, how many times would the capacitor have needed to be discharged to produce the same temperature rise? Explain.

9.1 9.2 18.1 18.2

22.7 It is possible to draw an analogy between what happens when a force is exerted on a spring and what happens when a p.d. is applied to a capacitor.

(a) For a spring $F = kx$ and for a capacitor $Q = CV$, where the symbols have their usual meanings, copy and complete the table to show which quantities for a capacitor are analogous to F, k and x for a spring.

spring	capacitor
F	
k	
x	

(b) Write down an expression for the energy stored in a stretched spring in terms of k and x.
(c) Write down an analogous expression for the energy stored in a charged capacitor.

3.1 4.3 12.1 12.4

Data speed of light $c = 3.00 \times 10^8$ m s^{-1}, mass of electron $m_e = 9.11 \times 10^{-31}$ kg, the Planck constant $h = 6.63 \times 10^{-34}$ J s

22.8 When a photon collides with an electron which is loosely bound to an atom it changes direction. The collision is elastic because there are no other forms of energy into which the kinetic energy of the photon may be changed. Before the collision the electron may be considered to be stationary. Because the photon behaves like a particle we can say that momentum and kinetic energy are conserved. (This scattering of photons is called the *Compton effect*, and is strong evidence for the particle nature of light.)

(a) Why must the electron have some kinetic energy after the collision?
(b) What can you say about the energy of the scattered photon, compared with the energy of the incident photon?
(c) Consider a beam of photons of frequency 2.1×10^{20} Hz. For each photon what is **(i)** the wavelength **(ii)** the energy **(iii)** the momentum?
(d) In which part of the electromagnetic spectrum are these photons?
(e) Suppose one photon strikes an electron head-on, and bounces back along its original path. Its new frequency is found to be 0.50×10^{20} Hz. What is the gain of kinetic energy of the electron?
(f) Use k.e. = $\frac{1}{2}mv^2$ to try to calculate the speed of the electron. Why cannot this be the correct way to calculate its speed?
(g) The electron is moving so fast that its mass has increased. The correct expression for the k.e. is $(m-m_0)c^2$ where m is its mass, m_0 is its mass when it is stationary, and c is the speed of light. Use this equation to calculate its mass.
(h) Calculate the momentum of the photon after the collision and hence deduce the momentum of the electron.
(i) The momentum of the electron is given by $p = mv$ where m is its mass. What is the speed of the electron?

5.2 15.1 17.3 21.6

Data: $g = 9.81$ N kg^{-1}, electronic charge $e = 1.60 \times 10^{-19}$ C

22.9 One end of a length of copper wire of diameter 0.50 mm is clamped to a laboratory bench and the other end runs over a pulley. The distance from the clamp to the pulley is 1.80 m. A mass of 2.0 kg hung at this end of the wire keeps it in tension. The density of copper is 8930 kg m^{-3}.

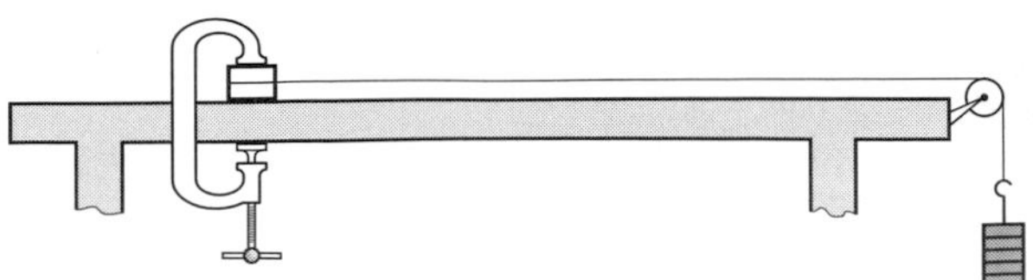

(a) What is the mass per unit length, μ, of the wire?

(b) The speed c of transverse waves on a wire is given by $c = \sqrt{(T/\mu)}$ where T is the tension in the wire. What is the speed of waves on this wire?
(c) What is the lowest frequency at which stationary waves can be set up on the wire? Draw a diagram to illustrate this, marking the positions of nodes and antinodes.
(d) Suppose you are provided with an alternating power supply of variable frequency, a U-magnet, and a means of connecting the power supply to the wire. Describe how you would generate stationary waves in the wire.

The same wire now carries a steady (direct) current of 3.0 A. The number of conduction electrons per unit volume in copper is $1.0 \times 10^{29}\ m^{-3}$.
(e) What is the drift speed of the electrons?
(f) If, instead, the wire carries an alternating current whose r.m.s. value is 3.0 A, what is the maximum speed of the electrons during their oscillations?
(g) Draw a graph to show how the *velocity* of the electrons during their oscillations varies with time if the frequency is 50 Hz. Label the axes and give numerical values.
(h) What is the amplitude of oscillation of the electrons?

12.1 13.3

22.10 **(a)** Under what circumstances would you expect the intensity of radiation from a small source to vary inversely with the square of the distance from the source (i.e. so that doubling the distance reduces the intensity to a quarter as much)?
(b) Discuss to what extent you would expect such a variation to apply in the following cases: **(i)** a light source in clear air **(ii)** an α-particle source in air **(iii)** a β-particle source in a vacuum **(iv)** a β-particle source in air **(v)** a γ-ray source in air.
(c) You are given a γ-ray source which is enclosed in a sealed container, and you are asked to show how the intensity of radiation A varies with distance r from the source. You cannot be sure of how far the source is from the edge of the container. Explain whether you would plot **(i)** A against r^2 **(ii)** A against $1/r^2$ **(iii)** $\sqrt{A}$ against r **(iv)** $\sqrt{A}$ against $1/r$. Draw a diagram to show the sort of graph you would expect to obtain.

1.3 4.3 5.1 5.4 7.3 12.1 20.2

Data: electronic charge $e = 1.60 \times 10^{-19}$ C, mass of electron $m_e = 9.11 \times 10^{-31}$ kg

22.11 The diagram shows some of the parts of an oscilloscope. The final anode (A_3) is at zero potential and the filament F is at a potential of –800 V. Anode A_2 is at a slightly lower potential than A_3, and anode A_1 is at a slightly lower potential than A_2. The grid G is at a slightly lower potential than F.

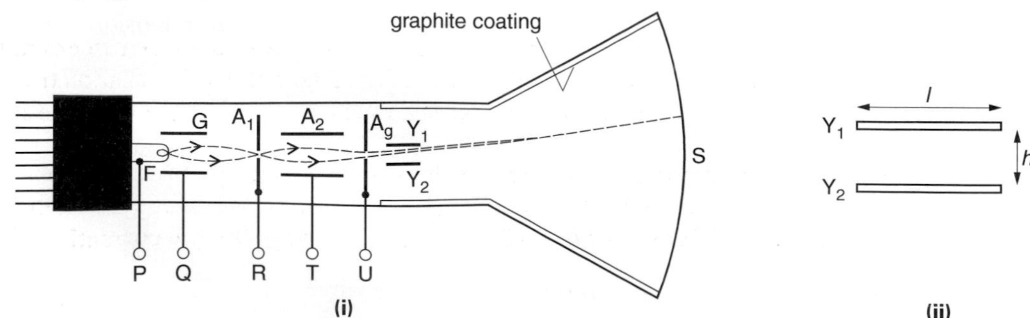

(a) Draw a diagram to show how you would connect power supplies to the terminals P, Q, R, T and U to achieve these potential differences.
(b) As shown, the inside of the tube is coated with graphite, and this is connected to terminal T. Mark on your diagram a point where you could place an ammeter to measure the current in the tube between A_3 and the screen S.
(c) How much electrical potential energy, in J, does an electron lose when it moves between F and A_3?
(d) With what speed does each electron pass through A_3?
(e) If the screen is at zero potential, what is the speed of an electron when it reaches the screen?
(f) At one setting of the brilliance control, 1.0×10^{16} electrons are leaving the cathode each second. What is the electric current between F and A_1?
(g) If the number of electrons passing A_3 is 90% of the number of electrons passing A_1, and the process of converting kinetic energy to light when the electrons hit the screen is 30% efficient, what is the rate of emission, in W, of light from the screen?
(h) Part **(ii)** of the diagram shows the deflector plates Y_1 and Y_2. A potential difference is applied to these plates to deflect the electron beam vertically. To create a uniform field between the plates, should the ratio h/l be small or large?
(i) To deflect the beam upwards, as shown, which of the plates must be at the higher potential?
(j) If the deflecting p.d. is 100 V, explain whether you would make the plates **(i)** +50 V and –50 V or **(ii)** +100 V and 0 V or **(iii)** 0 V and –100 V, or does it not matter?
(k) If the p.d. creates a uniform electric field between Y_1 and Y_2, what is the shape of the curve in which the electrons move while they are between the plates?
(l) If the length l of the plates is 30 mm, for what time is each electron moving between the plates?
(m) If the distance apart h of the plates is 10 mm, what is **(i)** the electric field strength between the plates **(ii)** the force on an electron **(iii)** the acceleration of an electron **(iv)** its vertical displacement as it leaves the plates?

5.1 18.1 20.3

Data: electronic charge $e = 1.60 \times 10^{-19}$ C, $\varepsilon_0 = 8.85 \times 10^{-12}$ F m^{-1}

22.12 In a small Van de Graaff generator the rubber belt is 5.0 cm wide. Positive electric charge is sprayed on to one side of it by a metal comb at the bottom end, and deposited on the dome at the top end. The density of charge on the belt is 8.0 μC m^{-2}. The belt moves at a steady speed of 0.40 m s^{-1}.
(a) In one second what area of belt moves past any point?
(b) What is the vertical electric current?
(c) Is its direction upwards or downwards?
(d) At intervals of 10 s the charge on the dome has increased to such an extent that it is discharged by means of a spark to a smaller metal dome a few centimetres away, which is earthed. How much charge is there on the dome just before the spark occurs?
(e) If the discharge takes 20 μs, what is the average current in the discharge?
(f) If the belt is stopped 5.0 s after a spark has occurred, the charge gradually leaks off the dome. If half leaks off in 40 s, what is the average leakage current?

(g) Draw a diagram of the dome and the smaller dome, and sketch the electric field lines around the two domes. Is the electric field strength in the gap larger or smaller than it would have been with the smaller dome absent?

(h) Now consider the large dome without the smaller dome present. Suppose the dome discharges when the electric field strength at its surface rises to $2.0 \times 10^6\ \text{V m}^{-1}$. Treating the dome as a perfect sphere of diameter 0.24 m, what would be the charge on it when the field strength reached this value?

(i) What would be the potential of the sphere?

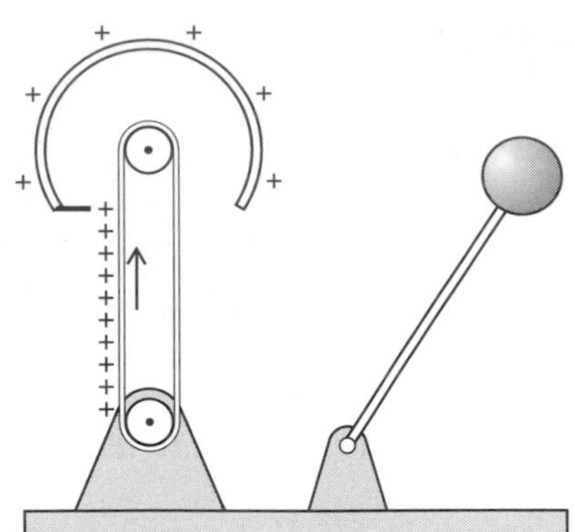

4.3 5.4 19.2

Data: electronic charge $e = 1.60 \times 10^{-19}$ C, mass of electron $m_e = 9.11 \times 10^{-31}$ kg

22.13 A cyclotron is a machine which uses the same p.d. many times to accelerate charged particles. The figure shows the principle: the ions are generated at S and are accelerated through a p.d. between the two hollow metal dees D_1 and D_2. In the dees they are moving at right angles to a magnetic field, so they move in a semicircle. When they return to the edge of the dee, the p.d. is in the opposite direction and accelerates them again across the gap to the other dee, where they again move in a semicircle. The process is repeated many more times than can be indicated in the figure.

(a) The p.d. between the dees might be 150 kV, and the ions might make 100 crossings of the dees. What difficulty might there be in trying to accelerate ions in a straight line through a single p.d. of 100 × 150 kV?

(b) Find the speed reached by a singly charged ion which has a mass of 1.7×10^{-27} kg if it starts from rest and is accelerated once through the p.d. of 150 kV.

(c) How long will it then take the ion to travel round the semicircular dee at this speed, if the magnetic flux density is 1.6 T?

(d) It will then again be accelerated through 150 kV. How long will it take to travel through the next semicircle: a shorter time, or the same time as before?

(e) The p.d. between the dees has to change direction so that it always accelerates the ions, from whichever direction they are coming. For this cyclotron, what would need to be the frequency of this alternation?

(f) Why would it be more difficult to construct a machine like this if the time for each ion to traverse each dee was not always the same?

(g) What difficulty might arise if the cyclotron were used to accelerate electrons?

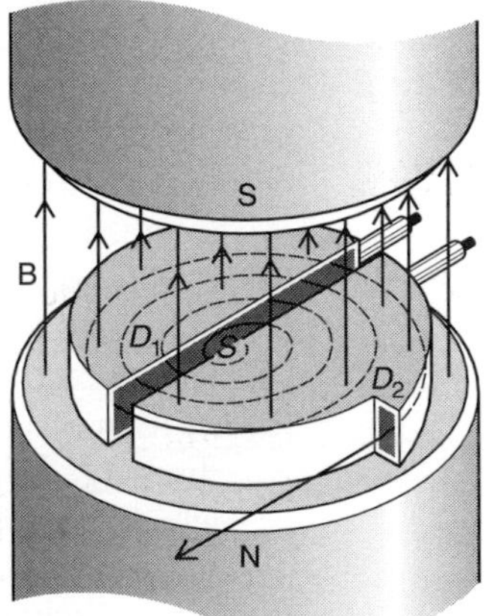

3.1 4.3 13.2 14.2

Data: 1 u ≡ 931.5 MeV

22.14 A nucleus of radon-220 ($^{220}_{86}Rn$) decays by α-particle emission.

(a) Write down the equation for this reaction given that the element with atomic number 84 is polonium (Po).

The masses of the atoms are given in the table:

nuclide	radon-220	helium-4	polonium-216
mass/u	220.011 40	4.002 604	216.001 92

(b) What is the decrease in mass in this reaction?

(c) How much kinetic energy (in MeV) will the two products share?

(d) The mass of the α-particle is 54 times less than the mass of the polonium nucleus. Explain why the speed of the α-particle is 54 times greater than the speed of the polonium nucleus.

(e) Hence explain why the kinetic energy of the α-particle is 54 times greater than the kinetic energy of the polonium nucleus.

(f) What is the kinetic energy (in MeV) of the resulting nucleus?

1.4 13.3 19.2

Data: electronic charge $e = 1.60 \times 10^{-19}$ C, mass of α-particle $= 6.69 \times 10^{-27}$ kg

22.15 A student tries to demonstrate the magnetic deflection of α-particles in a school laboratory. She has a strong permanent magnet which produces a uniform *B*-field of 220 mT over a distance of nearly 10 cm.

(a) Why must the experiment be done in an evacuated tube?

(b) How might she try to obtain a beam of α-particles?

(c) The α-particles she is using each have an energy of 4.8 MeV. What is their speed?

(d) Calculate the radius of the circle which they follow in the field when moving at right-angles to it.

(e) Discuss whether or not the demonstration will be effective.

19.2 20.2

Data: electronic charge $e = 1.60 \times 10^{-19}$ C

22.16 A stream of electrons is moving along the axis of an evacuated tube. The beam enters a region where there is a horizontal magnetic field at right angles to the beam.

(a) Draw a diagram to show the directions of the velocity, the field and the magnetic force.

(b) If the speed of the electrons is 2.00×10^6 m s^{-1} and the magnetic flux density is 30 mT, calculate the size of the magnetic force.

An electric field in the same region is then switched on.

(c) Mark on your diagram the direction of the field if the resultant force on the electrons is to be zero.

(d) What should be the size of the electric field strength?

(e) If the strengths of the two fields are *B* and *E*, and the velocity of the electrons is v, derive a relationship between *B*, *E* and v.

(f) Suppose the fields have been adjusted so that together they create zero resultant force on the electrons. Show on your diagram the direction in which the electrons would be deflected if they were then made to move a little faster.

(g) This arrangement of crossed fields is sometimes said to be a *velocity selector*. Why do you think it is given this name?

13.3 20.3

Data: electronic charge $e = 1.60 \times 10^{-19}$ C, $\varepsilon_0 = 8.85 \times 10^{-12}$ F m^{-1}

22.17 When an α-particle moves directly towards a nucleus it is pushed back by the positive charge on the nucleus. At some point the α-particle stops. At that point its initial kinetic energy has been transferred to electrical potential energy. The diagram shows this happening. This distance of closest approach gives us some idea of the size of the nucleus.

(a) Suppose an α-particle whose initial kinetic energy was 6.28 MeV is fired at some gold foil. What is this kinetic energy in joules?

(b) Suppose r is the closest distance of approach of the α-particle. The atomic number of gold is 79. Write down an expression for the electric potential energy of the α-particle at this point.

(c) Hence find the closest distance of approach by one of these α-particles to the centre of a gold nucleus.

(d) Repeat **(c)** for an α-particle whose initial k.e. was 4.63 MeV.

(e) Draw a diagram which shows the path of an α-particle which was approaching a gold nucleus, but not striking it head-on.

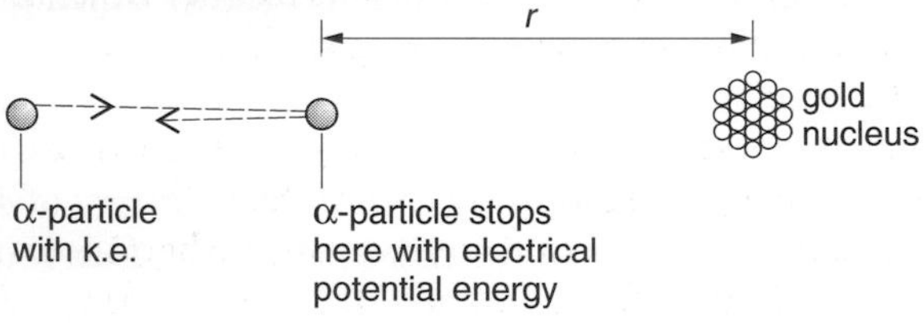

8.1 10.3 19.3

22.18 The photograph shows a metal ring supported by a magnetic force from the coil at the base of the stand.

(a) Why must the current in the coil be alternating?

(b) Why must the retort stand rod be made from iron or steel?

(c) Why is aluminium a good choice of metal for the ring?

(d) What can you deduce about the directions of the current in the coil and the ring at any moment?

(e) Why does the ring move upwards if the current in the coil is increased?

(f) The coil and ring act as a step-down transformer. However, it is not an ideal transformer, because the ring is so far from the coil. Assume its efficiency is 2.0%. If the p.d. supplied to the coil is 12 V, and the current in the coil is 2.0 A, what is the power transferred in the ring?

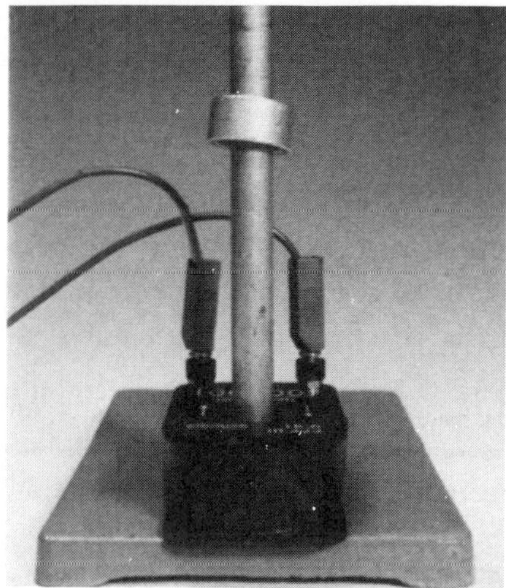

(g) The ring has an average radius of 6.0 mm, a thickness of 2.0 mm and a height of 15 mm. The density of aluminium is 2710 kg m^{-3}. What is the mass of the ring?

(h) The s.h.c. of aluminium is 880 J kg^{-1} K^{-1}. What is the initial rate of rise of temperature of the ring, in K min^{-1}?

2.4 19.2

Data: electronic charge $e = 1.60 \times 10^{-19}$ C, mass of electron $m_e = 9.11 \times 10^{-31}$ kg

22.19 The photograph shows the track of a charged particle in a bubble chamber. The track is curved because the particle is moving at right angles to a magnetic field, whose direction is into the paper.

(a) The particle carries a positive charge. Is it moving clockwise or anti-clockwise?

(b) What is happening to the radius of the path?

(c) The particle leaves a track in the bubble chamber because it is ionising the hydrogen. What must therefore happen to the kinetic energy of the particle and its speed?

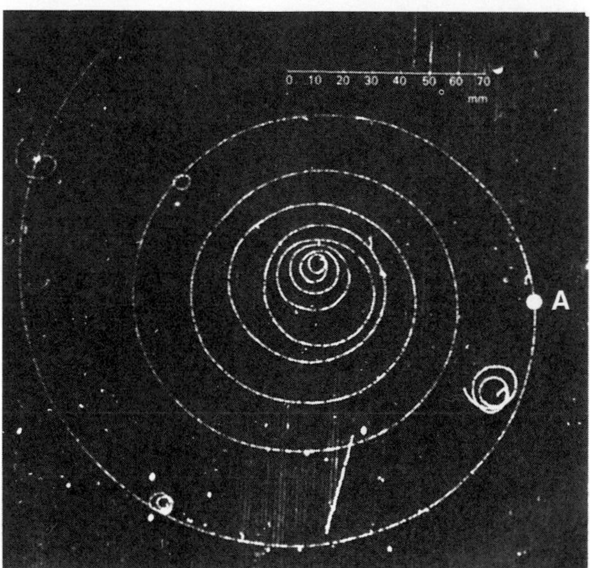

(d) Derive an expression for the momentum mv of the particle in terms of the magnetic flux density B, the charge Q on the particle and r, the radius of its path.

(e) Use the scale shown on the photograph to estimate the radius of the particle's path at point A.

(f) If the particle is a positron, and the magnetic flux density is 0.50 T, what is its momentum at point A?

(g) Particles in bubble chambers are often moving at speeds very close to the speed of light, so that in accordance with the ideas of relativity, their mass is much greater than their rest mass. Assuming that the speed of this positron is equal to the speed of light, what is its mass?

(h) How many times greater is this than its rest mass?

4.3 4.4 11.2 13.3

Data: electronic charge $e = 1.60 \times 10^{-19}$ C, $k = 1.38 \times 10^{-23}$ J K^{-1}

22.20 The neutrons emitted in the fission of a uranium nucleus typically have energies of 1 MeV and therefore speeds of 1.4×10^7 m s^{-1}. They need to be slowed down if there is to be a good chance of them being captured by another ^{235}U nucleus and so cause a chain reaction.

(a) α-particles slow down when they pass through any matter. Why is this mechanism not available to slow down neutrons?

(b) What happens when a neutron collides head-on with a stationary proton? Assume that the proton and the neutron have the same mass and that the collision is elastic.

(c) Unfortunately the neutron may be captured by the proton to form deuterium, so to slow down the neutrons a different element must be used. The diagram shows the head-on collision between a neutron and a carbon nucleus. Their masses have been written as m and $12m$, and the velocities before and after the collision are shown. Assume that the collision is elastic and use two different conservation laws to show that **(i)** $u = v + 12w$ **(ii)** $u^2 = v^2 + 12w^2$.

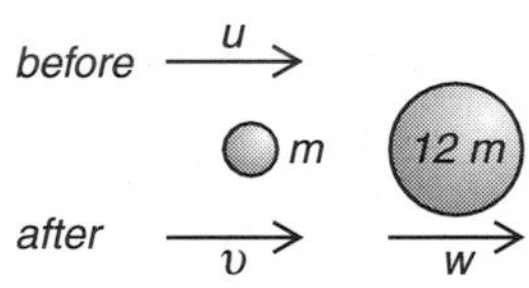

(d) Show that a solution of these equations is $v = -0.846u$ and $w = +0.154u$.

(e) What fraction of its energy does the neutron keep in this collision?

(f) When the neutron does not collide head-on with a carbon nucleus it does not lose so much energy. In a glancing collision it will keep almost all of it. What is the average amount of energy which the neutron keeps?

(g) The neutron will stop slowing down when its kinetic energy is roughly equal to the kinetic energy of the moderator atoms. Use $E = \frac{3}{2}kT$ to find the kinetic energy of the moderator atoms if the temperature of the core is 500 K. Give your answer in eV.

(h) How many collisions must the neutrons make, on average, before their k.e. is reduced from 1 MeV to your answer to **(g)**?

12.4 16.4

Data: mass of electron $m_e = 9.11 \times 10^{-31}$ kg, speed of light $c = 3.00 \times 10^8$ m s^{-1}

22.21 When light of wavelength λ strikes a circular target of radius r it is diffracted and there is a minimum intensity at an angle θ given by $\sin\theta = 0.611\lambda/r$. Similarly electrons approaching a nucleus are found to be diffracted. In one experiment electrons with an energy of 420 MeV were used.

(a) Show that if you could use the equation $E_k = \frac{1}{2}mv^2$ to calculate the speed of electrons with an energy of 420 MeV you would find that their speed was apparently 1.2×10^{10} m s^{-1}. Why is this an impossibility?

(b) At speeds approaching the speed of light, the Newtonian laws of mechanics cannot be used: at the speeds of these electrons the relationship between energy E and momentum p is (approximately) $E = pc$, where c is the speed of light. Use this relationship to calculate the momentum of electrons which have an energy of 420 MeV.

(c) What is the wavelength of these electrons?

(d) It was found that the first minimum of intensity when the electrons were scattered from carbon nuclei occurred at an angle of 50°. What does this information tell us about the radius of the carbon nucleus?

(e) In the earliest experiments at the Stanford Linear Accelerator, electrons of energy 6.0 GeV were used to probe the nucleus. For these electrons what is **(i)** the momentum **(ii)** the wavelength?

(f) Using the equation $\sin\theta = 0.611\lambda/r$ explain why using these electrons might be expected to give us information about the structure of protons and neutrons.

3.2 12.1 20.3 20.4

Data: $g = 9.81$ N kg^{-1}, speed of light $c = 3.00 \times 10^8$ m s^{-1}, radius of the Earth = 6.37 Mm, the Planck constant $h = 6.63 \times 10^{-34}$ J s

22.22 The Hubble space telescope is in orbit around the Earth at a height of 600 km. Its mass is 11 tonnes.

(a) What is the gravitational acceleration at that distance?

(b) How fast is it moving?

(c) What is the pull of the Earth on it?

It has solar panels to supply it with electrical energy. The intensity of radiation is 1.4 kW m^{-2}, and the mean wavelength of the radiation from the Sun is 550 nm. The telescope needs 5.0 kW of power. The solar panels convert energy with an efficiency of 25 %. Assume that the panels are able to face the Sun for 80 % of the time.

(d) What is the minimum area for the solar panels?

(e) What is the energy of a photon of wavelength 550 nm?

(f) The momentum p of a photon is given by the equation $p = E/c$. What is the momentum of one of these photons?

(g) When the panel is facing the Sun, how many photons arrive each second on each square metre of the solar panels, assuming that the radiation strikes them normally?

(h) Find the pressure exerted by the radiation on the solar panels, assuming that it is totally absorbed.

12.1 13.3 13.4 20.3

Data: Avogadro constant $L(=N_A) = 6.02 \times 10^{23}\ mol^{-1}$, A_r for plutonium = 238, A_r for oxygen = 16

22.23 The Cassini space probe was launched from Cape Canaveral on 6 October 1997. It was sent in an orbit that passed close to Venus twice, then past the Earth again, and then on the path shown in the diagram. The distances of the Earth, Jupiter and Saturn from the Sun are 1.50×10^{10} m, 7.78×10^{10} m and 14.27×10^{10} m, respectively.

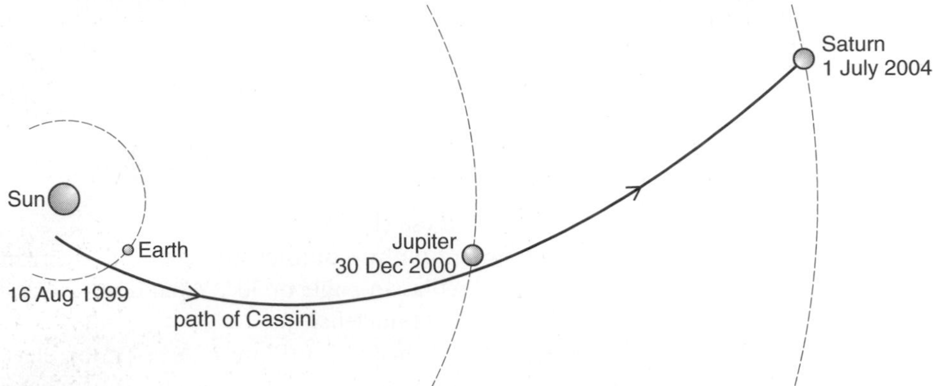

(a) Use the diagram to estimate the approximate distances that Cassini travelled between **(i)** the Earth and Jupiter **(ii)** Jupiter and Saturn.

(b) What were its average speeds on these two parts of its journey?

(c) Why did Cassini slow down, and why was its path curved?

(d) The power for Cassini was provided by 32.6 kg of plutonium dioxide (PuO_2). What was **(i)** the initial mass of plutonium **(ii)** the number of atoms of plutonium?

(e) The half-life of ^{238}Pu is 86 years. What is **(i)** the decay constant, in s^{-1} **(ii)** the activity, in Bq, of this source?

(f) The α-particles it emits each have an energy of 5.5 MeV. What is its rate of generation of internal energy?

(g) Why is it convenient to use an α-emitting source in this application?

(h) The internal energy is converted into electrical energy by means of a thermopile with an efficiency of 6%. What is the initial rate of production of electrical energy?

(i) $^{238}_{94}$Pu decays into $^{234}_{92}$U, which decays by α-emission into $^{230}_{90}$Th, which itself decays by α-emission. The half-lives of these last two decays are 2.5×10^5 years and 6.7×10^4 years respectively. Do they make any significant contribution to the activity of the source?

(j) What will be the rate of production of electrical energy when Cassini reaches Saturn?

(k) At the Earth's surface the intensity of radiation from the Sun is 1.4 kW m^{-2}. What is the intensity of the Sun's radiation in Saturn's orbit?

(l) How large would solar panels have to be to provide the same power as the decay of plutonium? Assume that the efficiency of conversion of the radiant energy into electrical energy is 25%.

17.1 20.1 20.3 20.4

Data: $g = 9.81\ \text{N kg}^{-1}$, $G = 6.67 \times 10^{-11}\ \text{N m}^2\ \text{kg}^{-2}$, radius of Earth $R = 6.37$ Mm

22.24 The diagram **(i)** shows a hollow spherical shell of matter. Its total mass is m. Two point masses m_1 and m_2 are placed inside and outside the shell. Sir Isaac Newton proved the resultant gravitational pull of the shell on m_1 is zero and the resultant gravitational pull of the shell on m_2 is as if all the mass of the shell were at its centre.

(a) Give qualitative, or partly quantitative, arguments for thinking that these statements are true.

(b) Now consider a solid sphere of mass M, as shown in **(ii)**. A point mass m is placed somewhere inside the sphere, at a distance r from the centre. Use Newton's results to explain why the resultant gravitational pull of the sphere on the point mass m is the same as if it was at the surface of a sphere of radius r. [Hint: think of the solid sphere as divided up into concentric spherical shells: the point mass m is inside some of these, and outside the others.]

(c) Explain why the result in **(b)** would still be true even if the sphere were not of uniform density but (like the Earth) made up of concentric spherical shells each of which had the same density.

Arguments like those above can be used to show that the gravitational acceleration g inside a sphere of uniform density is proportional to its distance from the centre.

(d) Sketch a graph to show this variation, labelling your axes with numerical values.

(e) Suppose a tunnel could be drilled diametrically through the Earth and evacuated, and that a mass m were let go at one end of the tunnel, as shown in part **(iii)** of the diagram. Describe what happens to the mass as it moves through the tunnel to the other side of the Earth.

(f) In terms of g_0, the gravitational acceleration at the Earth's surface, what would be the size of its acceleration **(i)** initially **(ii)** halfway to the centre **(iii)** at the centre **(iv)** halfway to the other surface **(v)** when it reaches the other surface?

(g) Show that the acceleration a is given by the equation $a = -(g_0/R)r$ where R is the radius of the Earth and r is the distance of the mass from the centre, and explain why this shows that the motion of the mass is simple harmonic.

(h) What is the period, in minutes, of the s.h.m.?

Suppose a satellite could be sent round the Earth so that it just skimmed the surface. (Air resistance, and mountains, would of course make this impracticable.)

(i) What is the acceleration of the satellite?

(j) What is its speed?

(k) What is its period?

(l) Explain why your answers to **(h)** and **(k)** are the same.

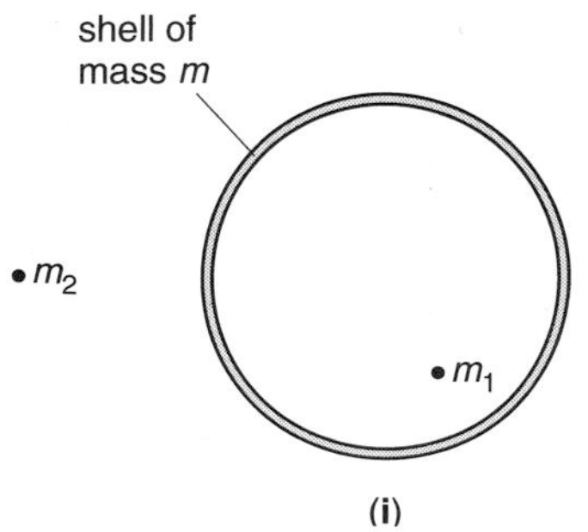

(i)

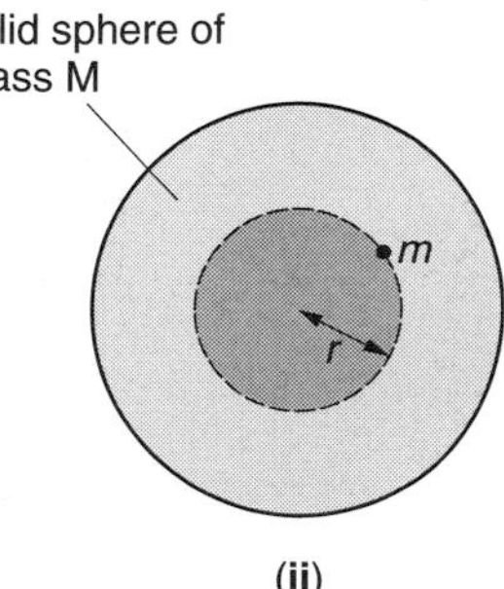

(ii)

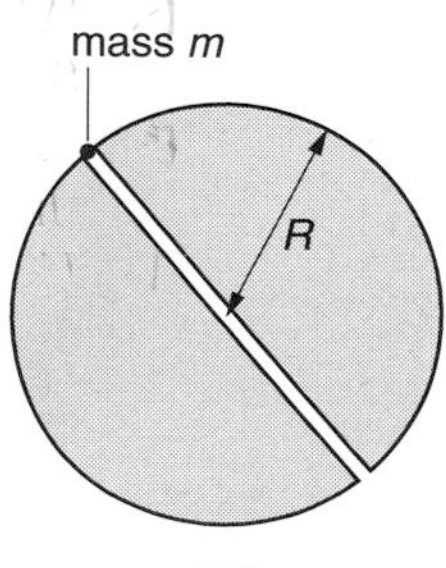

(iii)

11.3 14.2 20.3

Data: electronic charge $e = 1.60 \times 10^{-19}$ C, $\varepsilon_0 = 8.85 \times 10^{-12}$ F m^{-1}, the Boltzmann constant $k = 1.38 \times 10^{-23}$ J K^{-1}, 1 u ≡ 931.5 MeV

22.25 In the interior of the Sun, nuclear reactions take place whose net result is the conversion of hydrogen into helium with the release of energy. The three stages in this proton-proton cycle are as follows:

(i) ${}^1_1\text{H} + {}^1_1\text{H} \rightarrow {}^2_1\text{H} + {}^0_1\beta + \nu$

(ii) ${}^1_1\text{H} + {}^2_1\text{H} \rightarrow {}^3_2\text{He} + \gamma$

(iii) ${}^3_2\text{He} + {}^3_2\text{He} \rightarrow {}^4_2\text{He} + 2{}^1_1\text{H}$

(a) The masses of the nuclei are: ${}^1_1$H: 1.007 28 u, ${}^2_1$H: 2.013 55 u, ${}^3_2$He: 3.014 93 u; ${}^4_2$He: 4.001 51 u and the mass of the positron is 0.000 55 u. Calculate the binding energy for each of these reactions.

(b) For there to be two ${}^3_2$He nuclei available for reaction **(iii)**, the first two reactions must take place twice. What is the total binding energy in the creation of a single helium-4 nucleus?

(c) What is the total energy released in the creation of a single helium-4 nucleus?

(d) The first stage in the process is the fusion of two hydrogen nuclei. Why is energy needed to do this?

(e) When hydrogen is raised to high temperatures it dissociates into atoms and is then ionised. The ions have high speeds and are therefore able to fuse when they come together. Use $E = \frac{3}{2}kT$ to find the temperature at which the hydrogen ions would have the 1 MeV which they would need in order to be able to fuse.

(f) Why are these reactions sometimes called thermonuclear reactions?

(g) At what distance from the centre of one proton would another have an electrical potential energy of 1.00 MeV?

(h) At this distance what is size of the electric force which one proton exerts on the other?

Answers

1 Linear and circular motion

1.1 **(a)** 3.3 m s^{-1} **(b)** zero
1.3 **(a)** **(i)** 1.6 mm **(ii)** 0.18 mm from A towards B
(b) **(i)** 79 mm s^{-1}
(ii) 8.8 mm s^{-1} from A towards B
1.4 **(a)** 9 m s^{-1}
(b) –21 m s^{-1} **(c)** –12 m s^{-1}
(d) 20 m s^{-1} west
1.5 **(a)** 3.0 m s^{-1} **(b)** 11 m s^{-1}
1.7 **(b)** **(i)** 1.5 m s^{-1}
(ii) 0.80 m s^{-1} **(iii)** 0.50 m s^{-1}
(c) about 0.9 m s^{-1}
1.8 **(b)** **(i)** 23 m s^{-1} **(ii)** 11 m s^{-1}
(c) 35 ms^{-1}
1.9 35 ms
1.10 7 m
1.11 **(a)** 15.40 s **(b)** 64.92 m s^{-1}
1.13 P takes 50 s, Q takes 33 s, R takes 20 s
1.14 **(a)** 72 m **(b)** 72 m
1.17 **(a)** 57 m s^{-2} **(b)** –190 m s^{-2}
1.18 **(a)** +2.5 m s^{-2}, zero, –5.0 m s^{-2}
1.19 **(a)** 3.0 m s^{-1} **(b)** 6.0 m s^{-1}
(c) 1.2 m s^{-2}
1.20 **(a)** 33 s **(b)** 1.9 km
1.21 400 m
1.23 **(a)** 4.1 m s^{-2}
(b) just > 30 m
1.24 **(a)** **(i)** 0.33 m **(ii)** 0.81 m
(b) **(i)** 1.65 m s^{-1}
(ii) 4.05 m s^{-1}
(c) 0.60 s **(d)** 4.0 m s^{-2}
1.25 **(a)** 800 m s^{-2}
(b) **(i)** 0.07 s **(ii)** 30 mm
1.26 **(a)** 9.0×10^{12} m s^{-2}
(b) 2.3 m
1.27 **(b)** **(i)** 0.15 km **(ii)** 0.49 km
(c) **(i)** 0.96 m s^{-2} **(ii)** 3.0 m s^{-2}
1.28 **(a)** 21.3 m.p.h.
(b) **(i)** 8.0 m s^{-2} **(ii)** 1.5 m s^{-2}
1.29 **(b)** 45 m, 20 m, 5 m
1.30 **(a)** $v^2 = u^2 + 2as$
(b) $s = ut + \frac{1}{2}at^2$
1.31 **(a)** 26 m s^{-1} **(b)** 57 s
(c) 0.93 m s^{-2}
1.32 0.93 m s^{-2}
1.33 **(a)** 20 m s^{-1} **(b)** 0.18 s
(c) 220 m s^{-2}
1.34 220 m s^{-2}
1.35 **(a)**, **(b)** and **(c)** all 400 m
1.36 **(a)** 9.8 m s^{-1}, 0, –9.8 m s^{-1}, –19.6 m s^{-1}
(b) 19.6 m, 0 **(c)** 39.2 m
1.38 **(a)** 1.51 m s^{-1} **(b)** 3.03 m s^{-1}
(c) 10.1 m s^{-2}
1.40 **(a)** **(i)** 0.45 s **(ii)** 0.64 s
1.41 44 m
1.42 **(b)** 10.0 m s^{-1} each time
(c) 0.8 m **(i)** 0.45 m
(ii) 0.29 m
1.43 1.8 m
1.44 1.5 s
1.45 0.025 s, i.e. flashing at 40 Hz
1.46 **(a)** **(i)** 1.2 m **(ii)** 1.4°
(b) greater
1.47 **(a)** 12 m s^{-2} **(b)** 1.2 s
(c) 7.3 m **(d)** 51 m
1.48 **(a)** 9.4 m s^{-1} **(c)** 16°
1.49 **(b)** 2.3 s **(c)** 0.44 s
(d) 23 m s^{-1} **(e)** 8.6°
1.51 **(a)** **(i)** 15° s^{-1} **(ii)** 0.26 rad s^{-1}
(b) **(i)** 0.10° s^{-1}
(ii) 1.7×10^{-3} rad s^{-1}
1.52 470 m s^{-1}
1.53 12.5 m
1.54 **(a)** 21.7 rad s^{-1}, 207 rev min^{-1}
(b) 51.0 rad s^{-1}, 487 rev min^{-1}
1.55 **(a)** 10 m s^{-1} **(b)** 10 m s^{-1}
(c) 0 **(d)** 6.4 m s^{-1}
(e) 9.1 m s^{-1} SE
(f) 20 m s^{-1} NW
(g) 14 m s^{-1} SW
1.56 430 m s^{-2}
1.57 **(a)** 7.3×10^{-5} rad s^{-1}
(b) 3.1 km s^{-1} **(c)** 0.22 m s^{-2}
1.58 22 m s^{-1}
1.59 **(a)** 1.1 m s^{-2} **(b)** 4.4 m s^{-2}
1.60 **(a)** 13 s

2 Balanced and unbalanced forces

2.1 35 N
2.2 U = 22 N, F = 15 N, T = 29 N
2.5 **(c)** 850 N
2.6 **(b)** **(i)** 700 N **(ii)** 50 N
(iii) zero
2.7 **(b)** P = 6000 N, Q = 2000 N
2.8 **(a)** 41 N **(b)** 51 N
2.9 **(a)** 10 N **(b)** 17 N **(c)** 17 N
2.10 **(b)** 49 N, R = 46 N, F = 17 N
2.11 6.1°
2.12 19°
2.13 **(a)** 710 N **(b)** 69 kg
2.14 **(a)** 43° **(b)** 30 N
(c) 77°, 136 N
2.15 **(a)** 9.4 N **(b)** 18 N
2.16 **(a)** –1200 N m **(b)** 0
(c) 600 N
2.17 **(a)** –2000 N m, –600 N m
(b) 1300 N
2.18 **(a)** front 120 N, back 80 N
(c) 2.0 m
2.19 **(a)** 0.22 MN
(b) 8.7 m from A
2.20 **(a)** 140 N **(b)** 38 N
2.21 **(a)** about 20 cm **(b)** 3.3 N m
(c) 5 cm **(d)** 67 N **(e)** 50 N
2.22 **(b)** 3.0 kg
(c) xF = 1.93 N m for all x, hyperbola
2.24 **(b)** 17 N **(c)** 147 N, 17 N
(d) **(i)** 148 N **(ii)** 6.5°
2.25 **(b)** 28 N **(c)** 85 N
2.26 **(b)** 230 N
(c) X = 120 N, Y = 100 N

2.28 **(a)** 5.0 m s^{-2} **(b)** 4.0 m s^{-2} **(c)** 7.5 m s^{-2} **(d)** 1.2 m s^{-2} **(e)** 5.0 m s^{-2}
2.29 **(a)** −24 N **(b)** −10 N **(c)** −2.2 N **(d)** +26 kN **(e)** -1.8×10^{-9} N
2.30 0.55 m s^{-2}
2.31 **(a)** **(i)** 21 kN **(ii)** 8 kN **(b)** both 26 times
2.32 1200 N
2.33 **(a)** 3.0×10^3 m s^{-2} **(b)** 6.0×10^8 N
2.34 2.0×10^4 N
2.36 **(a)** no movement **(b)** no movement **(c)** $a = +2.0$ m s^{-2}
2.37 **(a)** **(i)** 340 N **(ii)** 730 N **(b)** **(i)** 5.2 m s^{-2} **(ii)** 11 m s^{-2}
2.38 **(a)** **(i)** 6.0 kN **(ii)** 5.2 kN
2.39 **(a)** 4.8 m s^{-2} **(b)** 0.29 kN forward **(c)** 0.29 kN
2.41 69 N
2.42 **(a)** **(i)** F/kN = 0.83, 0.67, 0.34, 0.085 **(ii)** F/kN = 0, 0.16, 0.49, 0.75
2.43 **(a)** 12 kN **(b)** 14 kN **(c)** 8.2 kN **(d)** 15 kN
2.44 **(b)** **(i)** 3.1 kN **(ii)** 1.7 kN
2.45 6.2 kN
2.46 0.40 N
2.47 **(a)** **(i)** 32 N **(ii)** 6.4 N
2.48 **(b)** 7.8 N **(c)** 3.9 kg
2.49 **(a)** thread at 4.7° to vertical
2.50 **(a)** 0.61 m s^{-2} **(b)** 134 N, 86 N
2.52 0.44 kN
2.54 **(a)** 0.16 kN
2.55 3.2 kN
2.56 **(b)** **(i)** 11 m s^{-2} upwards **(ii)** 0.63 kN
2.57 **(b)** 620 N **(c)** **(i)** 620 N **(ii)** 1.1 kN
2.58 **(a)** 12 m s^{-1} **(b)** **(i)** 3.9 kN **(ii)** 3.9 kN
2.59 **(b)** 3.0 kN **(c)** 22 m s^{-1}
2.60 **(a)** 733.5 N **(c)** 731.0 N **(d)** 731.0 N
2.61 **(a)** **(i)** 0.15 m **(ii)** 0.91 m s^{-1} **(iii)** 5.7 m s^{-2} **(b)** **(i)** 1.1 N **(ii)** 2.2 N
2.62 **(a)** **(i)** 8.7 kN **(ii)** 38 m s^{-1}

3 Linear momentum

3.1 **(a)** **(i)** 150 kg m s^{-1} north **(ii)** 150 kg m s^{-1} south **(b)** 2.0×10^4 kg m s^{-1} east **(c)** 5.0×10^9 kg m s^{-1} west
3.4 **(a)** +8.0 m s^{-1} **(b)** +1.0 m s^{-1} **(c)** −1.0 m s^{-1}
3.5 **(a)** moves away from bank at 3.0 m s^{-1} **(b)** no
3.6 0.28 kg
3.7 **(b)** 3.0 m s^{-1}
3.8 2.0 kg
3.9 0.52 m s^{-1} in direction of faster player
3.10 32 kg
3.11 24 m s^{-1}
3.13 3.4×10^5 m s^{-1}
3.14 **(a)** **(i)** -3.0×10^4 kg m s^{-1} **(ii)** $+3.0 \times 10^4$ kg m s^{-1} **(b)** **(i)** A: +3.0 m s^{-1}, B: +0.5 m s^{-1} **(ii)** A and B: +1.5 m s^{-1} **(c)** 75 000 kg m s^{-1} at all times
3.15 **(a)** 0.67 m s^{-1}
3.16 +8.7 m s^{-1}
3.20 **(a)** 3.0×10^3 N s east **(b)** 100 N s down **(c)** 480 N s down
3.21 **(a)** 45 N s **(b)** 45 N s
3.22 2.7 m s^{-1}
3.23 **(a)** 40 N s **(b)** 140 N s **(c)** 75 N s
3.24 **(a)** 7.4 kN
3.25 **(a)** about 2.5 N s **(b)** about 40 m s^{-1} **(c)** 15 000 m s^{-2} or 15 m s^{-1} per millisecond
3.26 **(b)** 90 N s $\Rightarrow$ 1.1 m s^{-1}
3.28 **(a)** acceleration **(b)** **(i)** 83 kN **(ii)** 83 kN
3.30 **(a)** 1.2 m s^{-1} **(c)** 1.8×10^4 N, 1.8×10^4 N
3.31 50 ms
3.33 **(a)** 4.2×10^4 N s **(b)** 4.2×10^4 N **(c)** 1.7×10^5 N
3.34 **(a)** 5.3×10^{-23} N s **(b)** 1.9×10^{21} **(c)** 100 kPa

4 Work and energy

4.1 **(a)** 8000 J **(b)** −8000 J
4.2 **(a)** parallel 35 kN, perpendicular 8.1 kN **(b)** **(i)** 140 MJ **(ii)** zero
4.3 **(a)** **(i)** 2.3 kJ **(ii)** 1.4 kJ **(b)** barbell has some k.e.
4.4 **(a)** 240 J, −240 J **(b)** 190 J, −190 J
4.5 **(b)** 24 J, −16 J, zero, zero
4.6 **(a)** 6.2 kJ, zero, −1.1 kJ, −5.1 kJ
4.7 zero
4.8 **(a)** 0.59 N **(b)** **(i)** zero, 12 mJ **(ii)** zero, −12 mJ
4.9 **(a)** **(i)** 0.80 J **(ii)** 3.2 J
4.10 **(b)** **(i)** 2.0 J **(ii)** 6.0 J **(iii)** 100 J **(iv)** 140 J
4.11 about 6.4 GJ
4.12 **(a)** about 10 MJ **(b)** about 2 kW
4.13 **(a)** 35 MJ **(b)** **(ii)** 100p
4.14 **(a)** kg m^2 s^{-3} **(b)** kg m^2 s^{-3}
4.15 **(a)** 27 W **(b)** 7.1 kW
4.16 98 W
4.17 **(a)** 44 s **(b)** 180 kJ
4.18 about 80 W
4.19 **(a)** 65 kJ **(c)** 1.1 kW
4.21 50 kN
4.22 **(a)** **(i)** 79% **(ii)** 100% **(b)** **(i)** 330 N **(ii)** 250 N
4.24 v should be v^2
4.25 **(a)** 17×10^6 m^2 or 17 km^2 **(b)** 0.006% **(c)** 0.4%
4.26 **(a)** 0.31 MJ **(b)** 1.7 kJ
4.28 **(a)** 3.2 J **(b)** 2.7 J **(c)** stored elastic energy **(d)** 0.5 J
4.29 9.3 μW
4.30 **(a)** 0.15 MJ **(b)** 830 W
4.31 **(a)** 9.0 kJ **(b)** 1.5 kJ
4.32 **(a)** 130 W **(b)** 56 W
4.33 **(a)** 1600 N **(b)** 800 m
4.34 **(a)** 23 m **(b)** 46 m **(c)** 103 m
4.35 **(a)** 16 kJ **(b)** 4.1 kN

4.36 a bouncing ball
4.37 **(a)** 25 m s^{-1} **(b)** 31 J **(c)** 20 m s^{-1}
4.38 0.88 mW
4.39 **(b)** 50 g mass gains k.e.
4.40 **(a)** 7.0 m s^{-1} **(b)** 0.41 J **(c)** 0.17 N
4.41 **(a)** 1.6 J **(b)** 1.6 J **(c)** speed = 4.4 m s^{-1}, it breaks
4.42 **(b)** **(i)** 7.7 m s^{-1} **(ii)** 8.3 m s^{-1}
4.44 **(a)** 7.2 m s^{-1} **(b)** 1.5 m above his take–off centre of gravity
4.45 gain of k.e. = 71 kJ, loss of g.p.e. = 103 kJ, push of air = 210 N
4.46 **(a)** **(i)** 75.6 J **(ii)** 75.5 J **(b)** 1.0 m s^{-1}
4.47 **(a)** 580 × 10^6 kg **(b)** 580 × 10^3 m^3, 1.6 m
4.48 **(a)** 46 s **(b)** 36 m s^{-1} **(c)** 3.7 kN
4.49 **(a)** –0.80 J **(b)** +0.40 J **(c)** +0.40 J **(d)** –1.6 J **(e)** +1.6 J, the lump has no k.e.
4.51 **(a)** 2.2 m s^{-1} **(b)** 150 kJ
4.52 **(a)** 0.96 J **(b)** 19 N
4.53 250 J
4.54 **(b)** 2.0 J
4.55 **(b)** 0.99 m s^{-1} **(c)** 250 m s^{-1}
4.56 93 m
4.58 0, 2υ

5 Electric currents and electrical energy

5.4 **(a)** 0.20 A **(b)** 0.20 A **(c)** 0.60 A **(d)** 0.40 A
5.5 C 0.20 A, D 0.15 A
5.6 5.0 A
5.7 **(a)** 144 C **(b)** 9.0 × 10^{20}
5.8 1.1 s
5.9 36 C
5.10 0.50 kA
5.11 20 kA
5.12 **(b)** **(i)** 6.0 kC **(ii)** 0.60 kC **(iii)** 13.2 kC
5.13 **(a)** 30 C **(b)** 10 mC **(c)** 20 μC
5.14 **(c)** 0.36 kC **(d)** 6.3 kC
5.16 8.2 mm s^{-1}
5.17 1.7 mm s^{-1}
5.18 about 15 h
5.20 **(a)** 1:4 **(b)** thinner wire
5.22 31 m s^{-1}
5.23 **(a)** 18 kC **(b)** 20 g **(c)** 1.1 × 10^{23} **(d)** 1.6 × 10^{-19} C
5.24 **(a)** 15 J **(b)** 45 J **(c)** 30 J
5.25 **(a)** 150 J **(b)** less chemical energy, more electrical energy
5.26 **(a)** 18 J **(b)** 36 J **(c)** 72 J
5.27 **(a)** 7.5 V **(b)** 4.5 V **(c)** 1.5 V
5.29 **(a)** 5 cells **(b)** 3.6 V
5.30 **(a)** – + – + + + **(b)** **(i)** 0 **(ii)** 2.0 V **(iii)** 2.0 V **(iv)** 0
5.31 **(a)** **(i)** 4.0 V 0.40 A **(ii)** 2.0 V 0.20 A **(b)** **(i)** 3.2 V 0.29 A **(ii)** 2.0 V 0.20 A
5.35 0.54 kJ
5.36 **(a)** 3.6 kJ **(b)** 25 h
5.37 **(a)** 7.2 kC **(b)** 11 kJ
5.38 **(a)** 0.97 kJ **(b)** **(i)** 2.9 kJ **(ii)** 0.97 kJ
5.39 **(a)** 28 kJ, 43 kJ, 86 kJ **(b)** 0.36 kJ g^{-1}, 0.43 kJ g^{-1}, 0.66 kJ g^{-1}
5.41 A & B, B & C, C & D, E & F, F & G, G & H
5.42 **(a)** 4.0 V **(b)** 4.0 C **(c)** 16 J **(d)** 4.0 V **(e)** 50 mA **(f)** 0.20 A **(g)** 0.80 W
5.43 **(a)** 0.60 kC **(b)** 7.2 kJ **(c)** 12 V
5.44 **(a)** all zero **(b)** 0.79 V
5.45 **(a)** A 1.46 V, B 1.46 V, C 0.67 V, D 0.67 V, E 0.00 V
5.46 **(a)** 0.15 m **(b)** 0.90 m
5.47 **(a)** +6.0 V, +4.0 V, 0.0 V **(b)** 0.0 V, –2.0 V, –6.0 V **(c)** +2.0 V, 0.0 V, –4.0 V
5.49 all 2.89 V
5.50 **(a)** 2.45 V **(b)** 3.41 V
5.51 **(a)** **(i)** 0.45 V **(ii)** 1.35 V **(b)** **(i)** 0.15 V, B **(ii)** 0.30 V, F
5.52 **(a)** **(i)** A –1.5 V, B +1.5 V, D 0 **(ii)** A 1.5 V, B –1.5 V, D –1.5 V **(b)** **(i)** no effect **(ii)** current from C to D
5.56 **(a)** 4 × 10^9 J **(b)** about 11 000
5.57 0.12 kW
5.58 4
5.59 96 W
5.60 0.14 MJ
5.61 54 kJ
5.62 0.42 A, 0.36 MJ
5.63 0.42 A, 0.25 A
5.64 **(a)** 2.5 mC **(b)** 0.25 kA **(c)** 60 kW
5.65 **(a)** 0.42 A **(b)** 0.63 A **(c)** 120 V **(d)** 0.50 A **(e)** 1.54 A **(f)** 0.37 kW
5.66 **(d)** 24 W **(e)** 10 A
5.67 **(a)** 20 kA **(b)** 4.3 × 10^{13} J **(c)** 2.6 × 10^{13} J

6 Electrical resistance

6.1 25 Ω
6.2 8.6 × 10^{10} Ω
6.3 50 V
6.4 **(b)** carbon **(c)** 91 kΩ **(d)** 99 kΩ **(e)** 83 kΩ
6.5 **(a)** 0.20 Ω **(b)** 0.12 V **(c)** 0.14 W
6.6 larger
6.9 **(a)** 0.12 A **(b)** 40 Ω, 0.15 A **(c)** 30 Ω, 0.20 A; 20 Ω, 0.30 A; 10 Ω, 0.60 A
6.10 90 W
6.11 R_1, 3.6 W
6.12 **(a)** second, 2.6 W **(b)** first, 11 W
6.13 **(a)** 2.4 Ω **(b)** 15 W **(c)** decreases, larger
6.14 0.51 m
6.15 **(b)** **(i)** 12 V **(ii)** 0.10 A **(iii)** 0.10 A
6.16 25 Ω, 6.0 Ω
6.17 **(i)** 2.50 Ω **(ii)** 5.00 Ω **(iii)** 8.33 Ω **(iv)** 6.67 Ω
6.18 **(a)** 50 Ω **(b)** 150 Ω **(c)** 50 Ω **(d)** 25 Ω **(e)** 75 Ω **(f)** 25 Ω

6.20 0.20 A, 0.30 A, 0.60 A

6.21 **(a) (i)** 6.0 V **(ii)** 3.0 V **(iii)** 2.4 V **(iv)** 3.0 V **(b) (ii)** zero **(iii)** 3.6 V

6.22 0.042 Ω

6.23 **(a)** 5.0 Ω **(b)** 1.0 Ω

6.24 **(a)** $<$ 1.00 Ω **(b)** $<$ 1.00 Ω

6.25 **(a)** 0.990 Ω **(b)** 0.989 Ω

6.28 **(a)** four **(b)** five

6.29 larger

6.30 **(a)** 2.3 V **(b)** 7.4 V **(c)** 23 V

6.31 **(a)** bd **(b)** bc **(c)** ac **(d)** ad

6.32 93 W

6.33 **(b)** 68 Ω

6.34 **(a) (i)** 0.96 kΩ **(ii)** 2.4 Ω **(b)** 20 **(c)** both dim **(d)** first nearly normally bright, second dark

6.37 **(c)** 132 mA **(e)** 8.2 V

6.38 **(c)** A **(d)** 1.5 Ω **(e)** 13.3 V **(f)** B

6.39 **(b)** tungsten 60 W, carbon 0.11 kW **(c) (i)** 0.77 A **(ii)** 165 V **(d) (i)** 0.19 A **(ii)** 370 V

6.40 **(a) (i)** infinite (ii) 29 Ω **(iii)** 4 Ω **(b) (i)** infinite **(ii)** almost zero

6.42 **(a)** A **(b)** B

6.44 0.34 Ω

6.45 24 μm

6.46 second wire

6.47 **(a)** 240 mm **(b)** 16

6.48 1.5 Ω

6.49 **(a)** 13 mm^2 **(b)** 2.5 Ω **(c)** 0.42 Ω

6.51 2.8 m

6.52 **(a)** 0.10 Ω **(b)** 3.0 A **(c)** 0.30 V **(d)** 12.6 V

6.53 **(a)** 1.1 mm **(b)** 0.8 kW

6.54 **(a)** 8.0 MΩ **(b)** the same

6.55 890°C

6.56 **(a)** 0.32 kΩ **(b)** 12

6.58 **(a)** copper **(b)** iron

6.60 **(a)** 0.101 A **(b)** 1.42 V **(c)** 1.42 V

6.61 **(a)** 1.488 V **(b)** 1.500 V

6.62 **(a)** 3.00 V, 1.00 Ω **(b)** 1.50 V, 0.25 Ω

6.63 **(b)** 1.5 Ω **(c) (i)** 0.30 W **(ii)** 0.24 W **(iii)** 0.060 W

6.64 **(a)** 0.27 kJ **(b)** 0.25 kJ

6.65 **(a)** 2.4 V **(b)** 2.4 V **(c)** 2.3 V, 2.2 V

6.66 5.2 V

6.67 **(a)** 68.9 mA **(b)** 1.52 V **(c)** 1.52 V **(d)** 107 mW **(e)** 104 mW

6.68 **(a)** 8.3 Ω **(b)** 12 Ω **(c)** 1.7 Ω

6.69 **(a)** 12.0 V **(b)** 11.9 V **(c)** 8.0 V

6.71 **(a)** 3.8 A **(b)** 19 h **(c)** 1.1 MJ **(d)** 27%

6.73 **(b)** 4

6.74 **(a)** 40 μA **(b)** 120 μA **(c)** 120 μA

6.76 0.66 kJ g^{-1}, 0.61 kJ g^{-1}, 0.54 kJ g^{-1}, 0.49 kJ g^{-1}

6.77 **(a)** 0.223 W, 0.275 W, 0.360 W, 0.500 W, 0.558 W, 0.563 W, 0.555 W, 0.527 W **(b)** both zero **(d)** 1.0 Ω

6.78 **(a)** 0.50 Ω **(b)** 1.50 V

7 Circuits and meters

7.1 **(a)** 2.4 V **(b)** 10 Ω

7.2 **(a)** 2.7 V **(b)** 0.23 kΩ

7.3 **(a)** 1.25 V **(b)** 4.7 V

7.4 **(a) (i)** 1.2 V **(ii)** 1.7 V **(b)** 55 mA **(c)** 17/12 **(d)** 0.19 A

7.5 A goes out, B gets brighter

7.6 **(a)** 6.0 Ω **(b) (i)** 2.5 V **(ii)** 3.2 V **(c) (i)** 0.53 A **(ii)** 0.32 A **(iii)** 0.21 A

7.7 **(a)** 12.5 Ω **(b)** 48 mA **(c)** larger

7.8 **(a)** 6.0 V, 6.0 V **(b)** 6.0 V, 6.0 V **(c)** 2.4 V, 9.6 V **(d)** 7.1 V, 4.9 V **(e)** 9.1 V, 2.9 V

7.9 **(a)** 6.0 V **(b)** 2.4 V **(c)** 7.1 V **(d)** 9.1 V

7.10 **(a) (i)** 4.0 V **(ii)** 8.0 V **(b) (i)** 8.0 V **(ii)** 4.0 V

7.11 **(a)** 4 times greater than in **(b)**

7.12 **(a)** 0.96 kΩ, 48 Ω **(b)** 0.24 A; 55 W, 2.8 Ω **(c)** lamp slightly dimmer; heater cold **(d)** smaller; the heater most

7.15 **(b)** 0.12 kΩ

7.16 **(a)** 0.10 A, 0.10 A **(b)** 0.10 A, 0.10 A **(c)** –0.10 A, 0.10 A **(d)** 0, 0

7.17 **(a)** all 2 A **(b)** all 1 A **(c)** all 2 A **(d)** all 3 V **(e)** all 2 V

7.19 **(c)** 2.1 V **(d)** 0.13 A **(e) (i)** 46 mA **(ii)** 86 mA

7.21 choose **(a)**

7.22 **(a) (i)** both 3.0 V **(ii)** both 3.0 V **(iii)** both 4.0 V **(b)** $P/Q = R/S$ **(c)** an ammeter or voltmeter between X and Y **(d)** 46 Ω.

7.23 **(c)** from X to Y

7.24 1.1 MΩ, 0.95 kΩ

7.25 **(a)** 0.11 V **(b)** decreases **(c)** 85 kΩ

7.27 **(c)** approx 95 Ω **(d)** approx 25 Ω **(f)** approx 110°C

7.28 **(a)** 0.50 mA **(b)** 76 mV

7.29 1.0 kΩ

7.31 **(a)** 0.69 V **(b)** 0.89 V **(c)** 0.69 V

7.32 **(a)** 60 μA, 6.0 V **(b)** 60 μA, 6.0 V **(b)** 40 μA, 6.0 V

7.34 **(b)** 28 mV **(c)** 81 mA

7.35 **(a)** +1.0 div **(b)** +1.0 div **(c)** +1.0 div **(d)** zero **(e)** +1.5 div **(f)** +0.5 div **(g)** –0.5 div

7.36 **(a)** 40 Hz **(b)** 33 Hz **(c)** 29 Hz

7.38 **(b)** 6.8 mV **(c)** 1 mV div^{-1} **(d)** 0.25 V, 100 mV div^{-1} **(e)** 63 mV **(f)** 6.3 mA

8 Density, pressure and flow

8.1 **(a)** 2.7×10^3 kg m^{-3} **(b)** 3.5×10^2 kg **(c)** 7.8×10^{-3} m^3
8.2 5.5×10^3 kg m^{-3}
8.3 17 kg
8.4 **(a)** 3.9×10^5 N m^{-2} **(b)** 0.28 m^2 **(c)** 1.3 m^2 **(d)** 40 N m^{-2} **(e)** 4.0×10^2 N
8.5 **(a)** 1.6×10^3 Pa, 1.6 kPa **(b)** 2.3×10^5 Pa, 2.3×10^2 kPa **(c)** 5.6×10^4 Pa, 56 kPa
8.6 1.1 kPa, 1.6 kPa, 3.3 kPa
8.7 **(a)** 27 MPa **(b)** 0.53 MPa **(c)** 16 kPa
8.9 10 m
8.11 8 km
8.12 **(b)** 101 kPa
8.13 2.0 kPa
8.14 **(a)** 4.7 MN **(b)** 0.25 MN **(c)** 0.19 MN
8.16 **(a)** 56 kN, at an angle of 50° with the horizontal **(b)** 56 kN
8.17 **(a)** A **(b)** 103.0 kPa **(c)** 179 mm
8.19 **(a)** 16 kPa, 11 kPa **(b)** 100 mmHg **(c)** 56 mmHg, 100 mmHg, 144 mmHg, 188 mmHg **(d)** 168 mmHg
8.20 **(b)** 0.27 m
8.21 70 m
8.23 **(a)** 1.75 MN **(b)** **(i)** 0.119 MN **(ii)** 0.238 MN **(c)** 1.63 MN, 1.51 MN
8.24 **(a)** 2.0 kN
8.26 0.14 m^3
8.28 **(a)** **(i)** 0.381 mN **(ii)** 2.43 g **(b)** **(i)** 0.443 mN, 23.87 g **(c)** 1.6%, 0.19%
8.31 **(a)** **(i)** 8.0×10^{-5} m^3 s^{-1} **(ii)** 8.0×10^{-5} m^3 s^{-1} **(b)** 0.80 m s^{-1}
8.32 **(a)** 1.8 m s^{-1} (b) 2.7 m s^{-1} **(c)** 0.98 cm
8.37 **(c)** **(i)** 0.36 kN **(ii)** 0.71 kN
8.43 0.25 MN

9 Mechanical properties of matter

9.1 5000 N
9.2 **(a)** 59 N **(b)** 59 N
9.3 **(a)** 59 N **(b)** 69 N **(c)** 64 N
9.4 **(a)** 100 N m^{-1}, 40 N m^{-1}, 10 N m^{-1} **(b)** **(i)** 12 N **(ii)** 4.8 N **(iii)** 1.2 N **(c)** **(i)** 0.16 m **(ii)** 0.040 m
9.5 **(a)** 6.0 N **(b)** 30 N
9.6 **(a)** 6.0 N **(b)** 12 N
9.7 **(a)** 0.20 m **(b)** 0.10 m, 200 N m^{-1} **(c)** 0.40 m, 50 N m^{-1} **(d)** **(i)** $2k$ **(ii)** $\frac{1}{2}k$
9.8 **(a)** 10 N m^{-1} **(b)** 60 N m^{-1} **(c)** 60 N m^{-1}
9.9 **(a)** no **(b)** yes
9.10 no
9.11 **(a)** 1.01×10^5 N m^{-2} **(b)** 2.7×10^5 N m^{-2} **(c)** 3.5×10^7 N m^{-2} **(d)** 2.8×10^9 N m^{-2} **(e)** 2.35×10^{11} N m^{-2}
9.12 **(a)** 0.60 GPa **(b)** 6.0 GPa **(c)** 2.0 kN **(d)** 5.0 N **(e)** 0.50 mm^2
9.13 **(a)** 0.18 mm **(b)** 1.0×10^{-7} m^2 **(c)** 3.4×10^7 N m^{-2} = 34 MPa
9.14 **(a)** 3.4×10^8 N m^{-2} **(b)** 0.59 kN **(c)** 1.7×10^{-6} m^2 **(d)** 1.5 mm
9.15 **(a)** 2.4×10^7 N m^{-2} **(b)** 1.6×10^8 N m^{-2} **(c)** 1.1×10^7 N m^{-2}
9.16 **(a)** 2.0×10^{-3} **(b)** 2.5 **(c)** 3.0 mm **(d)** 18 mm **(e)** 3.6 m
9.17 **(a)** 5.0×10^{-5} **(b)** 2.0
9.18 **(a)** 50 N, 50 MPa **(b)** 50 N, 100 MPa
9.19 **(a)** 83 GPa **(b)** 2.0 GPa **(c)** 12 MPa **(d)** 1.0×10^{-3} **(e)** 1.0×10^{-2}
9.20 **(b)** 0.19 kg mm^{-1} **(c)** 1.9 N mm^{-1} **(d)** 1.9×10^3 N m^{-1} **(e)** 1.3×10^{-7} m^2 **(f)** 3.0×10^{10} N m^{-2} (= 30 GPa) **(h)** 3.0 mm
9.21 **(a)** 0.50×10^{-3} or 5.0×10^{-4} **(c)** nickel **(d)** 200 GPa, 130 GPa, 70 GPa
9.22 **(a)** 100 MPa **(b)** 7.7×10^{-4} **(c)** 0.92 mm
9.23 **(a)** 1 **(b)** no
9.25 **(a)** **(i)** no **(ii)** yes **(iii)** yes **(iv)** yes **(b)** stress, strain
9.27 **(a)** **(i)** 8 **(ii)** 8 **(iii)** 4 **(iv)** 2
9.28 **(a)** 0.51 mm **(b)** 1.6 mm
9.29 **(a)** 37 kN **(b)** 120 MPa **(c)** 5.9×10^{-4}
9.30 **(a)** 4 **(b)** 4 **(c)** 8
9.31 **(a)** A, twice **(b)** B
9.32 **(a)** tensile force **(b)** tensile strain, extension, tensile stress
9.33 **(a)** tungsten **(b)** copper
9.34 **(a)** true **(b)** true **(c)** true
9.35 **(a)** 28 mm **(b)** 55 MPa **(c)** 0.41 MN
9.36 **(e)** 77 GPa **(h)** 150 MPa
9.43 aluminium
9.45 **(a)** 0.15 J **(b)** 0.60 J **(c)** 1.4 J
9.46 **(b)** **(i)** 1.0 J **(ii)** 3.0 J **(iii)** 9.0 J
9.47 **(a)** 1.7 mm **(b)** 39 mJ
9.49 **(a)** 50 mJ **(b)** 0.20 J **(c)** 0.45 J
9.50 **(a)** 65 mJ **(b)** 33 mJ, 32 mJ
9.51 1.5 J
9.53 **(a)** high tensile steel **(b)** 2.5 J, 8.1 J **(c)** mild steel
9.54 **(a)** upper **(b)** hysteresis
9.55 8.1×10^5 N m^{-1}
9.56 **(b)** 0.34 kJ, 0.19 kJ **(c)** **(i)** 0.34 kJ, 0.15 kJ
9.57 **(a)** 1.4 m s^{-1} **(b)** 20 cm from O, between O and B

10 Thermal properties of matter

10.3 **(a)** 19.4°C **(b)** 19.00°C
10.7 **(a)** 1.6215, 1.6235 etc. **(c)** 1.6295 **(d)** 445.11 K

10.13 **(a)** +0.20 MJ **(b)** –0.080 MJ **(c)** +0.12 MJ
10.14 **(a)** +, –, + **(b)** 0, –, +
10.15 3.2 GJ
10.16 £1
10.17 10 s
10.18 7.0 kW
10.19 1.3 K
10.20 29 K min^{-1}
10.21 **(a)** 167 ml **(b)** 2.6 kW
10.22 about 6 kJ must be removed
10.23 7.1 K min^{-1}
10.24 about 4 min
10.25 130 K
10.26 **(a)** 0.30 K
10.27 0.12 K
10.29 **(b)** 90°C
10.30 424 J kg^{-1} K^{-1}
10.31 **(a)** 23.1°C **(b)** 1.0 kJ kg^{-1} K^{-1}
10.34 **(a)** 28 W **(c)** 19 W
10.35 113 W
10.37 **(a)** 4.5 MJ **(b)** 0.17 MJ
10.38 **(a)** 38 min **(b)** 74 min
10.40 13 g
10.41 15.4 W
10.42 23 g
10.48 **(a)** 0.19 K min^{-1} **(b)** 0.12 K min^{-1} **(c)** 13 g min^{-1} **(d)** 0.17 K min^{-1}

11 The ideal gas

11.1 **(a)** 1.7×10^{-3} m^3 **(b)** 6.5×10^{-6} m^3 **(c)** 3.4×10^{-9} m^3
11.4 148 kPa
11.5 **(a)** 17 mol **(b)** 1.0×10^{25} **(c)** **(i)** 33 g **(ii)** 465 g
11.6 **(a)** 4.13×10^{22} **(b)** 1.88×10^{21}
11.7 **(a)** no **(b)** T/K
11.8 164 kPa
11.9 39.9°C
11.10 **(a)** 569 cm^3 **(b)** 531 cm^3
11.11 0.364 m^3
11.12 381 kPa
11.13 **(a)** 85 kPa **(b)** 0.16 MPa **(c)** 0.16 MPa **(d)** 0.19 MPa **(e)** 45 kPa
11.14 **(a)** 5.7 MPa
11.15 1.7×10^{-4} m^3
11.17 **(a)** 300 N m **(b)** 0.12 mol **(c)** 773 K
11.18 **(a)** **(i)** 10 **(ii)** 5.0 **(iii)** 6.3 **(b)** greatest **(i)**, least **(ii)**
11.19 **(a)** hydrogen 1.0 mol, helium 2.0 mol **(b)** 3.0 mol **(c)** 40 kPa
11.22 **(a)** 1.0×10^{-3} m^3 **(b)** 2.5×10^{22} **(c)** 4.0×10^{-26} m^3 **(d)** 1.1×10^{-28} m^3 **(e)** 350
11.23 **(a)** 2.0 m s **(b)** 5.0×10^{-23} kg m s^{-1} **(c)** 2.5×10^{-20} kg m s^{-2} **(d)** 1.25×10^4 N **(e)** 104 kPa
11.24 **(a)** 104 kPa **(b)** 167 kPa **(c)** 150 kPa
11.26 **(a)** $\frac{3}{2}nRT$ **(b)** $\frac{3}{2}kT$ **(d)** R/N_A
11.30 **(a)** 32.3 m s^{-1} **(b)** 33.6 m s^{-1}
11.31 **(a)** 432 m s^{-1} **(b)** 432 m s^{-1}
11.32 4.8×10^2 m s^{-1}
11.33 **(a)** **(i)** 6.0×10^{-21} J **(ii)** 4.1×10^2 m s^{-1} **(b)** **(i)** k.e. the same **(ii)** 5.1×10^2 m s^{-1}
11.34 **(a)** 300 K **(b)** 600 K
11.35 all 5.96×10^{-21} J
11.36 1.9 km s^{-1}, 0.51 km s^{-1}, 0.21 km s^{-1}
11.39 **(a)** lower graph, 600 K **(b)** **(i)** 600 **(ii)** 1200 **(c)** yes, number of molecules **(d)** **(i)** about 150 000 **(ii)** about 65 000 **(e)** **(i)** 400 m s^{-1} **(ii)** 560 m s^{-1} **(iii)** 1.4 **(f)** **(i)** 483 m s^{-1} **(ii)** 683 m s^{-1} **(iii)** $\sqrt{2} = 1.41$
11.44 **(a)** 15 kJ **(b)** 30 kJ
11.45 **(a)** 0.50 kN **(b)** 2.5 J
11.46 **(b)** **(i)** – **(ii)** + **(iii)** 0
11.47 **(a)** 2.4 mol **(b)** 200 kPa **(c)** 12×10^{-2} m^3
11.49 **(a)** **(i)** 0 **(ii)** + **(iii)** + **(b)** **(i)** + **(ii)** + **(iii)** –
11.50 **(a)** +2.0 kJ **(b)** –3.0 kJ **(c)** –5.0 kJ
11.51 **(b)** $\Delta U = +0.77$ kJ, $\Delta Q = +0.77$ kJ, $\Delta W = 0$ **(c)** $\Delta U = +0.77$ kJ, $\Delta Q = +1.28$ kJ, $\Delta W = -0.51$ kJ
11.52 **(a)** adiabatic **(b)** isothermal **(c)** constant volume
11.53 **(a)** yes **(b)** yes **(c)** no
11.54 **(a)** p/MPa = 0.20, 0.17, 0.14, 0.12, 0.11; $\Delta W = -5.9$ kJ
11.55 **(b)** $\Delta W = -5.2$ kJ **(c)** smaller **(d)** $\Delta Q = 0$, $\Delta U = -5.3$ kJ
11.56 **(b)** 0.38
11.57 **(a)** 0.45 **(b)** 0.54 **(c)** 0.60
11.59 **(a)** b: 200 kPa, 1.0 m^3 c: 200 kPa, 2.0 m^3 d: 100 kPa, 2.0 m^3
(c) to **(f)**

path	ΔQ/kJ	ΔW/kJ	ΔU/kJ
ab	+150	0	+150
bc	+500	–200	+300
cd	–300	0	–300
da	–250	+100	–150
$abcda$	+100	–100	0

(i) 650 kJ **(j)** 0.15
11.60 **(c)** to **(e)**

path	ΔW/kJ	ΔU/kJ	ΔQ/kJ
ab	+7.3	0	–7.3
bc	+15	+15	0
cd	–13.4	0	+13.4
da	–15	–15	0

(f) 6.1 kJ, 13.4 kJ **(g)** 0.45 **(i)** 0.45; the same; it is a cycle of maximum efficiency **(j)** increasing the difference between the temperatures
11.61 **(a)** 3.5 times
11.63 **(b)** £172.80 **(c)** 40 kW **(d)** the surrounding air **(e)** £115.20
11.64 **(a)** 5.0 MJ **(b)** 1.5
11.66 **(b)** 0.60

12 Photons and electrons

12.1 **(a)** 6.0 J **(b)** 1.6×10^{-19} J **(c)** 1.3×10^{-16} J
12.2 **(a)** 1.6×10^{-19} J **(b)** 800 eV
12.3 **(a)** 3.2×10^{-19} J **(b)** 3.2×10^{-18} J **(c)** 1.6×10^{-13} J **(d)** 8.0×10^{-13} J
12.4 **(a)** 2.0 eV **(b)** 20 eV **(c)** 1.0 MeV **(d)** 5.0 MeV
12.5 **(a)** 1.0 eV **(b)** 2.0 eV **(c)** 2.0 MeV
12.6 **(a) (i)** 3.6×10^{6} m s^{-1} **(ii)** 4.2×10^{6} m s^{-1} **(b) (i)** 8.5×10^{4} m s^{-1} **(ii)** 9.8×10^{6} m s^{-1}
12.7 **(a)** 200 keV **(b)** 3.2×10^{-14} J
12.8 $\sqrt{2}$:1
12.9 **(a)** 3.0×10^{-19} J **(b)** 4.6×10^{-19} J
12.10 **(a)** 1.33×10^{-19} J **(b)** 3.64×10^{-19} J **(c)** 5.45×10^{-19} J **(d)** 1.29×10^{-15} J **(e)** 8.6×10^{-14} J
12.11 **(a) (i)** 2.4×10^{14} Hz **(ii)** 1.2×10^{-6} m **(b)** infrared **(c) (i)** 4.8×10^{14} Hz **(ii)** 6.2×10^{-7} m **(d)** red
12.12 **(a) (i)** light **(ii)** X-rays in TV tube **(iii)** infrared, fire **(iv)** microwave oven **(v)** radio waves **(b) (i)** 2.5 eV **(ii)** 4.1 keV **(iii)** 0.62 eV **(iv)** 1.0×10^{-5} eV **(v)** 8.3×10^{-10} eV
12.13 **(a)** 5.18×10^{14} Hz **(b)** 3.43×10^{-19} J **(c)** 2.14 eV
12.15 **(a)** 3.14×10^{-19} J **(b)** 2.2×10^{15} s^{-1}
12.16 **(a)** 4.6×10^{19} Hz **(b)** 0.19 MeV
12.17 **(a)** about 0.10 W m^{-2} **(b)** 3.6×10^{-19} J **(c)** about 1.8×10^{16} s^{-1}
12.18 **(a)** 3.37×10^{-19} J **(b)** 4.4×10^{20} s^{-1} **(c)** 4.8×10^{-3} W m^{-2} **(d)** 1.4×10^{11} s^{-1} **(e)** 2.2 mm
12.19 **(a) (i)** 100 keV **(ii)** 1.6×10^{-14} J **(b) (i)** 2.4×10^{19} Hz **(ii)** 1.2×10^{-11} m
12.20 **(a)** 4.8×10^{18} Hz **(b)** 62 pm upwards
12.22 **(a) (i)** 10.2 eV **(ii)** 1.63×10^{-18} J **(iii)** 2.47×10^{15} Hz **(iv)** 1.21×10^{-7} m **(v)** ultraviolet **(b) (i)** 1.90 eV **(ii)** 3.04×10^{-19} J **(iii)** 4.61×10^{14} Hz **(iv)** 6.51×10^{-7} m **(v)** red **(c) (i)** 0.662 eV **(ii)** 1.06×10^{-19} J **(iii)** 1.60×10^{14} Hz **(iv)** 1.87×10^{-6} m **(v)** infrared
12.23 **(a)** 103 nm, ultraviolet **(b)** 122 nm, ultraviolet **(c)** 650 nm, red
12.25 6
12.26 ultraviolet
12.27 **(b)** $f_2 - f_1$
12.29 **(a)** 2.19×10^{6} m s^{-1} **(b)** 91.4 nm
12.30 **(a)** 13.60 eV **(b) (i)** nothing **(ii)** elastic collision, or electron loses 1.9 eV to the atom **(c)** 4.09×10^{-6} m **(d)** 9.75×10^{-8} m
12.32 **(a)** 50.0 eV **(b)** 58.9 nm, 53.6 nm, 51.8 nm
12.33 **(a)** Balmer **(b)** Paschen **(d)** 31.967×10^{14} Hz, 32.208×10^{14} Hz **(e)** 6 **(f)** 122 nm
12.34 **(a)** 2.5×10^{-7} m, ultraviolet
12.35 **(a)** nothing **(b)** atom excited, electron leaves with k.e. of 0.39 eV **(c)** atom excited
12.40 **(a)** 2.3 eV **(b)** 0.41 eV **(c)** 3.8×10^{5} m s^{-1}
12.41 **(a)** 1.00×10^{15} Hz, ultraviolet **(b)** 0.81 eV
12.42 **(c)** about 4.2×10^{-15} V s **(e)** 6.7×10^{-34} J s **(f)** A: –1.5 V; B: –2.7 V **(g)** A: 1.5 eV; B: 2.7 eV
12.43 0.20 μA
12.44 **(a)** 1.31 eV **(b)** –1.31 V
12.45 **(a) (i)** 8.3×10^{-10} eV **(ii)** 0.025 eV **(iii)** 6.2×10^{5} eV
12.46 **(b)** 1.6×10^{-18} J, 1.7×10^{-24} N s, 3.9×10^{-10} m **(c)** 1.6×10^{-17} J, 5.4×10^{-24} N s, 1.2×10^{-10} m
12.47 **(a)** 5.4×10^{-21} N s, 1.2×10^{-11} m **(b)** 2.3×10^{-21} N s, 2.8×10^{-13} m **(c)** 4.7×10^{-21} N s, 1.4×10^{-13} m
12.49 **(a)** 4.4×10^{-24} N s **(b)** 1.1×10^{-17} J **(c)** 67 V
12.50 **(b)** 11 kV

13 Radioactivity

13.3 **(a)** about 2000 **(b)** about 300 000
13.4 **(a)** 12, 13, 14 **(b)** 2.32×10^{-26} kg
13.5 **(a)** 3.27×10^{-25} kg **(b)** 5.90×10^{28} **(c)** 1.70×10^{-29} m^3 **(d)** 2.57×10^{-10} m; for aluminium: **(a)** 4.47×10^{-26} kg **(b)** 6.04×10^{28} **(c)** 1.65×10^{-29} m^3 **(d)** 2.55×10^{-10} m
13.6 **(a)** 7.6×10^{22} **(b)** 1.1×10^{25}
13.8 63.6
13.9 28.1
13.10 23.0 g
13.11 **(a)** 6.0×10^{23} **(b)** 3.0×10^{23} **(c)** 2.0×10^{23}
13.12 1.00, 1.14 etc.
13.14 neodymium-143

13.15 **(a)** positron
(b) 27p, 29n, 27e **(c)** $^{56}_{28}Ni$
13.16 **(a)** $^{140}_{57}La \rightarrow ^{136}_{55}Cs + ^{4}_{2}He$
13.18 **(a)** $^{26}_{13}Al + ^{0}_{-1}e \rightarrow ^{26}_{12}Mg$
(b) $^{26}_{13}Al \rightarrow ^{26}_{12}Mg + ^{0}_{1}\beta$
13.19 **(a)** no **(b)** yes
13.20 **(a)** no, yes **(b)** no, yes
(c) yes, yes
13.21 The recoiling nucleus, and the neutrino which is emitted
13.23 **(a)** 0.492 MeV, 0.452 MeV, 0.432 MeV, 0.327 MeV, 0.287 MeV, 0.040 MeV
(b) 3.1×10^{-11} m
13.28 **(a)** 2.1×10^5 **(b)** 1.6×10^{-19} C
(c) 2.7×10^{-11} A
13.30 **(a)** 5.49 MeV, 5.44 MeV
(b) 0.05 MeV **(c)** between 2 and 3×10^{-11} m
13.33 **(d)** First three values of lg(count min^{-1}): 4.00, 3.94, 3.65
(e) about 800 mg/cm^2
13.36 **(a)** 74 **(b)** 3
13.37 800 Bq
13.38 **(a)** 768 kBq **(b)** 15.0 h
(c) 1.5×10^{10} **(d)** 192 kBq
13.39 **(a)** 5.6×10^{-7} s^{-1}
(b) 1.8×10^{10} **(c)** 9.5×10^{-13} g
13.40 **(a)** 1.7×10^{-17} s^{-1}
(b) 3.8×10^{24} **(c)** 4.5×10^{20}
(d) 7.6 kBq
13.41 5.8×10^4 Bq
13.42 **(a)** $^{99}_{44}Ru$ **(b)** 6.0×10^{23}
(c) 2.2×10^{-13} s^{-1}
(d) 9.9×10^4 y
13.43 **(a)** 7.3×10^{-10} s^{-1}
(b) 2.7×10^{14} **(c)** 6.2×10^{-8} g
13.44 **(a)** 0.897 **(b)** **(i)** 527 **(ii)** 473
(c) 729
13.45 **(a)** 2.9×10^{-7} s^{-1}
(b) 7.0×10^{11}
(c) 1.5×10^{-10} g **(d)** 26%
13.46 **(a)** 100 kBq **(b)** 100 kBq
13.47 6.9×10^9 Bq
13.48 4.0 μg
13.49 **(a)** B **(b)** 53 s **(c)** 0.40
13.50 **(a)** 4.0 min
(f) -0.173 min^{-1}, 4.0 min
13.51 **(b)** A/Bq = 116, 58, 29, 14, 7.2, 3.6
13.52 **(d)** 40 minutes
13.53 **(b)** α 3, β 2 **(c)** 52 s
(c) products of Rn-220 decay, emitting α–particles
13.55 **(a)** 0.025 y^{-1} **(b)** 4.2×10^{10}
(c) 186 y
13.56 200 litres
13.57 **(a)** activity decreases by 0.02% **(b)** 1300 cm^3
(c) 3.0 litres
13.58 160 y
13.60 **(b)** **(i)** depth of water
(ii) cross-sectional area of pipe
(d) X
13.61 **(a)** 1.25×10^{11}
(b) 3.84×10^{-12} s^{-1}
(c) 389 **(d)** 2750 y
13.62 **(a)** $^{14}_{7}N$ **(b)** 3.84×10^{-12} s^{-1}
(c) 3.91×10^{13} atoms
(d) 0.62 kg

14 Nuclear power and nuclear matter

14.4 **(a)** **(i)** 0.087 d^{-1} **(ii)** 0.046 d^{-1}
(b) 0.133 d^{-1} **(c)** 5.2 days
14.5 **(a)** about 13%
(c) **(i)** 5%
(ii) 40%
14.9 **(a)** 1.008 67 u
(b) 0.000 548 598 u
(c) 12.0000 u
14.10 **(a)** $1.672\,51 \times 10^{-27}$ kg
(b) $1.501\,91 \times 10^{-10}$ J
(c) 938.7 MeV
14.11 **(a)** 1.49×10^{-10} **(b)** 931.5
(c) 1.07×10^{-3}
(d) 1.78×10^{30}
14.12 **(a)** 8.77×10^{-13} J
(b) **(i)** 9.74×10^{-30} kg
(ii) 0.005 87 u
(c) 1.5×10^{-3}
(d) 10.7
14.13 **(a)** 1.94×10^{-8} kg
(b) 3.9×10^{-12}
14.14 **(a)** 15.125 00 u
(b) 0.114 40 u
(c) 107 MeV
14.15 **(a)** 15.124 16 u, 15.123 32 u
(b) 0.124 06 u, 0.120 25 u
(c) 116 MeV, 112 MeV
(d) nitrogen-15
(e) nitrogen-15
14.16 **(a)** Al: 225.0 MeV, 8.332 MeV;
Fe: 492.3 MeV, 8.791 MeV;
Ba: 1158 MeV, 8.395 MeV;
U: 1802 MeV, 7.570 MeV
14.17 **(a)** 5.79 MeV
(b) the radon nucleus, which recoils
14.18 **(a)** 1.02 MeV
(b) 1.22×10^{-12} m
(c) 0.511 MeV
14.19 **(b)** 1.02 MeV
(c) 1.24×10^{20} Hz
14.20 **(a)** 1.64 MeV
(b) 4.24×10^7 m s^{-1}
14.21 **(d)** and **(e)** are possible
14.22 β^-
14.23 **(a)** charged: Ra^{2-}, He^{2+}
(b) Ra $+2m_e$, He $-2m_e$
14.24 **(a)** charged: +1 **(b)** $-m_e$
14.25 **(b)** 2
14.27 **(a)** $-0.005\,27$ u
(b) 4.91 MeV
14.28 **(b)** 14.003 074 u
(c) 0.000 168 u
(d) 0.156 MeV
14.29 31.9739 u
14.30 **(a)** $^{15}_{8}O \rightarrow ^{15}_{7}N + ^{0}_{1}\beta$
(b) 15.001 024 u
(c) two nuclei, positron, neutrino **(d)** 1.74 MeV
14.31 **(a)** $^{37}_{18}Ar + ^{0}_{-1}e \rightarrow ^{37}_{17}Cl$
(b) 0.81 MeV **(c)** recoil of chlorine nucleus
14.32 **(a)** less stable **(b)** 50–60
14.33 **(a)** more **(c)** 7.6 MeV, 8.6 MeV, 8.3 MeV
(d) **(i)** 1786 MeV
(ii) (783 + 1170) MeV
(e) 173 MeV
14.34 **(a)** 0.1859 u **(b)** 173 MeV
(c) 4.25 mol **(d)** 2.56×10^{24}
(e) 4.43×10^{26} MeV

(f) (i) 7.09×10^{13} J
(ii) 1.97×10^{7} kW h
(g) 19.7 h

14.35 (a) 2 (b) 4

14.36 (a) neutron-rich
(b) Ba 1.52; Kr 1.56; Ru 1.59, Sr 1.47, Cs 1.67; Rb 1.57 (e) 3, 4

14.40 (a) hydrogen
(b) ${}^{1}_{1}\text{H} + {}^{1}_{0}\text{n} \rightarrow {}^{2}_{1}\text{H} + \gamma$, deuterium
(c) carbon, graphite

14.41 (a) to absorb thermal neutrons
(b) boron, cadmium
(c) ${}^{11}_{5}\text{B} + {}^{1}_{0}\text{n} \rightarrow {}^{12}_{5}\text{B}$

14.47 (a) 2500 MW
(b) 2.81×10^{20}
(c) 0.467 mol
(d) 0.11 kg

14.48 (c) $5.7 \text{ x } 10^{-10}$ s^{-1}
(d) 2.2×10^{20} (e) 88 mg

14.49 (b) 5 MeV per nucleon
(c) more

14.50 (a) (i) 0.003 50 u
(ii) 3.27 MeV
(b) 3.82×10^{20} s^{-1}
(c) 0.22 kg

14.51 (a) 4.0×10^{26} W
(b) 4.4×10^{9} kg s^{-1}

14.54 (a) 2.1×10^{-4}
(b) 2.4×10^{18} kg m^{-3}

14.56 (b) 1.0×10^{-15} m
(c) (i) 2.8×10^{-15} m
(ii) 4.7×10^{-15} m

14.57 (a) about 0.5 mm
(b) about 5 m

14.58 (a) A (b) Am (c) $A(\frac{4}{3}\pi r_0{}^3)$
(d) $3m/4\pi r_0{}^3$
(f) 2×10^{17} kg m^{-3}

14.59 (a) 2×10^{17} kg m^{-3}
(b) about 30 km

14.60 (d) uud (e) udd

14.61 (a) (i) positron
(ii) anti-proton
(iii) anti-neutrino
(b) (i) mass (ii) charge
(c) annihilation (d) yes

14.62 (a) electron, anti-neutrino
(b) u changes to d (c) yes

14.63 (b) yes (c) weak
(d) e^{+}, positron (e) all +1
(f) all −1

14.64 (a) yes (b) yes (c) yes (d) yes

14.65 (a) yes (b) yes (c) yes (d) yes

14.66 (a) (i) an electron is repelled by another electron; electromagnetic; photon
(ii) an up quark is scattered by a down quark; strong nuclear, gluon (iii) a neutron decays to a proton, an electron and an anti-neutrino; weak nuclear
(iv) a massive particle attracts another massive particle; gravitational
(b) yes

14.67 (a) hadrons (b) (i) 3 (ii) 2

14.68 (b) (i) e (ii) $-e$ (iii) 0 (iv) 0
(c) (i) e (ii) $-e$ (iii) 0 (iv) 0
(e) mesons
(g) charge not multiple of $\pm e$

14.69 $u\bar{d}$, $\bar{u}d$, $u\bar{u}$ or $d\bar{d}$, $u\bar{s}$, $\bar{u}s$, $d\bar{s}$, $\bar{d}s$

14.70 (a) hadrons (b) (i) baryons
(ii) mesons

14.71 (a) (i) yes (ii) yes
(iii) no: B not conserved
(iv) yes (v) neither Q nor B conserved
(b) protons had additional energy

14.72 (b) $K^{-} + p \rightarrow \Sigma^{+} + \pi^{-}$
(c) 1, 1, −1
(d) it has no charge
(e) neutron

14.73 (a) $K^{-} + p \rightarrow \Sigma^{-} + \pi^{+} + \pi^{+} + \pi^{-}$ (b) −1, 1, −1

14.74 uus, dds

14.76 (a) leptons, quarks
(b) leptons
(c) all charged particles

14.77 (a) yes
(b) it may change by 1
(c) both possible
(d) (i) decreases by 1
(ii) increases by 1
(iii) decreases by 1

15 Waves

15.1 (b) 47 ms, 53 ms

15.2 1.5×10^{8} km

15.3 (a) 20 m (b) 2.8 h

15.4 620 km

15.6 15 mm to 6 mm

15.7 (a) 6000 m (b) 125 mm

15.10 (a) (i) 0.25 m (ii) 1.5 m
(b) (i) 2.0 m s^{-1} (ii) 1.3 Hz
(iii) 0.75 s

15.13 (b) 9.8 N

15.20 (a) 0.40 mW m^{-2}
(b) 0.10 mW m^{-2}
(c) 0.044 mW m^{-2}

15.21 0.24 mW

15.22 32 m

15.23 (a) 28 mW m^{-2}
(b) 55 mW m^{-2}

15.24 1.3 m

15.25 3.8×10^{20} W

15.26 (a) 24 W m^{-2}
(b) 0.11 W m^{-2}

15.27 7.4 km

15.28 (b) (i) 10 mm (ii) 14 mm

15.30 6.6×10^{6} m s^{-1}

15.31 (a) 3.2×10^{-12} m

15.32 (b) (i) 1.2×10^{7} m s^{-1}
(c) 5.5×10^{24} m

15.34 (a) 41.5° (b) 39.3° (c) 39.3°

15.35 4.3°

15.36 (a) 0.97° (b) 0.50° (c) 0.23°

15.39 4.6×10^{-8} s

15.40 (a) (i) 49.81° (ii) 47.29°
(iii) 40.37° (b) 2.419

15.41 (a) 0.49°

15.42 4.935 μs, 4.945 μs, 10 ns

15.43 (b) 9.4°

15.44 (a) 81.28° (b) 14.73 μs
(c) 14.56 μs

15.48 (a) 62

15.49 (b) 3.1 kHz, 1080 kHz
(c) 250

15.50 (a) greater than 3.6 kHz
(b) 0.23 ms

15.52 (b) 1.0 kHz

15.53 (d) 30 kbit s^{-1}

15.54 (b) 30 kHz (c) 15 kHz

15.56 (a) 8.0 μs (b) 117 μs

16 Interference patterns

16.6 **(b)** 44 m s^{-1}
16.8 nodes every 2 s
16.9 **(b)** 0.43 kHz, 0.85 kHz **(d)** 0.21 kHz, 0.64 kHz
16.12 **(a)** 104 m **(b)** maximum every 3.5 s
16.13 **(a)** 32 mm
16.16 **(c)** 11.0 mm
16.19 **(a)** 590 nm
16.20 **(a)** 0.59 mm, 0.71 mm
16.21 **(a)** **(ii)** 0.0015 mm
16.22 0.50 mm
16.23 **(b)** 28 mm
16.24 **(a)** 400 nm **(b)** 100 nm
16.25 **(a)** 20λ, 24λ **(b)** antiphase
16.26 **(b)** 510 Hz
16.29 5.2×10^{-7} m
16.31 beam width is 0.27 m at the flag
16.32 **(a)** 0.11° **(b)** 14 μm
16.33 **(b)** 0.28 mm **(c)** 0.070 mm
16.34 **(a)** **(i)** 1.57×10^{-6} m **(ii)** 6.36×10^{5} m^{-1} **(b)** 647 nm
16.35 7
16.36 **(a)** 10.5°

17 Oscillations

17.4 **(a)** **(ii)** 45 ks **(c)** **(i)** 256 Hz
17.5 **(a)** s^{-2} **(b)** $f_A = 3f_B$
17.6 **(a)** zero **(b)** –16.0 cm **(c)** zero **(d)** 11.3 cm
17.7 **(a)** 4.0 cm, 2.8 Hz **(c)** about 0.7 m s^{-1}
17.8 **(a)** **(i)** 1.6 m **(ii)** 0.67 s **(b)** **(i)** zero **(ii)** 15 m s^{-1} **(iii)** zero
17.10 **(a)** 0.13 m s^{-1} **(b)** 29 m s^{-2}
17.11 $f < 460$ Hz
17.12 0.16 m s^{-1}
17.13 170 m s^{-1}
17.14 2.05 p.m.
17.15 **(a)** 0.99 m **(c)** **(i)** 0.49 m s^{-2} **(ii)** 0.99 m s^{-2}
17.16 **(a)** **(i)** 98 mm **(ii)** 0.50 Hz **(b)** 0.31 m s^{-1}
17.19 0.79 s
17.20 upward acceleration of 0.87 m s^{-2}
17.22 **(a)** 2 **(b)** $4m$
17.23 66 kg
17.24 **(b)** **(i)** 0.79 m s^{-1}, 0.39 m s^{-1} **(ii)** the maximum slopes of the graphs
17.25 13 kN m^{-1}
17.26 **(a)** 720 N m^{-1} **(b)** 0.31 Hz
17.27 **(c)** 0.38 kN m^{-1} **(d)** 0.94 s **(e)** 0.53 m s^{-1}
17.29 **(a)** at $x = 12$ cm, $E_p = 0.22$ J
17.30 **(c)** 4.0 m s^{-1}, 3.5 m s^{-1}
17.32 **(a)** at $t = 0.6$ s, $x = -0.88$ m and $E_p = 32$ J
17.37 **(a)** 2 **(b)** 0.5 **(c)** 4 **(d)** 2

18 Capacitance

18.1 ±12 mC
18.2 0.25 kV
18.3 2nd column 2.0 V, 3rd column 7.1 μC
18.4 **(a)** ±20 mC **(b)** 5.0 mA
18.5 **(a)** 14 C **(b)** 71 kA
18.6 2.2 μA
18.7 200 μF
18.9 3.4 ms
18.10 **(a)** 0.70 s **(b)** 18 V **(c)** 26 V s^{-1} **(d)** steady 1.2 mA
18.11 **(a)** **(i)** 0 **(ii)** 12 V **(iii)** 12 μA **(iv)** 0 **(b)** **(i)** 12 V **(ii)** 0 **(iii)** 0 **(iv)** ±0.26 mC **(c)** 22 s
18.12 **(a)** 9.5 V **(b)** 2.5 V **(c)** ±55 μC
18.14 4.0 μF
18.15 0.54 mA
18.16 **(a)** **(i)** 850 V **(ii)** 560 V **(b)** 8×10^{-14} A **(c)** 9×10^{15} Ω
18.17 **(a)** 11 μF **(b)** 1.0 μF
18.19 **(b)** –60 μC, +60 μC **(c)** 2.0 V, 1.0 V **(d)** 3.0 V **(e)** 20 μF
18.21 **(a)** 0.40 μF **(b)** 80 μC $V_A = 160$ V, $V_B = 40$ V
18.23 for A **(a)** 6 V **(b)** 18 μC for B **(a)** 3 V **(b)** 4.5 μC for C **(a)** 3 V **(b)** 13.5 μC
18.24 **(a)** 0.28 mC **(b)** 0.28 mC **(c)** 69 μF **(d)** 4.1 V **(e)** 0.19 mC, 0.090 mC
18.25 **(a)** 120 nC **(c)** 1.0 nC **(d)** 0.83%
18.26 **(a)** 1.8 J **(b)** 1.8×10^{5} W
18.27 **(a)** 0.47 μF **(b)** for $q = 3$ μC, $W = 9.8$ nJ
18.28 **(a)** ±2.4 mC, 0.18 J **(b)** **(i)** 24 μF **(ii)** 100 V **(iii)** 0.12 J
18.29 **(a)** 4.0 mC **(b)** 2.0 J **(c)** 1.0 J
18.30 **(a)** 0.44 J **(b)** 0.22J
18.31 **(a)** 0.29 mJ **(b)** 0.032 mJ
18.34 5.9×10^{-10} F
18.36 29 mm square
18.37 2.4×10^{-12} F
18.38 **(a)** 7.6 mF **(b)** 1.4×10^{12} J
18.39 **(a)** 0.35 nF **(b)** 0.18 μC **(c)** 0.44 mJ **(d)** 10 kV **(e)** 0.88 mJ
18.40 **(a)** 13 μA **(b)** 8.6 μA
18.42 **(a)** 29 s
18.44 **(a)** 0.28 mC **(b)** 6.0 μA, 47 s **(c)** 47 s
18.45 **(a)** 430 s **(b)** 4.3 MΩ
18.47 5
18.48 **(a)** **(i)** 3.0 V s^{-1} **(ii)** 0.66 mA **(b)** 0.11 mA
18.49 **(a)** 1.5 s **(b)** 2.2 s
18.51 **(a)** **(i)** 88 μC **(ii)** 1.0 μA **(iii)** 88 s **(b)** 89 s **(c)** 1.3%
18.52 **(a)** 0.67 μA **(b)** 21 mV s^{-1} **(c)** 32 μF **(d)** **(ii)** 48 s **(e)** 32 μF

19 Electromagnetism

19.1 **(b)** zero at some point
19.5 **(a)** 21 μT **(b)** 45 μT
19.6 **(b)** $B_{centre} = 2B_{ends}$
19.7 9.0 mT
19.8 1.2 kA
19.10 **(a)** 0.33 mT **(b)** 0.060 mT
19.12 0.20 m
19.14 **(a)** 59 μT **(b)** two cables
19.15 4.6 kA
19.16 **(b)** kg s^{-2} A^{-1}
19.17 **(a)** 75 mN
19.18 33 mT

19.19 **(a)** 9.6 N **(b)** 3.3 N
19.20 $F_{OB} = 75$ mN, $F_{OD} = 0$
19.21 **(b) (i)** $F_{PQ} = F_{RS} = 14$ mN, $F_{PS} = F_{RQ} = 9.0$ mN
(ii) torque from 14 mN forces = 2.2×10^{-3} N m, other torque is zero
19.22 **(c)** force = $\mu_0 I^2 l/2\pi r$
19.25 **(a)** 4.8×10^{-14} N
19.26 **(a)** 9.6×10^{-12} N
(b) circular arc
(c) 1.5×10^{15} m s^{-2}
19.27 **(a)** 0.40 mm s^{-1}
(b) 1.6×10^{-23} N
(c) 3.1×10^{22} **(d)** 0.50 N
19.28 **(a)** 8.0×10^{-15} N **(c)** 46 mm
(d) 14 ns **(e)** 91 mm
(f) 14 ns
19.29 **(a)** $r = mv/Bq$
(b) $T = 2\pi m/Bq$
19.30 1.1, $22m$
19.31 **(a)** 1.1 mT **(b)** 4.1 T
19.34 4.5 mV
19.37 **(b)** 0.70 V
19.38 **(a)** 2.8 m s^{-1}
(b) 7.3×10^{-5} V
19.39 0.38 Wb
19.40 1.6 T
19.41 **(a)** 15 mWb **(b) (i)** 0
(ii) 7.5 mWb **(iii)** 0
(iv) 7.5 mWb
(v) 15 mWb **(vi)** 0
19.44 **(a)** 17 μWb
(b) –17 μWb
(c) 12 μV
19.45 **(a)** 20 mV **(b)** 1.7 A
19.46 **(b) (i)** × 0.5 **(ii)** same
(iii) × 2
(c) (i) × 0.25 **(ii)** same
(iii) × 4
19.47 29 V
19.48 **(b)** 28 T s^{-1} **(c)** 14 V
19.51 **(a)** 240 V **(b)** 0.10 A
(c) 24 W
19.52 **(a)** 10 **(b)** 10.4 A **(c)** 13 A
19.53 **(a)** 3.6 W **(b)** 7.2 W
(c) primary
19.54 **(a)** 110 **(b)** 5.5 A **(c)** 45 mA

20 Inverse square law fields

20.1 **(a)** 3.3×10^{-8} N **(b)** same
20.2 **(a) (i)** 0.59 kN **(ii)** 0.59 kN
20.3 **(a) (i)** 0.50 kN **(ii)** 6.2 m s^{-2}
(b) (i) 23 N **(ii)** 0.23 m s^{-2}
(c) (i) 2.2×10^{20} N
(ii) 2.8×10^{-3} m s^{-2}
20.5 **(i)** 2.2×10^{-7} N
(ii) 3.5×10^{-6} N
20.6 0.12 kN
20.7 $m_E = 81 m_M$
20.8 **(a)** 5.97×10^{24} kg
(b) 5.52×10^{3} kg m^{-3}
20.9 **(a)** 1.2×10^{-11} N
(b) 1.2×10^{7} times
20.10 **(a)** 1.1×10^{-6} N
(b) at $r = 30$ mm, $F = 40 \times 10^{-6}$ N
20.11 0.66 kN
20.12 **(a)** 0.25 mN **(b)** 0.28 μC
20.13 **(b)** 0.10 m **(c)** 9.7 nC
20.14 **(a) (i)** 4.8×10^{-45} N
(ii) 5.8×10^{-8} N
(b) 1.2×10^{36}
20.15 **(a)** 74 N, 180 N, 460 N
(b) 120 J, 280 J, 740 J
20.16 **(c)** 5.9 J kg^{-1}, 14 J kg^{-1}, 37 J kg^{-1}
(d) 12 J kg^{-1}, 28 J kg^{-1}, 74 J kg^{-1}
20.17 **(a) (i)** +120 kJ **(ii)** –82 kJ
20.18 **(a)** 10 m **(b) (i)** 19.6 N
(ii) 19.6 N
(c) (i) 588 J **(ii)** 196 J
(d) 392 J
20.19 **(b)** 196 J kg^{-1} for each
20.21 **(b)** 5.3×10^{4} km
20.22 1st column 1.6×10^{-19}
20.23 1.4 nC
20.24 **(a)** 4.0×10^{-15} N
(b) (i) 3.3 m s^{-2} **(ii)** 16 m s^{-2}
20.26 **(a)** 5.0×10^{4} N C^{-1}
(b) (i) 1.0 μN **(ii)** 1.0 μN
20.27 **(a)** 2.0×10^{4} N C^{-1}
(b) 3.2×10^{-15} N
(c) 2.4×10^{-17} J
(d) 2.4×10^{-17} J
20.29 6.4×10^{-19} C
20.30 0.49 kV, upper
20.31 **(a) (i)** 1.57×10^{-16} J
(ii) 1.53×10^{-16} J
20.32 **(a) (i)** 3.0×10^{-11} F
(ii) 1.2×10^{-7} C
(iii) 3.0×10^{-6} C m^{-2}
(iv) 3.3×10^{5} V m^{-1}
20.33 **(a) (i)** 1.1 N kg^{-1}
(b) (ii) 8.1 N kg^{-1}
(c) (ii) 4.7×10^{6} m
20.34 **(a)** 9.81 N kg^{-1}, 9.72 N kg^{-1}
20.36 5.9×10^{13} N kg^{-1}
20.37 **(c) (i)** 4.2×10^{-2} m s^{-2}
(ii) 1.1×10^{-4} m s^{-2}
20.38 **(a)** at $r = 10 \times 10^{6}$ m, $F = 40$ N
(c) 1.1×10^{8} J
20.39 **(a) (i)** 3.0×10^{10} J **(ii)** same
(b) (i) 1.2×10^{10} J
(ii) 6.4×10^{10} J
(iii) 9.2 km s^{-1}
20.40 **(a)** $v_e = \sqrt{(2Gm_E/r_E)}$
(b) 11 km s^{-1}
20.41 **(a)** 2.4 km s^{-1}
20.42 **(b)** 3.3×10^{-7} C
20.43 **(a)** 5.1×10^{11} N C^{-1}
(b) 8.1×10^{8} N
(c) 9.0×10^{22} m s^{-2}
20.44 **(b)** 9.0×10^{9} N m^2 C^{-2}
(c) 2.2×10^{-18} J
20.46 **(a)** 110 kV
(c) 130 kV m^{-1}
20.47 **(a)** 4.0 μC **(b)** 1.6 MV m^{-1}
20.48 **(b)** 1.8 nC m^{-2}
20.49 **(a) (i)** 3.1 km s^{-1}
(ii) 0.22 m s^{-2}
(b) 0.22 N kg^{-1}
(c) (i) 0.89 m s^{-2}
(d) 4.3 km s^{-1}
(e) 3.1×10^{4} s
20.50 **(a) (i)** 8.9 N kg^{-1}
(ii) 8.9 m s^{-2}
(c) (i) 7.7 km s^{-1}
(ii) 5.5×10^{3} s (just over 90 minutes)
20.51 **(a)** 9.3 N kg^{-1} **(b)** 9.3 m s^{-2}
(c) 0.65 kN **(d)** 9.3 m s^{-2}
(e) 0.65 kN **(f)** no
(g) zero
20.52 **(a)** 2.6×10^{-3} m s^{-2}
(b) $g_M/g_0 = 2.7 \times 10^{-4}$
(c) $r_E/d_{EM} = 1.7 \times 10^{-2}$

20.53 **(a)** 1.80×10^6 m
(b) 360 km
20.54 **(b)** 2.2×10^{30} kg
20.55 **(b)** 380×10^3 km
(d) 8.7×10^{25} kg
20.56 **(a)** 4.22×10^7 m, about 7
(c) about 40°

21 Practising calculations

21.1 **(a)** 3.4×10^{-2} **(b)** 5.3×10^{-3}
(c) 1.5×10^{-1} **(d)** 6.7×10^2
21.2 **(a)** 2 **(b)** 3 **(c)** 2
21.3 **(a)** 10^9 **(b)** 10^6 **(c)** 10^3
(d) 10^{-9} **(e)** 10^3 **(f)** 10^6
(g) 10^7 **(h)** 10^{-3} **(i)** 10^6
21.4 **(a)** 10^{-2} **(b)** 10^{-4} **(c)** 10^2
(d) 10^3 **(e)** 10^{-8}
21.5 **(a)** 1.4×10^6 **(b)** 6.4×10^3
(c) 6.8×10^{-42}
21.6 **(a)** 1.3×10^2 **(b)** 3.6×10^7
(c) 1.2×10^{-19}
21.7 **(a)** 17 **(b)** 1.0×10^{-4} **(c)** 2.8
21.8 **(a)** 0.20 **(b)** 0.020
(c) 0.000 40 **(c)** 3.0
21.9 **(a)** 100% **(b)** 200%
21.10 **(a)** 90 **(b)** 120 **(c)** 132
(d) 360
21.11 **(a)** 10% **(b)** 87% **(c)** 230%
(d) 0.20%
21.12 **(a)** 124 **(b)** 244 **(c)** 1.30
21.13 v/m s^{-1} = 30.1, 32.7, 29.9
21.14 **(a)** 0.49 **(b)** 0.069 **(c)** 0.023
(d) 12 **(e)** 1.3 **(f)** 9.2×10^2
(g) 0.66 **(h)** 0.063 **(i)** 11
(j) 0.99
21.15 **(a)** 6.3×10^{-2} m
(b) 1.2×10^{-2} m
(c) 8.3×10^5 m
(d) 5.5×10^{-7} m
(e) 5.3×10^{-2} kg
(f) 5.0×10^5 kg
(g) 1.2×10^{-4} kg
(h) 2.3×10^{-6} kg
(i) 1.8×10^3 s
(j) 2.3×10^{-2} s
(k) 8.6×10^4 s
(l) 4.5×10^{-2} m^3
21.16 **(a)** 1.6×10^{-4} m^2
(b) 5.3×10^{-6} m^2
(c) 1.7×10^{-6} m^2
(d) 7.8×10^{-6} m^3
(e) 3.4×10^{-8} m^3

21.17 **(a)** 4.9 m^2 **(b)** 0.23 m^2
(c) 5.3×10^{-6} m^2
21.18 **(a)** 4.7×10^{-10} F
(b) 1.5×10^3 V
(c) 5.0×10^7 W
(d) 4.0×10^{-8} s
21.19 1.3×10^5 N
21.20 9.8×10^{-4} m^3
21.21 **(a)** $v = x/t$ **(b)** $h = V/bd$
(c) $r = \sqrt{(A/\pi)}$ **(d)** $h = V/\pi r^2$
(e) $V = m/\rho$ **(f)** $a = (v-u)/t$
(g) $a = (v^2-u^2)/2x$
(h) $T_1 = T_2/(1-\eta)$
(i) $I = \sqrt{(P/R)}$ **(j)** $R = V^2/P$
(k) $f = 1/T$ **(l)** $l = gT^2/4\pi^2$
(m) $r = (E-V)/I$
(n) $r = (3V/4\pi)^{1/3}$
21.22 5.5×10^{14} Hz
21.23 2.0×10^{-2} m
21.24 9.1×10^{-2} m
21.25 9.0×10^{-3} m
21.26 **(b)** 5.1×10^3 s
21.27 **(a)** m **(b)** kg **(c)** s **(d)** A **(e)** K
21.28 1 N = 1 kg m s^{-2}
21.29 1 Pa = 1 kg m^{-1} s^{-2}
21.30 1 J = 1 kg m^2 s^{-2}
21.31 1 W = 1 kg m^2 s^{-3}
21.32 1 C = 1 A s
21.33 1 V = 1 kg m^2 s^{-3} A^{-1}
21.34 1 Hz = 1 s^{-1}
21.35 1 F = 1 m^{-2} kg^{-1} s^4 A^2
21.36 **(a)** 30 m s^{-1}, speed
(b) 1.0×10^{-3} J, work or energy
(c) 29 Pa, pressure
(d) 76 m^3, volume
21.37 **(a)** 7.1×10^{-7} C **(b)** 2.0 J
(c) 2.8×10^6 V
21.39 1.4×10^3 Wm^{-2}
21.40 0.21 MN m^{-2}
21.41 2.47 acres
21.42 £1.37
21.43 **(a)** t is greater than 10 s
(b) t is greater than 20 s and less then, or equal to, 40 s
(c) the size of the difference between m_1 and m_2 is 0.35 g
(d) A is proportional to the square of r
(e) the change in x is 0.35 m

21.44 **(a)** $R \propto l/A$ **(b)** $F \propto v^2/r$
21.45 **(a)** $\Delta v = +1.4$ m s^{-1}
(b) $\Delta V = -47$ cm^3
21.46 If the current remains constant, and the temperature rises by 10 K, the potential difference falls by 2.0 V.
21.47 **(a)** yes **(b)** yes **(c)** yes **(d)** yes
(e) yes **(f)** no **(g)** yes **(h)** yes
(i) no **(j)** yes
21.48 **(a)** all **(b)** B
21.49 **(a)** all **(b)** B, C
21.50 **(a)** 2.0 m s^{-1}
(b) 5.0×10^{-4} C V^{-1} or 5.0×10^{-4} F
(c) 1.0×10^2 m s^{-2}
(d) 0.40 Ω
(e) 4.0×10^{14} m^3 s^{-2}
(f) 4.0×10^{15} V s
21.51 **(a)** 200 m **(b)** 100 m
(c) 0.15 J
21.52 **(a)** 62 m **(b)** 6.8×10^{-3} C
21.53 **(a)** W, x^2, $\frac{1}{2}k$ **(b)** E, $1/d$, V
(c) F, $1/r^2$, Gm_1m_2
(d) f, $1/\lambda$, c or λ, $1/f$, c
(e) C, A, $\varepsilon_r\varepsilon_0/d$
(f) C, $1/d$, $\varepsilon_r\varepsilon_0 A$
(g) T, $\sqrt{l}$, $2\pi/\sqrt{g}$
(h) T, $1/\sqrt{k}$, $2\pi\sqrt{m}$
(i) V_s, f, h/e
21.54 **(b)** yes **(c)** yes
(d) 6.5×10^3 N m^{-1}
(e) 1.3×10^{11} N m^{-2}
21.55 **(b)** yes **(c)** no **(d)** −0.60 Ω
(e) $m = -0.60$ Ω, $c = 1.52$ V
21.56 0.34 km s^{-1}
21.57 **(a)** 0.427 **(b)** 0.904
(c) 0.473
21.58 **(a)** 0 **(b)** 1 **(c)** 0
21.59 **(a)** 1 **(b)** 0
21.60 Because tan 90° is infinite.
21.61 **(a)** 0.5938 **(b)** 36.43°
21.62 **(a)** 0.8046, 36.43°
(b) 0.7380, 36.43°
21.63 **(a)** 11.2 m **(b)** 5.00 m
21.64 0.75 m
21.65 **(a)** **(i)** 60° **(ii)** 50° **(iii)** 65°
(iv) 40°

(b) (i) x: (2.6 m s^{-1})cos30°; y: (2.6 m s^{-1})cos60° (ii) x: (5.4 N)cos50°; y: (5.4 N)cos40° (iii) x: (2.6 N s)cos25°; y: (2.6 N s)cos65° (iv) x: (0.35 T)cos40°; y: (0.35 T)cos50°

21.66 **(a)** yes **(b)** yes **(c)** no **(d)** yes **(e)** yes

21.67 **(a)** 7.57 m **(b)** 5.54 m

21.68 **(a)** 6.32 cm **(b)** 0.622 cm

21.69 **(a)** 25 N **(b)** 16°

21.70 354 m s^{-1} in a direction N 8.13° E

21.71 **(a)** 5.73° **(b)** 57.3° **(c)** 115° **(d)** 360°

21.72 **(a)** $\frac{1}{2}\pi$ rad (b) $\frac{1}{3}\pi$ rad **(c)** $\frac{2}{3}\pi$ rad

21.73 **(a)** 1.50 m **(b)** 3.00 m **(c)** $s = r\theta$

21.74 **(a)** 1.05 rad **(b)** 2.62 m

21.75 0.660 m

21.76 **(a)** **(i)** 3.636×10^{-3}, 0.2083°, 3.636×10^{-3} rad **(ii)** 3.636×10^{-3} rad

21.77 19 m

21.78 **(a)** 2.98×10^{-5} rad **(b)** 0.102 minutes

21.79 **(a)** 3.09×10^{16} m **(b)** 3.67×10^{-6} rad = 0.758 seconds

21.80 **(a)** 0.767 rad = 43.9° **(b)** 41.9° **(c)** –4.7%

21.81 **(c)** +1, –1 **(d)** 0°, 180°, 360°

21.82 **(c)** +1, –1 **(d)** 90°, 270°

21.83 –1.2 m

21.84 –1.5 m

21.85 **(b)** 4.0 s **(c)** velocity **(d)** t = 1.00 s **(e)** 1.9 m s^{-1} **(f)** **(i)** 3.8 m s^{-1} **(ii)** 0.94 m s^{-1}

21.86 **(a)** 1.4 m s^{-1} **(b)** 3.8 km s^{-2}

21.87 **(a)** 3.1×10^{6} V s^{-1} **(b)** 5.0 μs **(c)** ±0.62 V

21.88 **(a)** 2×10^{-5} Hz **(b)** 2×10^{-3} m s^{-1}

21.89 **(a)** 1.40 **(b)** 2.40 **(c)** 3.40

21.90 **(a)** 3.22 **(c)** 5.52 **(d)** 7.82

21.92 **(a)** 1.00 **(b)** 2.00 **(c)** 3.00

21.93 **(a)** 1.00 **(b)** 2.00 **(c)** 3.00

21.94 **(a)** 2.00 **(b)** 3.00 **(c)** 2.50

21.95 **(a)** 2.00 **(b)** 3.00 **(c)** 2.50

21.96 **(a)** 6.7×10^{-2} **(b)** 2.0 **(c)** 0.47

21.97 **(b)** 1.5

21.98 **(a)** 40 y^{-1} **(b)** 100 y^{-1}

21.100 **(a)** **(i)** 640 **(ii)** 512 **(b)** 1250

21.101 **(a)** **(i)** 147 **(ii)** 89 **(iii)** 54 **(b)** 1.39 s **(c)** 9.21 s

21.102 **(b)** 31 s; the half-life **(c)** 62 s **(d)** 45 s

21.103 **(a)** 10.0, 8.18, 6.70, 5.49, 4.49, 3.68

21.104 **(a)** s **(b)** 0.33 s **(c)** **(i)** 0.33 s **(ii)** 0.66 s **(iii)** 0.99 s **(d)** 0.23 s **(e)** **(i)** 0.46 s **(ii)** 0.69 s

21.105 **(d)** 0.43 h^{-1} **(e)** 1.6 h

21.107 **(a)** 2100 s^{-1} **(b)** 7 **(c)** yes **(d)** number too small for laws to hold

21.108 **(a)** 0.14 h^{-1} **(b)** 1.3×10^{10}

21.109 **(a)** 1.25×10^{-4} m^{-1} **(c)** **(i)** 29 kPa **(ii)** 8.2 kPa

21.111 **(c)** 0.44 d^{-1} **(d)** 1.6 d

21.112 **(a)** second **(b)** **(i)** 1.97 V **(ii)** 3.16 V **(c)** 2.41 s

22 Synoptic questions

22.1 **(a)** 7.7 m s^{-1} **(c)** 17 kN **(d)** 34 kN **(e)** 70 MPa

22.2 **(c)** 3.9×10^{-3} deg^{-1} **(e)** $3\frac{1}{4}$ **(g)** **(i)** 14 N **(ii)** 20 N

22.3 **(a)** 0.13 mm^{2} **(d)** 4.4×10^{-4} **(e)** 4.4×10^{-4} **(f)** l increases, A decreases **(g)** from Y to X

22.4 **(c)** 1.5 m **(d)** 5.2×10^{3} m s^{-1} **(f)** 209 GPa **(g)** 72 GPa **(i)** 0.29 ms

22.5 **(b)** A **(c)** 362 m s^{-1} **(d)** 226 rev s^{-1} **(e)** 2.3 μm

22.6 **(a)** 2.0 g **(b)** 83 ms **(c)** 2.4 K **(e)** 4

22.8 **(c)** **(i)** 1.4×10^{-12} m **(ii)** 1.4×10^{-13} J **(iii)** 4.6×10^{-22} N s **(d)** γ **(e)** 1.1×10^{-13} J **(f)** greater than c **(g)** 2.1×10^{-30} kg **(h)** 5.7×10^{-22} N s **(i)** 2.7×10^{8} m s^{-1}

22.9 **(a)** 1.8×10^{-3} kg m^{-1} **(b)** 1.1×10^{2} m s^{-1} **(c)** 29 Hz **(e)** 9.5×10^{-4} m s^{-1} **(f)** 1.4×10^{-3} m s^{-1} **(h)** 4.3×10^{-6} m

22.11 **(c)** 1.3×10^{-16} J **(d)** 1.7×10^{7} m s^{-1} **(e)** 1.7×10^{7} m s^{-1} **(f)** 1.6 mA **(g)** 0.34 W **(h)** small **(i)** Y_1 **(j)** +50 V and –50 V **(k)** parabolic **(l)** 1.8×10^{-9} s **(m)** **(i)** 1.0×10^{4} V m^{-1} **(ii)** 1.6×10^{-15} N **(iii)** 1.8×10^{15} m s^{-2} **(iv)** 2.8 mm

22.12 **(a)** 0.020 m^{2} s^{-1} **(b)** 0.16 μA **(c)** up **(d)** 1.6 μC **(e)** 80 mA **(f)** 0.20 nA **(g)** larger **(h)** 3.2 μC **(i)** 0.24 MV

22.13 **(b)** 5.3×10^{-6} m s^{-1} **(c)** 21 ns **(d)** same **(e)** 24 MHz

22.14 **(b)** 0.006 876 u **(c)** 6.405 MeV **(f)** 6.28 MeV

22.15 **(c)** 1.5×10^{7} m s^{-1} **(d)** 1.4 m

22.16 **(b)** 9.6×10^{-15} N **(d)** 60 kV m^{-1}

22.17 **(a)** 1.00×10^{-12} J **(b)** (3.6×10^{-26} J m)/r **(c)** 3.6×10^{-14} m **(d)** 4.9×10^{-14} m

22.18 **(f)** 0.48 W **(g)** 3.1 g **(h)** 11 K min^{-1}

22.19 **(d)** $mv = BQr$ **(e)** 70 mm **(f)** 5.6×10^{-21} N s **(g)** 1.9×10^{-29} kg **(h)** 20

22.20 **(e)** 0.72 **(f)** 0.86 **(g)** 0.065 eV **(h)** 110

22.21 **(b)** 2.2×10^{-19} N s **(c)** 3.0×10^{-15} m **(d)** radius ≈ 2.4×10^{-15} m **(e)** **(i)** 3.2×10^{-18} N s **(ii)** 2.1×10^{-16} m

22.22 **(a)** 8.2 m s^{-2} **(b)** 7.6 km s^{-1} **(c)** 90 kN **(d)** 18 m^{2} **(e)** 3.6×10^{-19} J **(f)** 1.2×10^{-27} N s **(g)** 3.9×10^{21} s^{-1} m^{-2} **(h)** 4.7 μN m^{-2}

22.23 **(a)** **(i)** 7×10^{10} m
(ii) 8×10^{10} m
(b) **(i)** 1.6 km s^{-1}
(ii) 0.72 km s^{-1}
(d) **(i)** 28.7 kg **(ii)** 7.3×10^{25}
(e) **(i)** 2.6×10^{-10} s^{-1}
(ii) 1.9×10^{16} Bq
(f) 16 kW
(h) 0.98 kW
(i) no
(j) 0.94 kW
(k) 15 W m^{-2}
(l) 260 m^2

22.24 **(f)** **(i)** g_0 **(ii)** $\frac{1}{2}g_0$ **(iii)** 0
(iv) $\frac{1}{2}g_0$ **(v)** g_0
(h) 84 min
(i) 9.8 m s^{-2}
(j) 7.9 km s^{-1}
(k) 84 min

22.25 **(a)** 0.000 46 u, 0.005 90 u, 0.013 79 u
(b) 0.026 51 u
(c) 24.7 MeV
(d) 7.8×10^9 K
(g) 1.44×10^{-15} m
(h) 0.111 kN